Petroleum from Coal

History of Modern Science

VOLUME 5

The titles published in this series are listed at *brill.com/hims*

Petroleum from Coal

A Century of Synthesis

By

Anthony N. Strangës

BRILL

LEIDEN | BOSTON

Cover illustration:
Top left: Friedrich Bergius (see Fig. 2.1). Source: Courtesy of Johannes Bergius and Renate Bergius Burgess.
Top right: Franz Fischer (see Fig. 2.8). Source: Max-Planck-Institut für Kohlenforschung, Mülheim an der Ruhr.
Bottom left: Matthias Pier (see Fig. 3.1). Source: Werner Lang, "Matthias Pier, Ehrenbürger von Nackenheim," *Nackenheimer Heimatkundliche Schriftenreihe*, Heft 16, (Herausgegeben vom Verkehrsverein e.V., Nackenheim am Rhein, 1982). Pier Collection, BASF Archives, Ludwigshafen.
Bottom right: Hans Tropsch (see Fig. 2.8). Source: Max-Planck-Institut für Kohlenforschung, Mülheim an der Ruhr.

The Library of Congress Cataloging-in-Publication Data is available online at https://catalog.loc.gov
LC record available at https://lccn.loc.gov/2023056215

Typeface for the Latin, Greek, and Cyrillic scripts: "Brill". See and download: brill.com/brill-typeface.

ISSN 2352-7145
ISBN 978-90-04-54411-6 (hardback)
ISBN 978-90-04-69091-2 (e-book)
DOI 10.1163/9789004690912

This book is printed on acid-free paper and produced in a sustainable manner.

To my wife Sonya;
my family Krista, Victoria, Giovanna, Dominic Fazzino, and Zachary Dwyer;
and Kara, Michael, Miranda, and Madeline Holmgreen

Contents

Preface

In the first decades of the twentieth century, petroleum increasingly replaced coal as the major fuel consumed industrially and domestically. Germany, Britain, Japan, and other nations with little or no domestic petroleum found themselves at a serious disadvantage and sought ways to relieve or reduce their liquid fuel shortages. Success came first in Germany when Friedrich Bergius in 1913 and Franz Fischer and Hans Tropsch in 1926 invented laboratory processes for converting coal to petroleum and which on the eve of World War II were producing synthetic liquid fuel on an industrial (commercial) scale. No other nation matched Germany's synthetic fuel production. Britain, for economic reasons, confined its synthetic fuel development to a single industrial-size plant. The Japanese in their rush to industrialize their promising research program created a technological disaster. Canada restricted its synthetic fuel research to laboratory and pilot plants, the United States made little effort to develop a synthetic fuel program until shortly after World War II had ended.

In the ensuing decades, the synthetic fuel industry experienced a series of stops and starts largely determined by the economics and supply of natural petroleum and the lack of any long-term commitment from either industry or government. Its history has followed a cyclical rather than the hoped-for linear path of progressive development. After a nearly century's existence the time has come to analyze the ups and downs of the synthetic fuel industry and to provide a comprehensive and comparative history of its global development. *Petroleum from Coal: A Century of Synthesis* examines: (1) the invention and early development of Bergius's coal hydrogenation (liquefaction) process and the Fischer-Tropsch gas synthesis from 1910 to 1926; (2) Germany's industrialization of the two conversion processes from 1927 to 1945, particularly during the World War II years; and (3) the global transfer of the German technology to Britain, Canada, Japan, the United States, and other nations in the 1930s–90s. It will show why and how the two conversion processes arose first in Germany; explain why and how German scientists and engineers, such as Matthias Pier at Badische Anilin- und Soda-Fabrik (BASF) in Ludwigshafen (on 9 December 1925 a component of IG Farben, Interessengemeinschaft Farbenindustrie AG in Frankfurt), succeeded in industrializing the laboratory and pilot plant conversions during a time of social, political, and economic upheaval in Germany; and analyze the pre- and post-World War II vicissitudes of coal-to-oil conversion processes globally.

Most of the sources needed to research and write *Petroleum from Coal* are in the Texas A&M University archives. Members of the Allied investigative teams, such as the United States Technical Oil Mission (TOM), Combined Intelligence Operations Sub-Committee (CIOS), British Intelligence Operations Sub-Committee (BIOS), Field Intelligence Agency Technical (FIAT), and US Naval Technical Mission to Japan, who went to Germany and Japan during the closing months of World War II, captured and compiled information on the German and Japanese synthetic fuel industries in approximately 2,720 documents. Copies of these documents and related synthetic fuel documents were scattered around the United States in government repositories and archives, at the Commerce Department, Library of Congress, among TOM members, and in German and British archives.

Beginning in October 1975 Texas A&M's German Document Retrieval Project, of which I was a member, collected and brought many of the synthetic fuel documents to Texas A&M. In the ensuing years I have added to the collection a significant number of synthetic fuel papers that I obtained during visits to BASF's archives in Ludwigshafen, Theodor Goldschmidt archives in Essen, the Bundesarchiv in Koblenz, Imperial Chemical Industries (ICI, since 2008 AkzoNobel) archives in London, the National Diet Library in Tokyo, and the Central State Archives in Rome. The Texas A&M archival collection contains about 300,000 pages and 305 TOM microfilm reels of German documents, and numerous Japanese, American, British, and Canadian documents, making it one of the largest collections on the history of synthetic fuel production. The collection includes German synthetic fuel plant records on catalysts, plant-operating conditions, production figures, structural materials, and other technical details; United States (US) Bureau of Mines' translations of the German plant records and interviews (interrogations) of German scientists such as Matthias Pier, Heinrich Bütefisch, and Ernst Donath; documents and correspondence between the German and Japanese governments, synthetic fuel companies, and scientists; United States Naval Technical Mission to Japan Reports (185 reports on microfilm) which are the same as the reports in the National Diet Library in Tokyo (*Materials about Five Year Plan*, edited by the Commerce and Industry Section of South Manchuria Railway), and in the Records of General Headquarters/Supreme Commander for the Allied Forces (GHQ/SCAP, Record Group 331, National Archives and Records Administration, Washington, DC); ICI papers and correspondence dealing with its Billingham plant and with government officials; US Bureau of Mines reports on its Louisiana, Missouri, two demonstration plants that were operating in the late 1940s–early 1950s and reports to Congress.

As a member of the German Document Retrieval Project I also interviewed and corresponded with several TOM members, Vladimir Haensel, John Gordon Allen, William Appler Horne, and Hans Schindler, some of whom provided synthetic fuel documents that they possessed, and interviewed and corresponded with several German synthetic fuel scientists such as Ernst Donath and Hans Schappert, who came to the United States during Project Paperclip to work for the US Bureau of Mines. In the early 2000s I collaborated with Steve LeViness at Syntroleum (in 2007 Emerging Fuels Technology) in Tulsa, Oklahoma, to create the Fischer-Tropsch Archive and put online many of the synthetic fuel documents from Texas A&M's archives, including the original German documents, US Bureau of Mines publications, British and American investigative reports and contents of the microfilm reels that they produced, and numerous papers on the Fischer-Tropsch (F-T) synthesis published in recent years.

Other authors, such as the British economic historian William J. Reader and the German synthetic fuel scientist Ernst Donath, have discussed the British and the German synthetic fuel programs. They looked at the science-engineering, the economics, and the politics of coal-to-petroleum conversion, but no one yet has approached the subject comprehensively and comparatively and examined it from a global perspective. Its fragmented historical treatment has left unanswered important questions such as why Germany's synthetic fuel industry succeeded whereas Japan's did not, despite Japan's having access to the German synthetic fuel research and development; why the Germans constructed twenty-one plants whereas the British, who were as technologically advanced as the Germans, built only one plant; why Canada limited its investigations to the laboratory and pilot plant; and why the United States was the last of the industrialized nations to begin research and development on coal-to-petroleum conversion processes.

Petroleum from Coal addresses these essential issues. It examines synthetic fuel's one hundred year history and makes clear that the impetus for synthetic fuel's technological development often came too late, after the perceived energy crisis had abated or ended, and that a long-term commitment to its future might have ensured the industry's survival. Because synthetic fuel science-technology is complex, long-term research would have tested the scientific and engineering principles in the laboratory and in pilot and demonstration plants and solved the scientific and engineering problems before the crises arose.

In 1930s Germany and more recently in South Africa, long-term commitment permitted their synthetic fuel industries to survive. Conversely, the 1970s oil crises provide a classic example of government-industry non-commitment to establishing and supporting a long-term energy program, specifically synthetic fuel research and development, and we can make the same argument for other alternative energy programs. Future energy crises are unpredictable but probably inevitable; the future of natural petroleum, a finite energy source, is predictable.

I have not discussed the environmental impact of the emerging synthetic fuel industry in the 1930s–early 1950s, the pre-and post-World War II years and the synthetic fuel industry's invention and development years, because the German, British, Japanese, Canadian, and American scientists and engineers working on synthetic fuel processes during those years published nothing on the environmental impact of the synthetic fuel processes. The numerous synthetic fuel documents that they published and I examined say nothing about synthetic liquid fuel or fossil fuel (coal and petroleum) production and combustion polluting and warming the atmosphere. The synthetic fuel scientists had no interest in investigating atmospheric pollution or warming or other environmental issues that resulted from the combustion of synthetic liquid fuel or from coal and petroleum. Establishing a link between fossil fuel combustion and Earth's (global) warming and between fossil fuel combustion and the acidity of rain (acid rain) was not their objective. Nations, such as Germany, Britain, and Japan that had little, if any, naturally occurring petroleum, committed to synthetic fuel projects because they wanted to develop a domestic coal-to-oil synthetic fuel industry and not have to rely on petroleum imports. Nor did the greater scientific community of the 1930s–early 1950s, particularly its climate scientists with few exceptions, publish any research, either pro or con, on the environmental impact of synthetic and fossil fuel combustion. Serious concerns about industrial-size fossil fuel operations, such as Canada's oil sands-bitumen extraction process or Sasol's (South Africa) F-T process, polluting the environment and warming the globe, despite having done so for decades, came later in the 1950s–80s.

Why did the majority of the scientific community until the 1950s–80s dismiss, ignore or reject the few research publications linking fossil fuel consumption and global warming-environmental pollution? An examination of their papers and presentations shows that the majority remained convinced that little evidence supported a warming or an acidic polluting of Earth's atmosphere. If any increase of Earth's atmospheric temperature or acidity had occurred, the increase would be slow and insignificant. As the twentieth century passed through its early decades, from 1900 through the 1940s, and the evidence of any significant change in climate still lacking, research on global warming-environmental pollution gradually lost the scientific community's interest and faded from discussion. The successive tragedies of World War I, the Great Depression, and World War II pushed climate science studies further into the background.

But beginning in the 1950s, and particularly since the 1980s–early 1990s, as scientific evidence clearly established the atmosphere's warming and pollution, the overwhelming majority of informed and knowledgeable climate scientists have accepted (1) the reality of global warming and the necessity of reducing carbon dioxide and other greenhouse gas emissions resulting from fossil fuel combustion and (2) the existence of a clear connection between acid rain-atmospheric pollution and the increasing combustion of fossil fuels. Public opinion on linking fossil fuel combustion and global warming-environmental pollution paralleled the scientific community's increasing acceptance of a combustion-warming-pollution link.

A brief discussion on the evolution of scientific thinking on global warming and on the acid rain-fossil fuel combustion link follows in the Epilogue. It provides a summary of the advances that occurred in the climate sciences in the late 1800s–1900s and a comparison with the advances that occurred contemporaneously in synthetic fuel research and development.[1]

Energy-related developments and the impact they have on society are never far removed from academic, industrial, business, and general interest, and because of the interconnectedness of science, technology, and society, I remain confident that the historical development of the synthetic fuel industry that I have discussed in *Petroleum from Coal* will find a readership that includes historians of science, technology, and economics, scientists, engineers, social scientists, and economists.

Finally, I wish to thank those who in several ways provided valuable advice and assistance that enabled me to write a comprehensive history of the synthetic fuel industry, with a special thanks to Kurt Irgolic (1938–99), formerly at Texas A&M University, College Station, Texas; Steve LeViness formerly at Syntroleum in Tulsa; Ruth Fromm, archivist at BASF; Bergius's daughter Renate Burgess (1910–88) in London, his son Johannes Bergius (1916–88) in Bonn, his nephew and chemist Dietrich Stoltzenberg (1926–2008) in Hamburg; synthetic fuel scientists Ernst Donath and Hans Schappert formerly at BASF (IG Farben); Clive Weeks (1942–2016) at ICI in London; TOM members that I interviewed; Ken Agee, president, Emerging Fuels Technology (Syntroleum); and Burt Davis (1934–2018), formerly at the Center for Applied Energy Research, University of Kentucky, Lexington, Kentucky. I also wish to acknowledge the financial support I received from the National Science Foundation, Washington, DC; Center for Energy and Mineral Resources, Texas A&M University; Office of University Research Services, Texas A&M University; and the Canadian Studies Program, Academic Relations Office, Canadian Embassy, Washington, DC.

1 I have written a detailed account of the environment-energy issue in my history of science book, *Science Changed the World* (Dubuque, Iowa: Kendall Hunt, 2018).

Permission Acknowledgements

1. University of Chicago Press: permission to publish a modified version of my *Isis* article "Friedrich Bergius and the Rise of the German Synthetic Fuel Industry." *Isis* 75 (1984): 643–67.
2. Johns Hopkins University Press: permission to publish modified versions of my *Technology and Culture* articles "From Birmingham to Billingham: Synthetic Fuels in Great Britain, 1910–1945." *Technology and Culture* 26 (1985): 726–57 and "Canada's Mines Branch and Its Synthetic Fuel Program for Energy Independence." *Technology and Culture* 32 (1991): 521–54.
3. Taylor and Francis: permission to publish modified versions of my *Annals of Science* articles, "Synthetic Fuel Production in Prewar and World War II Japan: A Case Study in Technological Failure." *Annals of Science* 50 (1993): 229–65 and "The US Bureau of Mines Synthetic Liquid Fuels Programme, 1920–1950s: German Connections and American Advances." *Annals of Science* 54 (1997): 29–68.
4. Springer Nature: permission to publish my modified book chapter, "Germany's Synthetic Fuel Industry." In *The German Chemical Industry in the Twentieth Century*. Edited by John E. Lesch, 147–216. Dordrecht, The Netherlands: Kluwer Academic Publishers, 2000.

Illustrations and Tables

Illustrations

Tables

Abbreviations

Ber. deut. chem. Gesell.	*Berichte deutsche chemische Gesellschaft*
BIOS	British Intelligence Objectives Agency
Chem. Trade J.	*Chemical Trade Journal*
CIOS	Combined Intelligence Objectives Agency
C. R. Acad. Sci.	*Comptes rendus de l'Académie des Sciences*
FIAT	Field Intelligence Agency Technical
J. Chem. Soc.	*Journal of the Chemical Society*
J. Fuel Soc. Japan	*Journal of the Fuel Society of Japan*
J. Inst. Energy	*Journal of the Institute of Energy*
J. Inst. Fuel	*Journal of the Institute of Fuel*
J. Soc. Chem. Ind. Japan	*Journal of the Society of Chemical Industry, Japan*
J. Soc. Chem. Ind.	*Journal of the Society of Chemical Industry* (*Journal of the Society of Chemistry and Industry*)
J. Inst. Petroleum Technologists	*Journal of the Institute of Petroleum Technologists*
Proc. Royal Society	*Proceedings of the Royal Society*
Nav. Tech, Miss. Japan	US Naval Technical Mission to Japan
Z. angew. Chem.	*Zeitschrift für angewandte Chemie*
Z. anorg. Chem.	*Zeitschrift für anorganische und allgemeine Chemie*
Z. Elecktrochem.	*Zeitschrift für Elektrochemie*

CHAPTER 1

Overview of the Synthetic Fuel Industry

1 Introductory

Petroleum from Coal: A Century of Synthesis provides a global and comparative history of the twentieth-century coal-to-petroleum conversion processes or synthetic fuel industry. It examines (1) the invention and development of Bergius coal hydrogenation (direct coal liquefaction DCL) and Fischer-Tropsch (F-T) gas synthesis (indirect coal liquefaction ICL) during the period 1910–26;[1] (2) Germany's industrialization of the two synthetic fuel processes from 1927 to 1945, particularly during the National Socialist German Workers Party (NSDAP, Nazi)-World War II years; and (3) the global transfer of the German technology to Britain, Japan, Canada, the United States, South Africa, and other nations from the 1930s through the 1990s. It will show why the two synthetic fuel processes arose first in Germany and how the chemists Friedrich Bergius (1884–1949) in 1913 and Franz Fischer (1877–1947) and Hans Tropsch (1889–1935) in 1926 invented and developed their synthetic fuel processes; explain why and how industrial scientists and engineers, such as Matthias Pier (1892–1965) at BASF (IG Farben on 9 December 1925) and Otto Roelen

1 Scientists have used the following terms to describe the same processes.

a. Bergius high-pressure coal hydrogenation ≡ coal hydrogenation ≡ coal liquefaction ≡ direct coal liquefaction (DCL). Each refers to the process of converting coal or a coal paste to a petroleum-like liquid by adding hydrogen gas to the coal or paste.

$$\text{Coal} + H_2(g) \rightarrow \text{petroleum hydrocarbons}$$

b. Fischer-Tropsch process (F-T) ≡ F-T gas synthesis ≡ indirect coal liquefaction (ICL)

$$\text{Coal} + H_2O(g) \rightarrow CO(g) + H_2(g) \rightarrow \text{petroleum hydrocarbons}$$

c. Gas-to-Liquid (GTL)

$$CH_4\ (\text{methane gas}) + H_2O(g) \rightarrow \text{petroleum hydrocarbons}$$

In converting metric tons to barrels, the authors of the sources cited used slightly different conversion values: 1 metric ton = 7.2, 7.3, or 7.5 barrels. The values hardly differ, and I have left them unchanged in the chapters.

 | DOI:10.1163/9789004690912_002

(1897–1993) at Ruhrchemie in Oberhausen-Holten, succeeded in industrializing the laboratory and pilot plant syntheses during a time of social, political, and economic upheaval in Germany; and analyze the pre- and post-World War II vicissitudes of the synthetic fuel industry globally. No study yet has assessed historically the synthetic fuel industry's technological evolution, economics, and social impact. *Petroleum from Coal* addresses these important issues and evaluates the role synthetic fuel has played in our energy history.

Petroleum from Coal has two themes. (1) It examines whether an innovation's intrinsic scientific and engineering qualities, social-political-economic factors, or a combination of intrinsic qualities and external factors, determine an innovation's success or failure. For the Bergius coal hydrogenation and F-T synthesis, a combination of intrinsic qualities and external factors contributed to their success. In Germany, the processes succeeded technologically before the Nazi government rose to power in 1933 and proceeded to subsidize industrial-size development. Industrialization may not have occurred otherwise. Imperial Chemical Industries' (ICI) technologically successful coal hydrogenation process received similar incentives from the British government in the 1930s. The United States (US) Bureau of Mines' synthetic fuel program succeeded technologically, but because of unfavorable economics, an abundance of cheap petroleum in the early 1950s, and an end to the government's commitment to synthetic fuel research and development, coal hydrogenation and the F-T synthesis never achieved industrialization. In World War II Japan, the opposite happened. The Japanese government provided economic and social incentives, but because the Japanese leaped from the laboratory to industrial-size production without sufficient pilot plant experience, the two synthetic fuel processes were technological failures. In Canada, the Mines Branch's plan of long-term development of low-grade bitumen extracted from Canada's huge oil sands deposits and crude oil reserves met with considerable technological success in the pre- and post-World War II decades, but unfavorable economics resulted in a cautious and limited approach to industrialization.

(2) *Petroleum from Coal* makes clear that the impetus for the technological development of synthetic fuel usually came too late, after the perceived energy crisis had abated or ended, and that a long-term commitment to its future might have ensured the industry's survival. (The same argument holds for other alternative energies). Synthetic fuel science and engineering, such as the development of catalysts and environmentally friendly apparatus, are complex, and required long-term research to test the science-engineering principles of synthetic fuel production and solve the science-engineering problems in the laboratory and in pilot and demonstration plants before the crises arose.

In 1930s Germany and post-World War II South Africa, which are special cases, long-term industry and government commitment permitted their synthetic fuel industries to survive. Future energy crises remain unpredictable but probably inevitable, whereas we know the future of natural petroleum. It is a finite resource that faces eventual exhaustion that other available energy sources can only delay.

2 Global Energy Situation 1910–90s: Energy Crises and Revivals

An examination of the global energy situation during the decades from 1910 to the 1990s answers the issues raised above. Petroleum had become essential to the economies of industrialized nations by the 1920s. The mass production of automobiles, the introduction of airplanes and petroleum-powered ships, and the recognition of petroleum's high energy content compared to wood and coal, required a shift from solid to liquid fuels as a major energy source.

Industrialized nations responded in different ways. For Germany, Britain, Japan, and other nations that had little or no domestic petroleum, the shift to a petroleum-based economy presented a serious challenge for which only a few options existed. They could continue to import petroleum, which all of them did; they could acquire by force the petroleum resources of other nations, which Germany, Japan, and Italy did during their 1930s–40s wartime occupations in Europe and the Far East; or they could synthesize petroleum either from their domestic coal resources or from the coal resources they seized from occupied nations.

Germany was the first of the industrialized nations to succeed in synthesizing petroleum when Bergius in Rheinau-Mannheim in 1913 and Fischer and Tropsch at the Kaiser Wilhelm Institute for Coal Research (KWI) in Mülheim in 1926 invented processes for converting coal to oil. Their pioneering work enabled Germany to develop a technologically successful synthetic fuel industry that grew from a single coal hydrogenation plant constructed in Leuna, near Merseburg, in 1927 to twenty-one coal hydrogenation and F-T synthesis plants that in 1944 reached a peak production of 26 million barrels of synthetic fuel.

Britain's synthetic fuel program evolved from post-World War I laboratory and pilot plant studies that began at the University of Birmingham in 1920 on the F-T synthesis and in 1923 on coal hydrogenation. The Fuel Research Station in East Greenwich also began research on coal hydrogenation in 1923, and the program reached its zenith in 1935 when ICI constructed a coal hydrogenation plant at Billingham that had the capacity to synthesize annually 1.28 million barrels of petroleum. British research and development matched Germany's, but because of hydrogenation's high cost and the government's decision to rely on petroleum imports rather than price supports for an expanded domestic industry, Billingham remained Britain's only industrial-size synthetic fuel plant. F-T synthesis in the 1930s–40s never advanced beyond the construction of four small experimental plants: Birmingham, the Fuel Research Station's two plants that operated from 1935 to 1939, and Synthetic Oils Ltd. near Glasgow.

Britain and Germany had the most successful synthetic fuel programs. The others were either smaller-size operations, such as France's three demonstration plants (two coal hydrogenation and one F-T synthesis), Canada's bitumen hydrogenation pilot-demonstration plants, and Italy's two crude petroleum hydrogenating (refining) plants; or technological failures as were Japan's five industrial-size plants (two coal hydrogenation and three F-T) that produced only about 360,000 barrels of liquid fuel during the World War II years.

A much more favorable energy situation existed in the United States, and it experienced few problems in making an energy shift because it possessed large resources of both petroleum and coal. The US Bureau of Mines, at its Pittsburgh Experiment Station, had begun small-size research on the F-T synthesis in 1927 and coal hydrogenation in 1936, but did no serious study of either process until the government expressed considerable concern about the country's rapidly increasing petroleum consumption in the immediate post-World War II years. At that time the Bureau began a demonstration program, and from 1949 to 1953 when government funding ended, it operated a small 200–300 barrel per day (bbl//d) coal hydrogenation plant and a smaller 50 bbl/d F-T plant at Louisiana, Missouri. In addition to the Bureau's program, American industrialists constructed four synthetic fuel plants in the late 1940s and mid-1950s, none of which achieved full capacity before shutdown in the 1950s for economic and technical reasons. Three were F-T synthesis plants located in Garden City, Kansas; Brownsville, Texas; and Liberty, Pennsylvania. The fourth plant was a coal hydrogenation plant in Institute, West Virginia.

Following the plant shutdowns in the United States and until the global energy crises of October 1973–March 1974 and November 1979–January 1981, all major synthetic fuel research and development ceased except for the construction in March 1955 of the South African Coal, Oil, and Gas Corporation's (Sasol) $104.8 million F-T synthesis plant in Sasolburg, south of Johannesburg. South Africa's desire for energy independence and the low quality of its coal dictated the choice of F-T gas synthesis rather than coal hydrogenation. Its Johannesburg plant remained the only operational industrial-size synthetic fuel plant until the 1970s energy crises and South Africa's concern about hostile world reaction to its apartheid policy prompted Sasol to construct two more F-T synthesis plants in March 1980 and June 1982 in Secunda.[2]

The 1970s energy crises also revitalized synthetic fuel research and development in the United States and Germany and led to joint government-industry programs that quickly disappeared once the crises had passed. Gulf Oil, Atlantic Richfield, and Exxon in the United States; Saarbergwerke AG in Saarbrüken, Ruhrkohle AG in Essen, and Veba-Chemie AG (in 1978 Veba Oel AG, which in 2001 British Petroleum BP acquired) in Gelsenkirchen, Germany, constructed F-T synthesis and coal hydrogenation pilot plants in the 1970s and early 1980s only to end their operation with the collapse of petroleum prices in 1982/83–2001.

In the mid-1990s two developments triggered another synthetic fuel revival in the United States: (1) petroleum imports again reached 50% of total consumption, or what they were during the October 1973–March 1974 Arab petroleum embargo, and (2) an abundance of natural gas, equivalent to 800 billion barrels of petroleum, but largely inaccessible by pipeline, existed. Syntroleum in Tulsa, Oklahoma; Exxon in Baytown, Texas; and Atlantic Richfield in Plano, Texas, developed modified F-T syntheses that produced liquid fuels from natural gas (gas-to-liquid, GTL) and thereby offered a way of reducing the United States' dependence on petroleum imports. The Department of Energy (DOE) at its Pittsburgh Energy Technology Center through the 1980s–90s also continued small-size (pilot plant) research on improved versions of coal hydrogenation. DOE pointed out that global coal reserves greatly exceeded petroleum reserves, anywhere from five to twenty-four times, and that it expected petroleum reserves to decline significantly in 2010–30. Energy forecasters likewise predicted a United States petroleum shortfall of 1–3.5 million bbl/d in 2030.

3 Impact of Hydraulic Fracturing (Fracking) Revolution on Natural Petroleum and Synthetic Fuel Production

The imbalance between United States domestic production and consumption had persisted since the 1960s, but a 1980s–90s advance in petroleum recovery, combining horizontal drilling and hydraulic fracturing, revolutionized domestic petroleum and natural gas production and significantly reduced, nearly eliminated, the gap between production and consumption. Combined horizontal drilling and hydraulic fracturing (fracking revolution), which began in

2 Different sources give different dates for the start of Sasol's Secunda III F-T synthesis plant, May 1981 and 1984.

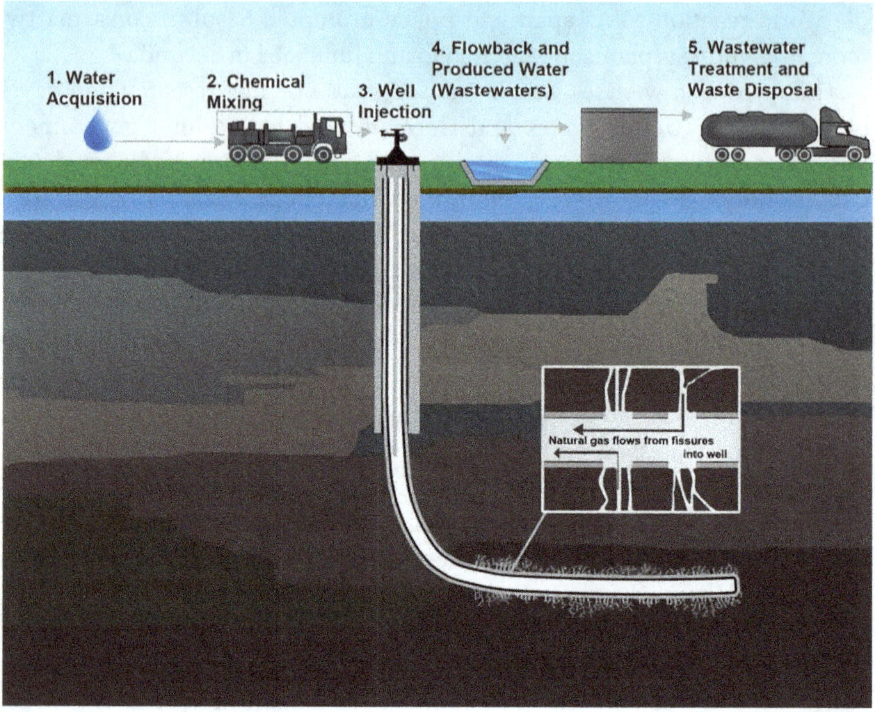

FIGURE 1.1 Hydraulic fracturing
SOURCE: ENVIRONMENTAL PROTECTION AGENCY.
https://land8.com/fracking-all-you-need-to-know/

1986–91 in the Barnett Shale (5,000–8,000 feet deep) in the Dallas-Fort Worth region of Texas (north Texas), enabled the United States to recover oil and natural gas from its extensive, but deep (1,000–14,000 feet), dense oil shale deposits. The fast growing application of the combined process to other shale deposits in North Dakota (Bakken Shale 7.4 billion barrels estimated); Pennsylvania (Marcellus Shale 47 billion barrels estimated and Utica Shale 7.9 billion barrels estimated); New York (Marcellus and Utica Shale), West Virginia (Marcellus Shale); Texas [Barnett (North), Eagle Ford (South), Haynesville (East) 7.9 billion barrels total estimated, Permian Basin (West), 15 billion barrels estimated]; and other states significantly reduced the United States' reliance on petroleum imports to meet its domestic demand. At the same time the fracking revolution led to growing concerns that the process has contributed to increased earthquake activity because of wastewater injected underground, pollution of drinking water resources, and methane gas (greenhouse gas) emissions.

Horizontal drilling and hydraulic fracturing have long histories. Horizontal drilling, previously called directional drilling, dates from 1934 in Conroe, Texas. It remained insufficiently developed until 1973 but has dominated drilling in the 2000s. Of the 1,756 rigs operating in 2013 in the United States, 1,100 were horizontal drilling rigs. Hydraulic fracturing (fracking), which began in 1947–49 in Oklahoma and Texas, recovers oil and natural gas trapped in dense oil shale formations by blasting at high pressure, a mix of water (90%), sand (about 9%), and chemicals (acids, salts, alcohols, gels, about 1%) into the formation to release the previously unrecoverable trapped oil and gas.

George Mitchell (1919–2013) in The Woodlands, Texas, founder of Mitchell Energy and Development Corporation, pioneered the use of hydraulic fracturing from 1946, developing new techniques in 1986 and 1997. He established his corporation in 1946 and 1949 under various names, and sold it for $3.5 billion in 2002 to Devon Energy in Oklahoma City. Mitchell combined his new-improved fracking process, in which he invested ten years and $6 million, with the advances in horizontal drilling, and successfully extracted the natural gas trapped in the Barnett Shale formation. His revolutionary technological innovation made possible economically the recovery of natural gas and oil from oil shale in the United States and globally, and the resulting increase (20–30% from 2008–12) in US domestic production of oil and natural gas.

The fracking revolution that followed the oil crises of October 1973–March 1974 and November 1979–January 1981 also shifted the construction of any potential industrial-size Bergius coal hydrogenation and F-T gas synthesis plants and gas-to-liquid plants from the United States and Europe to Asia, the Middle East, and Africa. At that time, the late 1990s, DOE switched its coal hydrogenation research to *standby*, ending its funding in 1999. Pilot plant studies in 1 L-30 bbl/d reactors that Hydrocarbon Technologies Incorporated in Lawrenceville, New Jersey, has carried out remain the only coal hydrogenation research project operating in the United States in the 1990s–2000s.[3] Syntroleum in Tulsa and Rentech, with headquarters in Los Angeles, California, were among the few American companies that carried out pilot plant F-T gas synthesis and gas-to-liquid studies in the late 1990s–early 2000s.[4]

---

3 Hydrocarbon Research Incorporated in Trenton and Lawrenceville, New Jersey, in 1943, became Hydrocarbon Technologies Incorporated in Trenton in 1995, and Headwaters Incorporated in Draper, Utah, in 2001.

4 Rentech switched from gas-to-liquid production of petroleum hydrocarbons to wood pellets production in 2013. Headwaters still licenses coal conversion and heavy oil upgrading but mainly manufactures building products for construction. Boral Limited, a building and construction materials company in Sydney, Australia, acquired Headwaters for $2.6 billion in 2017.

TABLE 1.1 International industrial-size synthetic fuel plants

Year	Company	Location	Plant cost	Type	Production
1993	Shell, Bintulu plant	Sarawak, Malaysia	$850 million	GTL	12,500 bbl/d 14,700 bbl/d (2010)
2007	Sasol-Qatar Petroleum, Orxy plant	Ras-Laffan, Qatar	$950 million	GTL	32,000 bbl/d
2008–2009	Shenhua Group Corporation, Hydrocarbon Technologies, Majata plant	80 miles south of Baotau, Inner Mongolia	$3.2 billion	DCL H-Coal process	22,000 bbl/d
2011	Shell-Qatar Petroleum, Pearl plant	Ras-Laffan, Qatar	$24 billion	GTL	140,000 bbl/d, world's largest GTL plant
2014	Chevron	Escravos, Nigeria	$10 billion	GTL	33,000 bbl/d
2016	Ningxia (subsidiary of Shenhua)	Yinchuan, in Hui Autonomous Region in Western China	$7.9 billion	ICL	100,000 bbl/d

SOURCE: COURTESY OF AUTHOR

The table above shows the industrial-size Bergius coal hydrogenation, Fischer-Tropsch gas synthesis, and gas-to-liquid plants that became operational outside the United States beginning in the 1990s.

Four factors, alone or in combination, therefore, have led industrialized nations at various times during the twentieth century to conclude that synthetic liquid fuel could contribute to their increasing liquid fuel requirements: (1) the shift from solid to liquid fuel as a major energy source, (2) the invention of the Bergius and F-T coal-to-petroleum and gas-to-liquid conversion or synthetic fuel processes, (3) recognition that global petroleum reserves were finite and much less than global coal reserves and that petroleum's days as a plentiful energy source were limited, and (4) the desire for energy independence.

With the exception of Sasol's three industrial-size F-T synthesis plants in South Africa constructed in the 1955, 1980, and 1982, and the few industrial-size F-T synthesis and coal hydrogenation plants constructed in the 1990s–early 2000s in China, Qatar, Malaysia, and Nigeria, the synthetic fuel industry, like most alternative energies, has endured a series of fits and starts that plagued it to near demise in the early 1980s. The history of the synthetic fuel industry shows that shortly after Bergius invented coal hydrogenation in 1913 the enthusiasm the process generated in petroleum-poor Germany during the early World War I years largely disappeared once Germany acquired Romania's oil fields. Similarly, in the United States the East Texas petroleum discoveries that began on 5 October 1930 and continued throughout the decade ended the US Bureau of Mines' Pittsburgh Depression-era small pilot plant program. The Bureau's promising resumption of research shortly after World War II also ceased within a decade with the discovery of new petroleum fields, which allayed all fears of an impending energy crisis, and the emphasis placed on developing nuclear energy. A combination of economics and politics almost totally ended the 1970s–80s synthetic fuel research and development in the United States and Europe. The later industrial-size developments occurred mainly in China which has huge coal fields and Qatar with its abundant sources of natural gas.

The historical record has demonstrated that after a century of research and development synthetic liquid fuel has not emerged as a significant alternative energy source. Despite the technological success of Bergius coal hydrogenation, Fischer-Tropsch gas synthesis, and gas-to-liquid processes, the synthetic fuel industry's lack of progress and seemingly cyclical history appear directly related to government's and industry's uninterest in making a firm and a long-term commitment to the economic survival of synthetic fuel research and development. The synthetic fuel industry experienced intermittent periods of intense activity internationally in times of crises, only to face quick dismissal as unnecessary or uneconomical upon disappearance of the crises. Even the synthetic fuel industry's argument that synthetic liquid fuels burn much cleaner than coal, and if substituted for coal they will reduce the emissions that have contributed to acid rain formation, greenhouse effect, and to an overall deterioration of air quality has failed to silence its critics. The hope of transforming the synthetic fuel industry's accomplishments at the pilot and demonstration plant stage into widespread or global industrial-size production has not materialized.

4 Scientific and Engineering Advances in Synthetic Fuel Research and Development

What twentieth-century scientific and engineering advances contributed to Bergius's and Fischer's invention of synthetic fuel processes and to the industrialization of these processes primarily in Germany, but also in Britain, Japan, and other nations?

In the pre-World War II period only Germany and Japan developed both Bergius coal hydrogenation and the F-T synthesis. The other nations regarded the F-T synthesis as technologically lagging and economically unfeasible, and they focused their research and development on Bergius coal hydrogenation as the better way to synthesize liquid fuels, especially gasoline. Bergius in Rheinau-Mannheim began the German drive for energy independence with his invention and early development of high-pressure coal hydrogenation in the years 1910–26. Bergius crushed and dissolved a coal containing less than 85% carbon in a heavy oil to form a paste. He reacted the coal-oil paste with hydrogen gas at high pressure (P = 200 atmospheres, atm) and high temperature (T = 400° C) and obtained petroleum-like liquids. Bergius sold his patents to BASF in July 1925, and from 1925 to 1930 Pier at BASF (IG Farben) made major advancements that significantly improved product yield and quality. Pier developed sulfur-resistant catalysts, such as tungsten sulfide (WS_2), and separated the coal-to-liquid conversion into two stages, a liquid stage and a vapor stage. By 1 September 1939 when World War II began seven coal hydrogenation plants (plus Ludwigshafen) were operating and beginning to contribute increasingly to Germany's domestic liquid fuel supply. When plant construction ceased in 1942 twelve coal hydrogenation plants converted coal and coal tar into gasoline, diesel oil, and other petroleum products. From the first coal hydrogenation plant that began operation at Leuna on 1 April 1927, the twelve coal hydrogenation plants in early 1944 reached a peak production of over 3,170,000 metric tons (21.6 million barrels) of synthetic fuel. In World War II these plants provided 95% of the German air force's aviation gasoline and 50% of Germany's total liquid fuel requirements.

The most important breakthroughs for the F-T synthesis occurred from 1926 to 1935 with Fischer's and Roelen's success in developing cobalt catalysts and establishing the optimum pressures (1 atm and 10–15 atm) for the synthesis. They made such significant progress in this period that at the start of World War II German industrial firms such as Ruhrchemie in Oberhausen (Sterkrade)-Holten and Braunkohle Benzin (Brabag) in Berlin had constructed nine F-T synthesis plants to complement the twelve coal hydrogenation plants that operated during the war. Although smaller than the coal hydrogenation plants, F-T plants nevertheless provided 12–15% of Germany's wartime liquid fuels production, reaching a maximum of 4.15 million barrels in 1944.

Japanese research on the F-T synthesis began in 1927 with the catalytic studies of Gen-itsu Kita (1883–1952) at Kyoto Imperial University. Kita's studies reached the semicommercial pilot plant stage in 1937 and continued until 1941. The Japanese established other F-T research laboratories at the Imperial Fuel Research Center in Kawaguchi, Tokyo Imperial University, and Mitsui Bussan KK in Rumoi during the period 1927–37, and in 1937 Mitsui acquired exclusive Japanese commercial rights to the F-T synthesis from Ruhrchemie which controlled F-T licensing. Mitsui paid ¥7.2 million (¥ yen) for the license and before the war's end constructed three plants, at Miike (pronounced Oyke), Amagasaki, and Takikawa, and had another two plants under construction. None of the plants approached design capacity, the best being Miike's 50% output in 1943. Synthesis gas shortages, low catalytic activity, and leaks in the cooling system presented major problems that Mitsui engineers in their wartime haste failed to eliminate at the pilot plant stage before moving to industrial-size operation. A similar set of circumstances led to Japan's failed coal hydrogenation program. Its leap from laboratory to industrial-size production without sufficient pilot plant experience produced disastrous results at its two hydrogenation plants in Fushun, Manchuria, and Agochi, Korea. Consequently, even after twenty years of research and development, the Japanese synthetic fuel program produced little liquid fuel and unlike the German program had no effect on the outcome of World War II.

All post-World War II F-T research and development including the US Bureau of Mines, South Africa's Sasol, and West Germany's Rheinbraun used the Ruhrchemie process, but the most significant improvement resulted from adoption of the fluidized-bed catalytic cracking process that Warren K. Lewis (1882–1975) and Edwin Gilland (1909–73) at MIT invented in late 1939 for the catalytic cracking of petroleum. In the fluidized-bed process used in the F-T synthesis plants to convert the synthesis gas reactant to liquid fuel products, the preheated synthesis gas reactant, when blown through a bed of finely divided powdered catalyst in a converter, surrounds the particles of the catalyst, transforming the bed into a mobile fluid that resembles a boiling liquid. The motion significantly increased contact between catalyst and reactant, greatly improving temperature control and the reaction rates obtained with the German converters, the US Bureau of Mines laboratory-developed converters, and all other early F-T converters.

An older German F-T synthesis plant of 10,000 bbl/d capacity required 128 small complicated and costly fixed-bed converters. The converters provided the large steel cooling surface areas that removed the heat of reaction and maintained the catalyst's activity within the narrow temperature range required for synthesis. A similar-sized plant using the Lewis-Gilland fluidized-bed catalytic process had only four converters and 1% of the German fixed-bed converters' cooling surface. Fluidization simplified the plant's engineering design and resulted in a considerable reduction in plant investment cost. It enabled later versions of the F-T synthesis to produce gasoline and diesel oil at a price near that of petroleum-derived fuels.

5 Economics and Politics of Synthetic Fuel Development

Despite the claims made and the optimism expressed throughout its history, the cost of synthesizing petroleum seldom approached the production cost of natural petroleum, sometimes reaching twice the production cost. The US Bureau of Mines' production costs obtained from the operation of its coal hydrogenation and F-T demonstration plants at Louisiana, Missouri, in the early 1950s were the lowest at 19.1¢/gal, comparing very favorably with the higher German and British costs of 20¢-30¢/gal but unfavorably with the 10.6¢/gal cost of natural petroleum at refineries. For this reason in every one of the pre- and post-World War II synthetic fuel programs, with the exception of the four synthetic fuel plants private industry operated in the United States in the late 1940s–early 1950s, federal governments provided support for the research and development through tariffs on imported petroleum, price supports for synthetic fuel, or by entering into joint projects with private industry.

In Germany, Britain, and Japan in the 1930s, tariff increases, subsidies, and government purchases went into effect. In Canada and the United States, government agencies, such as Canada's Mines Branch and the US's Bureau of Mines and DOE, from the 1930s and continuing to the 1990s directed most of the synthetic fuel programs for the duration of their existence or entered into joint agreements with industry, such as the DOE – Gulf Oil and DOE-Exxon projects. The industrial-size Sasol F-T synthesis plants in operation in South Africa are government-subsidized. *Petroleum from Coal* will discuss plant and production costs, tariffs, and licensing agreements for coal hydrogenation and F-T processes.

The social-political environment in which synthetic fuel development occurred sometimes has generated controversy, particularly the situation in Nazi Germany in the 1930s–40s, which ranks as the most controversial and requires clarification. Coal hydrogenation had achieved technological success by 1930 as a result of Pier's invention of poison-resisting tungsten catalysts and the two-stage, liquid phase-gaseous phase, hydrogenation that increased the yield of high-quality products such as gasoline. Officials at IG Farben, at that time the only company producing synthetic fuel, saw coal hydrogenation as a way of reducing Germany's dependence on imported petroleum and already had begun its industrialization with the construction of the Leuna coal hydrogenation plant in 1927. The Nazi government which came to power in January 1933 played no part in synthetic fuel's initial costly development. What the Nazis provided after 1933 was the economic assistance through loans for plant construction and government purchases for the German military that helped the synthetic fuel industry to survive and expand in the 1930s–40s.

A similar situation occurred with the F-T synthesis. Although not as well developed as coal hydrogenation, F-T synthesis also achieved technological success by the mid-1930s, and its industrialization and expansion benefited from the Nazi government's economic assistance. The synthetic fuel plants, like many other German industries, such as Daimler-Benz AG and Volkswagen AG, used forced and slave laborers to relieve worker shortages and reduce production costs which were not controversial issues in wartime Germany, but the Nazi-industry labor policy had nothing to do with the synthetic fuel industry's technological development.

Political controversy also surrounded ICI's government-supported program at Billingham in the 1930s and the US Bureau of Mines' demonstration program in Louisiana, Missouri, in the 1950s. The British concerns were with economics, with how much of an impact one synthetic fuel plant could make, and the safety of the plant from bombing during wartime. In the United States the issue was the government's role in the production of synthetic fuel. Oil companies, such as Standard Oil (New Jersey), and political conservatives in Congress questioned the wisdom of the government's supporting the Bureau's demonstration program. They championed the exclusive role of private industry and waged an extensive campaign to end the Bureau's program which probably would not have survived anyway because of the synthetic fuel's high production cost. Similar political controversy accompanied the later DOE-industry programs that the Carter administration (1976–80) initiated during the 1970s energy crises but ended a few years later during the Reagan administration (1980–88).

The stops and starts that plagued synthetic fuel development over several decades and slowed development of solar, wind, and other alternative energies, resulted mainly because government and industry did not commit long term to their success. In the 1990s–early 2000s synthetic fuel development went through another short-lived start period, propelled mainly by the availability of cheap natural gas. Whereas the previous programs focused on coal hydrogenation the latest programs mainly have adopted the gas-to-liquid process and F-T synthesis.

6 Chapter Titles and Summaries

Chapter 1: *Overview of the Synthetic Fuel Industry*

An overview of *Petroleum from Coal's* contents and makes clear the two themes developed.

Chapter 2: *Friedrich Bergius, Franz Fischer, and the Rise of the German Synthetic Fuel Industry*

What factors contributed to Bergius's invention of coal hydrogenation just before World War I, his technological successes and failures, and his constant struggle to secure funds to continue his research. Financial problems led Bergius to establish a complex network of companies and to enter into licensing agreements with several oil and chemical companies including IG Farben which purchased his patents in May 1925. For his important contributions to the high-pressure hydrogenation of coal, Bergius shared the 1931 Nobel Prize in chemistry with Carl Bosch (1874–1940), IG Farben's principal and in 1935 chairman of the Board of Directors.

The section on Fischer parallels that of Bergius, starting with Fischer and Tropsch's invention of the gas synthesis or F-T synthesis in 1926, their development of suitable cobalt catalysts and the licensing agreement with Ruhrchemie in 1934, and the World War II significance of the F-T synthesis.[5]

5 Ruhrchemie's 1934/35 *Annual Report* gives a summary of the 1934 financial agreement with the Kaiser Wilhelm Institute to license and develop the Fischer-Tropsch process and noted that the process in Ruhrchemie's experimental plant was satisfying enough to erect regular production plants. The report included an expenditure of RM 1,368,215.37 for constructing an experimental F-T plant and for improving and completing Ruhrchemie's nitrogen plant; RM 4,623,518.37 for the annual amortization cost of the F-T experimental plant (a fixed asset) and protective rights for the F-T process paid to the taxation office in 1934; and a total amortization cost of RM 21,385,001.59. See Ruhrchemie Aktiengesellscaft Oberhausen-Holten Geschäftsbericht für das Jahr 1934/35, pp. 58–59 (*Annual Report* 00007, online at Ruhrchemie AG, early documents and newspaper articles on Ruhr chemistry in the 20th century press kit of the ZBW-Leibniz Information Center for Economics). See also Manfred Rasch, *Kohlechemie im Revier. Zur Geschichte der Ruhrchemie AG 1927–1966* (*Coal Chemistry in the Area* (Münster, 2018) and *Geschichte des Kaiser-Wilhelm-Instituts für Kohlenforschung 1913–1943* (*History of the Kaiser Wilhelm Institute* (Weinheim, 1989).

Chapter 3: *Germany's Synthetic Fuel Industry 1927–45*

From 1927 to 1945, IG Farben, Ruhrchemie, and other German companies constructed 21 industrial-size synthetic fuel plants, 12 coal hydrogenation and 9 Fischer-Tropsch plants. After IG Farben obtained Bergius's patents, Pier, director of high-pressure research, introduced two major innovations in 1927 that made coal hydrogenation's industrialization technologically feasible.

Economically, the process remained expensive and non-competitive until 1933 when the Nazi government entered into agreements with IG Farben that guaranteed production costs and a reasonable profit. The twelve coal hydrogenation plants that emerged in the 1930s–40s provided 95% of Germany's aviation fuel and about 50% of Germany's total petroleum requirement. IG Farben also entered into international agreements prior to World War II. The most significant of these, the Standard Oil agreement in 1927 resulted in the construction of two industrial-size plants for the hydrogenation of heavy oil at Baton Rouge, Louisiana, and Bayway, New Jersey.

Ruhrchemie reached a licensing agreement with Fischer in 1934, and within the next five years nine F-T plants began operation. Although not as large as the hydrogenation plants, they contributed another 12–15% to Germany's wartime petroleum production. Allied wartime bombing damaged the synthetic fuel plants, some severely, and after the war the Allies ordered all the plants dismantled or converted to refining petroleum.

Chapter 4: *From Birmingham to Billingham: Synthetic Fuel in Great Britain*

Britain, like Germany, had plenty of coal but no petroleum. Researchers at Birmingham University and the government's Fuel Research Station in East Greenwich therefore investigated Bergius's coal hydrogenation process and followed his progress closely. In 1924, Bergius established a British Bergius Syndicate to handle licensing of his process in the British Empire. ICI in 1927 bought the rights from the syndicate and developed the process commercially at its Billingham plant in 1935. ICI's Billingham plant succeeded technologically and produced high-quality aviation fuel during World War II. But because of its high production cost and ICI's failure to secure a government subsidy, ICI's synthetic petroleum could not compete with natural petroleum and resulted in the Billingham plant's shutdown in 1958. Unlike the American members of the post-World War II investigative teams who visited the German synthetic fuel plants and interrogated the German fuel scientists, Germany's wartime progress failed to impress the British team members, who saw nothing new in the German synthetic fuel plants.

Chapter 5: *Japan's World War II Synthetic Fuel Industry: A Study in Technological Failure*

The German and British coal conversion programs proved that the processes had succeeded technologically but required subsidies for their economic success. In Japan the opposite happened. Like Germany and Britain, Japan had no petroleum. Once the military gained control of the government and began Japan's aggression and Asian expansionism in the early 1930s, the Japanese decided to satisfy some of their petroleum requirements with synthetic fuel. The navy took the lead, but its program failed mainly because it never developed good coal hydrogenation catalysts. By late 1944–early 1945, and the Japanese desperate for petroleum, the Japanese army finally entered into negotiations with IG Farben to purchase licensing rights to IG Farben's coal hydrogenation catalysts and other coal conversion expertise. Despite being wartime allies, neither nation trusted the other in the negotiations which came very late in the war and provided little incentive to the Japanese. The Japanese tried to leap from the laboratory to industrial-size production with insufficient pilot plant experience, and this clearly contributed to their failure.

The Japanese synthetic fuel program also provides an answer to the question of whether a new technology succeeds or fails because of its intrinsic quality or because of social-economic factors. The Japanese government provided every possible social and economic incentive to aid the development of a synthetic fuel industry, but its hasty action resulted in failure. In fact the Japanese program seems to support the often-made claim that Japanese industrialists lacked originality and sometimes succeeded only by copying a technological innovation. But the rushed 1940s wartime conditions drove Japan's hasty actions, thereby invalidating the Japanese lack of originality claim.

Chapter 6: *Canada's Mines Branch and Its Synthetic Fuel Program for Energy Independence*

The Canadian government's Mines Branch assumed the leadership role in developing Canada's synthetic fuel industry beginning in 1930. At first the Mines Branch tried to liquefy coal but gradually concentrated on hydrogenating the bitumen separated from Alberta's huge oil sands deposits. The Canadian program was never as large as those in Germany, Britain, Japan, or the United States, and remained at the laboratory or pilot plant stage of development. The Mines Branch made this choice because it adopted a long-term policy of carrying out basic research on developing Canada's low-grade fossil fuel reserves, particularly the oil sands, and not their industrial development. In the 1970s, the Mines Branch developed a new process called the canmet process that seemed to satisfy both the technological and economic criteria that determine the success or failure of an invention or new technology. Private industry's adoption of the canmet process resulted in commercial development of the oil sands as a source of liquid fuel.

Chapter 7: *Synthetic Fuel Developments in the United States*

Synthetic fuel investigations began in the United States in the 1920s at the US Bureau of Mines' Pittsburgh laboratory. Because of petroleum's abundance, industry and government lacked the incentive of those countries that had no domestic sources of natural petroleum. World War II, however, showed how quickly the United States could consume huge quantities of petroleum, and in 1944 before the war ended, the government launched a demonstration program at Louisiana, Missouri, where the Bureau constructed and operated a coal hydrogenation and a F-T synthesis demonstration plant from 1949 to 1953. The Allies confiscated almost the entire record of Germany's World War II synthetic fuel industry and brought seven German scientists to the United States to join the Bureau's demonstration program.

Technologically the Bureau's program succeeded, but economically synthetic fuel could not compete with natural petroleum. The high production cost and fierce opposition from Congress and some Bureau scientists forced the shutdown of the demonstration plants in 1953.

By the 1960s, the United States had become a net importer of petroleum, a situation that hardly anyone noticed. The real concern over petroleum imports occurred in the winter of October 1973–March 1974 with the Arab oil embargo and again with the Iranian hostage crisis of November 1979–January 1981. Prices jumped from about $2.00/bbl to over $40/bbl and led to a brief revival of interest in synthetic fuel. Once again the US government in cooperative efforts with Exxon, Gulf, and other petroleum companies supported synthetic fuel research. But the 1980–82 recession and the collapse of oil prices in the period 1982/83–2001 ended most of the technologically successful demonstration programs.

Chapter 8: *Epilogue: A Century of Global Synthetic Fuel Developments*

World War II and later synthetic fuel developments include the German, British, and Japanese wartime programs; the three post-World War II Sasol industrial-size F-T synthesis plants in South Africa; the 1970s and 1990s pilot and semicommercial coal hydrogenation, F-T synthesis, and gas-to-liquid plants in the United States; and the 1990s–2000s industrial-size coal hydrogenation, F-T synthesis, and gas-to-liquid plants operating in China, Qatar, Malaysia, and Nigeria.

Most of the European investigations were war-related and ceased once World War II ended in 1945 or shortly afterwards. In Italy, Benito Mussolini's state-supported Azienda Nazional Idrogenazione Combustibili (ANIC) operated two crude petroleum hydrogenation refineries at Bari beginning in 1938 and Livorno beginning in 1939. The three French plants, two coal hydrogenation in 1935 and one Fischer-Tropsch in 1937, were large demonstration or semicommercial plants located in the Pas-de-Calais region of northern France. South Africa's three government-subsidized Sasol F-T synthesis plants, the first of which Sasol constructed in Sasolburg (Sasol I) in 1955, now produces chemicals. The F-T synthesis plants constructed in Secunda in 1980 (Sasol II) and 1982 (Sasol III), and which the South African government considers vital to its energy independence, still produce synthetic fuel. The two Sasol plants provide about 50% of global synthetic fuel production.

In the United States and Europe the October 1973–March 1974 and November 1979–January 1981 oil crises revitalized synthetic fuel research and development and led to joint government-industry coal hydrogenation programs, but the research-development and programs quickly disappeared by the mid-1980s once the crises had passed.

Another synthetic fuel revitalization, mainly F-T and gas-to-liquid synthesis, occurred in the 1990s–early 2000s. Research and development continued in pilot and demonstration synthetic fuel plants in the United States. Outside of the United States, in China, Qatar, Malaysia, and Nigeria, industrial companies such as Shell, Chevron, and Sasol constructed and operate six industrial-size synthetic fuel plants. These six plants and Sasol's two Secunda F-T synthesis plants remain the world's only industrial-size operational synthetic fuel plants in the early 2000s.

7 A Note on Sources

Most of the information on the Fischer-Tropsch and coal hydrogenation processes has come from the Allied investigative teams that went to Germany during World War II's closing months. These teams, such as United States Technical Oil Mission (TOM), a part of the Combined Intelligence Objectives Subcommittee (CIOS); the British Intelligence Objectives Subcommittee (BIOS); and Field Intelligence Agency Technical (FIAT) examined the thousands of technical reports Allied troops captured at the synthetic fuel plants, interrogated (interviewed) many of the German synthetic fuel scientists, and sent their information to CIOS in London for translating and abstracting. CIOS prepared 141 microfilm reels, and after moving its operation to the United States produced another 164 reels. CIOS, BIOS, and FIAT also printed and released more than 2,720 comprehensive reports, unsurpassed in breadth and depth, on the German synthetic fuel plants, many of which are on TOM microfilm reels. The German scientists and engineers, some of whom spoke English, cooperated with the investigative teams, and the breadth and depth of the reports often depended on the plant officials' availability. Alternatively, most of the investigative teams' members could read and write German, which at that time was the physical scientists' second language, and from the published literature had considerable knowledge of Germany's synthetic fuel industry

In addition to the investigative reports, several exhaustive summaries of the reports are available. The most important of these are *Report on the Petroleum and Synthetic Oil Industry of Germany* (London: Ministry of Fuel and Power, 1947) and the Joint Intelligence Objectives Agency (JIOA), *High-Pressure Hydrogenation at Ludwigshafen-Heidelberg*, FIAT, *Final Report 1317* (9 vols., Dayton, Ohio: Central Air Document Office, March 1951). The Ministry's *Report* deals with the Fischer-Tropsch synthesis and the coal hydrogenation process whereas the Joint Intelligence's *High-Pressure* discusses only coal hydrogenation. A third comprehensive source is Henry H. Storch, Norma Golumbic, and Robert B. Anderson, *The Fischer-Tropsch and Related Syntheses* (New York: John Wiley & Sons, 1951). It also relies heavily on the captured German World War II synthetic fuel documents. These are the best and most comprehensive sources, and I have used them extensively. Other important secondary sources cited in this study are Peter Hayes, *Industry and Ideology, IG Farben in the Nazi Era* (New York: Cambridge University Press, 1987); Raymond G. Stokes, *Divide and Prosper: the Heirs of IG Farben under Allied Authority, 1945–1951* (Berkeley: University of California Press, 1988); John Gimbel *Science, Technology, and Reparations: Exploitation and Plunder in Postwar Germany* (Stanford: Stanford University Press, 1992); and Jeffrey A. Johnson, *The Kaiser's Chemists, Science and Modernization in Imperial Germany* (Chapel Hill: University of North Carolina Press, 1990).

In October 1975 after the Arab oil embargo and crisis of October 1973–March 1974 Richard Wainerdi (1931–2021) and Kurt Irgolic established the German Document Retrieval Project at Texas A&M University. They set as its objective the collecting, translating, and organizing of the thousands of German World War II documents and reports that the Allied intelligence teams had brought to the United States and now were scattered around the country in various government repositories, archives, and even with members of the TOM. The German Document Retrieval Project, of which I was a member, accomplished its objective, and as a result Texas A&M University's archives contain what is very likely the most comprehensive collection of information on Germany's

World War II synthetic fuel industry (Synthetic Fuel Collection). I have used this collection in *Petroleum from Coal* and in other papers I have written on the history of synthetic fuel. The citations on the plant descriptions are from CIOS, BIOS, and FIAT reports and the Ministry's and Joint Intelligence committees' reports, with other sources included when required for greater detail or clarification.

The German Document Retrieval Project also deposited in the Texas A&M University Archives the extensive collection of documents and reports on the Japanese synthetic fuel industry that the Naval Technical Mission to Japan assembled while in Japan from 4 September 1945 until 21 February 1946. The Naval team interrogated 14 leading Japanese fuel scientists, compiled the interrogations and related fuel documents, some of which the Japanese scientists translated into English, and wrote comprehensive reports of its investigations. The Naval team inspected Japanese fuel research institutes and fuel plants, and included descriptions and photographs in its reports. The documents and reports, visits to the synthetic fuel and coal conversion facilities and plants, and the Japanese scientists' pre-war publications provided a comprehensive record of Japan's coal-to-oil conversion program.

The Texas A&M Archives (Synthetic Fuel Collection) also contain documents on the British, Canadian, and American synthetic fuel projects. A brief history of the German Document Retrieval Project is online at the Fischer-Tropsch Archive (fischer-tropsch.org).

CHAPTER 2

Friedrich Bergius, Franz Fischer and Hans Tropsch: The Rise of the German Synthetic Fuel Industry

1 Introductory

Germany has few petroleum deposits. Prior to the twentieth century its lack of a liquid fuel was not a serious problem, because Germany possessed abundant coal reserves. Coal provided for commercial and home heating, it also fulfilled the needs of industry and the military, particularly the navy.

In the first decade of the twentieth century, Germany's energy requirements began to change. Two reasons were especially important. First, Germany became increasingly dependent on gasoline and diesel oil engines. The appearance of automobiles, trucks, and then airplanes made a plentiful supply of gasoline absolutely essential; moreover, ocean-going ships increasingly used diesel oil rather than coal as their energy source. Second, Germany's continuing industrialization and urbanization magnified the shortcomings of coal as an energy source. German scientists and engineers began to replace coal with smokeless liquid fuels, which burned more cleanly, handled more conveniently, and had a higher energy content.

Petroleum was clearly the fuel of the future, and to insure that Germany would never be without it, her scientists and engineers created a domestic source of petroleum. They synthesized petroleum from coal, a plentiful natural substance. Of the several processes the Germans used to convert coal into petroleum, high-pressure coal hydrogenation (liquefaction) was the most highly developed. Its history falls into two broad periods: 1910–26, during which time its inventor, Friedrich Bergius, advanced the process to the first stages of industrialization, and 1927–45 the period during which German industrial scientists and engineers carried out its industrial development. This chapter examines the first of these periods. It traces the growth of high-pressure coal hydrogenation from a laboratory process to a small plant operation and shows the difficulties Bergius faced in making the transformation. The three major problems he confronted and solved were how to (1) work at the high pressures required for converting coal into petroleum; (2) provide a sufficient supply of hydrogen gas for the conversion; and (3) regulate the reaction's temperature, because the reaction was exothermic and highly sensitive to temperature change.

 | DOI:10.1163/9789004690912_003

In spite of these successes, two important problems remained. Bergius did not investigate the influence of different catalysts on the reaction, nor did he explore alternatives to his process, in which hydrogenation of coal and splitting of the products occurred in one step. As a result, relatively less product split into gasoline-size molecules. In addition, the gasoline was of low quality, and without refining it could not compete with gasoline obtained from natural petroleum. In 1925–26 Matthias Pier at Badische Anilin- und Soda-Fabrik (BASF) in Leuna, near Merseburg, which merged with five other German chemical companies on 9 December 1925 to form IG Farben AG, remedied the defects of the Bergius process and undertook the industrial development of high-pressure coal hydrogenation in Germany.

The thirteen years Bergius spent investigating the conversion of coal to liquid fuels also exemplify how chemical technology (chemistry and chemical engineering) in the first two decades of the twentieth century underwent a transition from an empirical technology to a science-based chemical technology, dependent on scientific theories, laws, and principles. Theoretical, or physical chemistry, had emerged as a formal branch of chemistry in 1887. Its investigations included electrochemistry, the application of Henri Le Chatelier's principle to the study of chemical equilibria, catalysis, and thermodynamic studies at high pressures. Walther Nernst (1864–1941) at the University of Berlin and Fritz Haber (1868–1934) at the Technische Hochschule in Karlsruhe (1896–1912) and then director of the Kaiser Wilhelm Institute for Physical Chemistry and Electrochemistry in Berlin-Dahlem (1912–33) were two of the new physical chemistry's leaders. Their expertise in high-pressure studies of gaseous reactions caught Bergius's attention and led him in 1910 to study first the hydrogenation of heavy oils and then the hydrogenation of coal.[1]

The hydrogenation of heavy oil, Bergius said, was an *invention to order* because even though the scientific and technical knowledge available at that time made the reaction possible, the invention to a certain extent had its origin in a commercial and economic need. The conditions were entirely different for the hydrogenation of coal. Its invention, he said, resulted from pure scientific investigation on the formation of coal from vegetable matter that led to theories on the structure of coal. Coal hydrogenation was closer to an accidental discovery, and one more likely to have come from an independent investigator (such as Bergius) than from a highly organized industrial research laboratory whose scientists likely would not have the freedom to pursue research on the

1 Anthony N. Stranges, "Friedrich Bergius and the Rise of the German Synthetic Fuel Industry," *Isis*, 75 (1984): 643–67.

unimportant topic of how vegetable matter transformed into coal over millions of years. This is why Bergius believed coal hydrogenation was his most important contribution to chemistry and chemical engineering.[2] The scientific community agreed, and in 1931 Bergius and Carl Bosch, BASF's managing director (1919), IG Farben's principal (1925) and chairman of the Board of Directors (1935), shared the 1931 Nobel Prize in Chemistry for their research on coal hydrogenation.[3]

In 1926, thirteen years after Bergius patented his coal hydrogenation process, the chemists Franz Fischer and Hans Tropsch at the Kaiser Wilhelm Institute for Coal Research (KWI) in Mülheim, Ruhr, invented and developed a second important process that Germany used to synthesize petroleum from coal. Their process, the Fischer-Tropsch (F-T) synthesis, reacted coal with steam to give a mixture of carbon monoxide and hydrogen gases and then converted the gases to petroleum liquids. Scientists later called the two-stage F-T synthesis indirect coal liquefaction (ICL) versus the Bergius direct coal liquefaction (DCL). The most important breakthroughs for the F-T synthesis occurred from 1926 to 1935 with Fischer's and Otto Roelen's success in developing cobalt catalysts and establishing the optimum pressures (1 atm and 10–15 atm) for the synthesis. They made such significant progress in this period that prior to World War II German industrial firms, such as Ruhrchemie and Braunkohle Benzin (Brabag), constructed nine F-T plants to complement the twelve coal hydrogenation plants that operated during the war.

A major misconception in the history of synthetic fuel development is that the Bergius and Fischer-Tropsch processes were competitive. While this statement has some truth, a more accurate assessment makes clear that the two processes were complementary. Both could produce gasoline and diesel

---

2 Bergius included the synthesis of ammonia as an invention to order because he said "the dependence of European countries on supplies of saltpeter from Chile, and the necessity for intensifying their own agricultural industries, rendered necessary the development of a process for the manufacture of artificial manures. The thoughts of the inventor consequently were directed toward the problem and the solution was found. Many of the outstanding achievements of the dye industry belong, however, to the other class of inventions. They owe their origin to the scientific thirst for knowledge of the constitution and reactions of the compounds of organic chemistry." Friedrich Bergius, The Melchett Lecture. "An Inventor's Difficulties in Building Up a Large-Scale Industry" and "The Career of Dr. Bergius," *J. Inst. Fuel*, 8 (December 1934): 68–79, on 68–69.

3 Bergius, Melchett Lecture (note 2); Friedrich Bergius, "Chemical Reactions under High Pressure," Nobel Lecture, 21 May 1932, in *Nobel Lectures: Chemistry 1922–1941* (New York: Elsevier Inc., 1966), pp. 244–76; Carl Bosch, "The Development of the Chemical High Pressure Method During the Establishment of the New Ammonia Industry," Nobel Lecture, 21 May 1932, in *Nobel Lectures: Chemistry 1922–1941* (New York: Elsevier Inc., 1966), pp. 197–241.

fuel, but only coal hydrogenation produced high-quality aviation fuel. The F-T process was more suitable for the production of chemicals.

2 Hydrogenation Begins

Coal hydrogenation is a process that converts different kinds of coal into petroleum by reacting the coal with hydrogen gas at high pressure and high temperature. Distilling the synthetic petroleum yields its major components.

Marcellin Berthelot (1827–1907) at the Collége de France in Paris, succeeded in hydrogenating coal for the first time in 1869. He added finely powdered bituminous coal to a saturated hydroiodic acid (HI) solution and heated the reactants in a sealed glass tube from ten to twenty hours at 275° C. Berthelot combined the coal with twenty times its weight of acid and obtained a 60% conversion to liquid hydrocarbons, mainly saturated compounds such as hexane (C_6H_{14}), dodecane ($C_{12}H_{26}$), and cyclohexane (C_6H_{12}), but with some benzene (C_6H_6). The reaction also produced a solid residue (33%), and gases, chiefly hydrogen (7%). Berthelot's coal hydrogenation experiments were one part of his extensive investigation of hydroiodic acid's action on organic compounds. He successfully hydrogenated wood and partially carbonized charcoal but failed to hydrogenate pure carbon (graphite), coke, and a completely carbonized charcoal.[4]

Berthelot's experiments showed that hydrogenation occurred only if the carbon of the reactants is part of a compound and not present as the free element. This suggested that coal does not contain elemental carbon and revealed at least one aspect of coal's structure. Hydroiodic acid later became a widely used laboratory reagent for hydrogenating and reducing organic compounds, but because of its high cost Berthelot's "méthode universelle" had no potential industrial value. Coal hydrogenation remained a neglected reaction until the first decades of the twentieth century, when chemists began to study the hydrogenation process extensively.

Two factors contributed to the eventual success of the hydrogenation process: the introduction of metallic catalysts and the application of high pressure. Although a few publications on catalytic hydrogenation had appeared earlier in the nineteenth-century chemical literature, the real breakthrough

4 Marcellin Berthelot, "Methode universelle pour réduire et saturer d'hydrogéne les composés organiques," *Bulletin de la Société Chimique de France*, 11 (1869): 278–86. For conversion of pressure units 1 atm = 1.103×10^2 kPa. Berthelot applied no external pressure to the reaction.

came with Paul Sabatier's brilliant researches. Sabatier (1854–1941), Berthelot's assistant at the Collège de France and in 1882 professor of chemistry at the University of Toulouse, laid the foundation for the catalytic hydrogenation of organic compounds in the laboratory and in industry. For fifteen years, beginning in June 1897, Sabatier, his longtime collaborator the Catholic priest Jean Baptiste Senderens (1856–1936) at the Catholic University of Toulouse, and Sabatier's pupils Alphonse Mailhe (1872–1932), and Marcel Murat demonstrated the ability of finely divided metals, such as nickel, platinum, copper, cobalt, and iron, to act as catalysts in the hydrogenation of unsaturated organic compounds. Using glass reaction tubes heated to 150–200° C, they carried out numerous hydrogenations, among them ethylene to ethane, benzene to cyclohexane, phenol to cyclohexanol, and olein to stearin. By the end of 1900 Sabatier had established catalytic hydrogenation as a major new chemical process, and in 1912 he received the Nobel Prize in chemistry for his research with metallic catalysts.[5]

High-pressure hydrogenation began less than a decade after Sabatier initiated his catalytic studies. Working at the Mikhail Artillery Academy in St. Petersburg (Leningrad) from 1903 to 1914, Vladimir Ipatiev (1867–1952) hydrogenated aldehydes, ketones, benzene (C_6H_6), and phenol (C_6H_5OH) at high pressure. He reduced such compounds as isobutyric aldehyde to the alcohol, acetone to isopropyl alcohol, and naphthalene to tetrahydro- and decahydronaphthalene. Ipatiev used nickel or nickel oxides as catalysts, temperatures of 200–250° C, and pressures of 100–120 atm. Reaction times ran from six hours to three days. His hydrogenating chamber, or autoclave, was a soft malleable steel tube open at one end, with an inside diameter of 2.5 centimeters (cm) and a capacity of 25–27 cubic centimeters (cm^3). A steel cover fitted with a heat-treated red copper gasket and clamped to the tube with three retaining bolts enclosed the other end. Two short tubes extended from the cover. One of them led to a pressure gauge, the other allowed gases to enter or exit. He heated the autoclave electrically with nickel resistance wire.

Ipatiev's autoclave, later called the Ipatiev bomb, was a major breakthrough in high-pressure synthesis, for it withstood pressures of 400 atm at 600° C and

5 Paul Sabatier and J. B. Senderens, "Hydrogénations directes réalisées en presence du nickel réduit: Préparation de l'hexahydrobenzène," *Comptes rendus de l'Académie des Sciences*, 132 (1901): 210–12. Sabatier's book *La catalyse en chimie organique* (Paris: Librairie Polytechnique, 1913), trans. E. Emmett Reid, *Catalysis in Organic Chemistry* (New York: D. Van Nostrand, 1922), pp. 125–26, lists numerous references to his publications. See also Paul Sabatier, "The Method of Direct Hydrogenation by Catalysis," Nobel Lecture, 11 December 1912, published in *Nobel Lectures: Chemistry 1901–1921* (New York: Elsevier Inc., 1966), pp. 221–31.

1,300 atm at low temperatures. Prior to its construction, reaction pressures had never exceeded 20–25 atm. Now chemists could use high pressures to investigate reactions that took place very slowly or not at all at atmospheric pressure. This gave further impetus to the search for new and better catalysts,[6] and in 1907–08, five years after Ipatiev's pioneering high-pressure studies, Nernst successfully synthesized ammonia (NH_3) by hydrogenating nitrogen gas at high pressure. Nernst and Fritz Jost (b. 1885) at Berlin reacted the two gases over a platinum catalyst at 5–70 atm and 685° C, obtaining a 0.896% yield.[7]

$$N_2(g) + 3\,H_2(g) \rightarrow 2\,NH_3(g)$$

Nernst originally was investigating the thermodynamic equilibrium of nitrogen's oxides, but he, and other scientists, had another incentive to investigate the synthesis of ammonia. In 1898 William Crookes (1832–1919) in London had warned the British Association for the Advancement of Science that the world faced a serious food shortage unless scientists found a way to fix the atmosphere's nitrogen for fertilizer production. His warning coincided with relevant advances in applied electrical and mechanical engineering and with the advent of the new school of physical chemists who had begun to investigate gas reactions. Crookes' warning led Nernst to the ammonia synthesis, but he was not the first to investigate its synthesis. In 1895, before Crookes' warning, Adolph Frank (1834–1916) and Nikodem Caro (1871–1935) at Dynamit AG in Hamburg invented the cyanamide process for ammonia production. Four years later, in 1899, Frank and Caro, with financial support from German

6 Vladimir Ipatiev, "Catalytic Reductions at High Pressures and Temperatures" (in Russian), *Russkoe Fiziko-Kimischeskoe Obshchestvo*, 33 (1900–1902): 143–54, 496–506, 532–45, 632–43; 34 (1902): 182–95, 315–23, 351–56; 36 (1904): 786–835; 36 (1906–1907): 63–97, 1108–83; 39 (1907): 681–702. German translations of some of Ipatiev's articles appeared in *Ber. deut. chem. Gesell.*, 37 (1904): 2961–3005; 40 (1907): 1827–30; 41 (1908): 991–1007; 42 (1909): 2089–2102; 43 (1910): 3383–93, 3546–53. See also Vladimir Ipatiev, *Catalytic Reactions at High Pressures and Temperatures* (New York: Macmillan Publishers Ltd., 1936), pp. 31–33, 379–81, and Ipatiev, *The Life of a Chemist* (Stanford: Stanford Univ. Press, 1946), pp. 122–25.

7 Walther Nernst, "Über die Berechnung chemischer Gleichgewichte aus thermischen Messungen," *Z. Elektrochem.*, 12 (1906): 738–43; Walther Nernst and Fritz Jost, "Über des Ammoniakgleichgewicht," *Z. Elektrochem.*, 13 (1907): 521–24; Fritz Jost, "Über das Ammoniakgleichgewicht," *Z. anorg. Chem.*, 57 (1908): 414–30. See also Erwin Hiebert, "Walther Nernst," *Dictionary of Scientific Biography* 16 Vols. (New York: Charles Scribner's Sons, 1970–1980). Vol. 15, Supplement 1, pp. 432–53. Alfred von Nagel, *Stickstoff* (Ludwigshafen: Schriftenreihe des Firmenarchivs des Badischen Anilin- und Soda-Fabrik AG, 3, 1969) gives a good discussion of Nernst's work on the ammonia synthesis.

bankers and industrialists, founded Cyanidgesellschaft GmbH in Frankfurt to develop the cyanamide process.

$$CaO(s) + 3\,C(s) \rightarrow CaC_2(s) + C(s)$$
$$CaC_2(s) + N_2(g) \rightarrow CaCN_2(s) + C(s)$$
$$CaCN_2 + 3\,H_2O(l) \rightarrow CaCO_3(s) + 2\,NH_3(g)$$

A second nitrogen-related process emerged in 1903. Kristian Birkeland (1867–1913) at the University of Christiania (Oslo) and Samuel Eyde (1866–1940), general director of Norsk Hydro in Christiania, invented the Birkeland-Eyde electric arc process for the production of nitric acid (HNO_3).

$$N_2(g) + O_2(g) \rightarrow 2\,NO(g)$$
$$2\,NO(g) + O_2(g) \rightarrow 2\,NO_2(g)$$
$$3\,NO_2(g) + H_2O(l) \rightarrow 2\,HNO_3(l) + NO(g)$$

The most significant of the nitrogen-related syntheses followed in 1904. Fritz Haber in Karlsruhe was investigating, like Nernst, the direct synthesis of ammonia from its elements, and in 1904–05, two years before Nernst, he succeeded. Haber, with Gabriel van Oordt, carried out the reaction at 1 atm, 1020° C, iron catalyst, and obtained a 0.012% yield.[8]

Although Nernst's high-pressure synthesis was more successful than Haber's, by 1908 Nernst had all but abandoned ammonia's synthesis; the heat theorem had become his major scientific interest. Haber continued the research that eventually led to ammonia's industrial production. In 1908 Haber and the British chemist Robert Le Rossignol (1884–1976) synthesized ammonia in a quartz tube at 30 atm and 980° C, using a platinum foil catalyst. Their 6% volume yield showed clearly the advantage of higher pressure on the synthesis. By 1909 Haber and Le Rossignol had obtained an 8% volume yield of

---

8 Kristian Birkeland, French Patent 335,692 of 18 September 1903; US Patent 772,862 of 18 October 1904; Samuel Eyde, "Oxidation of Atmospheric Nitrogen and Development of Resulting Industries in Norway," *Industrial and Engineering Chemistry*, 4 (1912): 771–74, and *Chemical and Metallurgical Engineering*, 10 (1912): 617–19. Birkeland-Eyde process required 65,000 kilowatt-hours per ton of nitrogen fixed. Adolph Frank, "Über die direkte Verwertung des stickstoffs der Atmosphäre zur Gewinnung von Düngmitteln und anderen chemischen Produkton," *Z. angew. Chem.*, 19 (1906): 835–40; Nikodem Caro, "Über einheimische Stickstoffquellen," *Z. angew. Chem.*, 19 (1906): 1569–81; Caro, "Industrien des Kalkstickstoffs und verwandter Verfahren," *Z. angew. Chem.*, 22 (1909): 1178–82; Fritz Haber, *Thermodynamik technischer Gasreaktionen* (Munich: Oldenbourg, 1905), p. 187; Fritz Haber and G. Van Oordt, "Über Bildung von Ammoniak aus den Elementen," *Z. anorg. Chem.*, 43 (1905): 111–15; 44 (1905): 341–78; 47 (1905): 42–44.

ammonia, 80 grams per hour (g/hr), and they patented the process.[9] By 1913, employing steel autoclaves similar in design to Ipatiev's at 200 atm, 500° C, and with a variety of metallic catalysts (iron, tungsten, osmium), Haber and Le Rossignol more than doubled the yield.[10]

3 First Industrial Application of High-Pressure Hydrogenation

Haber served as a BASF consultant. He kept the company informed of his progress, and during the four-year period 1909–13 Bosch at BASF successfully transformed Haber's gas-phase ammonia synthesis from a laboratory process to a commercial industry. Bosch's success came after he solved three crucial problems: finding effective, inexpensive catalysts; constructing large, hydrogen resistant converters for the hydrogenation reaction; and producing the two reacting gases, hydrogen and nitrogen, economically. Bosch's colleague, BASF chemist Alwin Mittasch (1869–1953), solved the catalyst problem. For three years, 1909–12, Mittasch and his research assistants tested 2,500 substances in about 6,500 experiments. The experiments included testing metals such as iron, cobalt, and nickel combined with the oxides of aluminum, magnesium, and potassium, or with other metals such as molybdenum, tungsten, and calcium. Iron activated with aluminum oxide or other metallic oxides was the best catalyst.[11]

9 Fritz Haber and Robert Le Rossignol, "Über das Ammoniakgleichgewichte," *Ber. Deut. Chem. Gesell.*, 40 (1907): 2144–54; Haber and Le Rossignol, "Bestimmung des Ammoniakgleichgewichtes unter Druck," *Z. Elecktrochem.*, 14 (1908): 181–96, 513–14; Haber and Le Rossignol, German Patent 235,421 of 13 October 1908; US Patent 971,501 of 27 September 1910. Haber also received German Patents 229,126 of 29 November 1910 (filed 15 June 1909) and 238,450 of 28 September 1911 (filed 14 September 1909) and published "Über die Darstellung des Ammoniaks aus Stickstoff und Wasserstoff," *Z. Elektrochem.*, 16 (1910): 244–46.

10 Fritz Haber, "The Synthesis of Ammonia from its Elements," Nobel lecture, 2 June 1920, in *Nobel Lectures: Chemistry 1901–1921* (New York: Elsevier Inc., 1966), pp. 327–40; Fritz Haber and Robert Le Rossignol, "Über die technische Darstellung von Ammoniak aus den Elementen," *Z. Elektrochem.*, 19 (1913): 53–72, on p. 55.

11 BASF, German Patents 249,447 of 23 July 1912; 254,344 of 2 December 1912; 254,437 of 3 December 1912; Alwin Mittasch, *Geschichte der Ammoniaksynthese* (Weinheim: Verlag Chemie, 1951), pp. 93–116; A. Mittasch and W. Frankenburger, "Zur geschichtlichen Entwicklung und zur Theorie der Ammoniakkatlyse," *Z. Elektrochem.*, 35 (1929): 920–27; L. F. Haber, *The Chemical Industry 1900–1930* (Oxford: Clarendon Press, 1971), pp. 93–95; Eduard Faber, "From Chemistry to Philosophy: the Way of Alwin Mittasch," *Chymia*, 11 (1966): 157–78.

In spring 1910 Bosch began work on the converters for the reacting gases, nitrogen and hydrogen, in a small experimental plant in Ludwigshafen. Haber's original converter measured only 0.75 meters (m) high and 2 to 3 cm wide, and susceptible to corrosion from hydrogen gas. Bosch and his engineering group successfully increased the converter's height to 2 m, then to 4 m, and in 1911 they solved the problem of constructing a converter that resisted hydrogen attack by the introduction of double-tubing. The new converter consisted of an inner tube of low carbon steel or wrought iron enclosed in a tightly fitting outer tube of ordinary steel that contained a large number of small holes. At high pressures some of the hydrogen gas diffused through the inner tube, but it did little damage because the steel contained little carbon with which the hydrogen could react. When hydrogen came in contact with the outer tube, its pressure had fallen to atmospheric pressure, and at this low value hydrogen could not react with and thereby embrittle the carbon steel. It passed harmlessly through the outer tube's many small holes. When BASF's first industrial-size synthetic ammonia plant at nearby Oppau opened on 9 September 1913, eleven months before the outbreak of World War I, it had a hydrogen-resistant converter 8 m high and 45 cm wide, weighing 8.5 metric tons.[12]

An air-liquefaction unit that Carl von Linde (1842–1934) in Munich developed in 1895 supplied the nitrogen gas required for the synthesis. To produce hydrogen gas BASF at first used the Linde-Frank-Caro process that its three inventors had developed by 1911. By 1916 Bosch's group had perfected a new method for hydrogen's production, the catalytic decomposition of water gas, or Bosch-Wild process. In this process steam reacted with water gas (a mixture of hydrogen and carbon monoxide gases) in the presence of an iron (III) oxide (Fe_2O_3) or chromium (III) oxide (Cr_2O_3) catalyst (T = 450° C) to form carbon dioxide and hydrogen.

$$H_2(g) + CO(g) + H_2O(g) \rightarrow CO_2(g) + 2\,H_2(g)$$

Washing the products with water in scrubbing towers removed carbon dioxide. Compressing the gaseous mixture at 200 atm and then dissolving it in solutions first of tetraamine copper (II) formate [$Cu(NH_3)_4(HCOO)_2$] and then

12 BASF, German Patent 254,571 of 6 December 1912; Carl Bosch, US Patent 1,188,530 of 27 June 1916 (filed 5 February 1912); Haber, *The Chemical Industry* (note 11), p. 94; Farber, "From Chemistry to Philosophy" (note 11), p. 165; Walter Witschakowski, *Hochdrucktechnik* (Ludwigshafen: Schriftenreihe des Firmenarchivs des Badischen Anilin- und Soda-Fabrik AG, 12, 1974), pp. 10–15.

of sodium hydroxide (NaOH) removed carbon monoxide, leaving 97–98% pure hydrogen gas.[13]

The following year, in April 1917, BASF completed construction of its second synthetic ammonia plant at Leuna. At this time Bosch introduced a new, improved process for producing the reacting gases. Instead of first separating hydrogen from the water gas mixture and then having it react with nitrogen, Bosch prepared a cheaper form of water gas known as producer gas. This gas consisted of hydrogen, carbon monoxide, carbon dioxide, and nitrogen and resulted from heating incandescent coke with a mixture of air and steam. By adjusting the temperature and volume of air added, employing a catalyst such as iron (III) oxide, and removing the carbon dioxide present, Bosch obtained a nitrogen-hydrogen mixture in the correct proportion for synthesizing ammonia.[14] As a result of BASF's scientific and technological breakthroughs, Oppau and Leuna had a combined daily ammonia capacity of 620 metric tons before World War I ended on 11 November 1918.[15] Germany maintained its leadership in high-pressure technology throughout the 1920s, largely because of Bosch's efforts.

$$2\,C(s) + Air(N_2 + O_2) + H_2O(g) \rightarrow H_2(g) + CO(g) + CO_2(g) + N_2(g)$$

Haber's successful high-pressure ammonia synthesis, industrialized by 1913, marked the start of industrial high-pressure reactions. Soon after Haber's synthesis, Bergius in 1912–13 began to apply high-pressure science and engineering to a new hydrogen addition reaction, the hydrogenation of coal to produce synthetic petroleum.

13 Bosch and Wild, US Patents 1,113,096 and 1,113,097 of 6 October 1914. Wilhelm Wild was a BASF chemist; Badische Co., French Patent 459,918 of 2 July 1913; German Patent 297,258 of 11 September 1914; J. R. Partington and L. H. Parker, *The Nitrogen Industry* (London: Constable and Company Limited, 1922), pp. 160–67; T. P. Hilditch, *Catalytic Processes in Applied Chemistry* (New York: D. Van Nostrand, 1931), pp. 77–83. Another name for the Bosch-Wild Process is the BAMAG Process (Berlin Anhaltischer Maschinenbau Aktien-Gesellschaft), after the firm that built the equipment used in the process.

14 Bosch, "The Procedure of the Synthesis of Ammonia," *Chemical Age*, 29 (1921): 451–56; Bosch, "Wo ein Wille ist, ist auch ein Weg," *The Industrial Chemist*, 10 (1934): 90–94; Nagel, *Stickstoff* (note 7), pp. 41–44; Hilditch, *Catalytic Processes in Applied Chemistry* (note 13) p. 77; Fritz ter Meer, *Die IG Farben* (Düsseldorf: Econ-Verlag GMBH, 1953), p. 26; Gerhart Wolf, *BASF: Development of a World-Wide Enterprise* (Ludwigshafen: Schriftenreihe des Firmenarchivs Badischen Anilin- und Soda-Fabrik AG, 6, 1970), pp. 32–37; 43.

15 Witschakowski, *Hochdrucktechnik* (note 12), pp. 10–12; Partington and Parker, *The Nitrogen Industry* (note 13), p. 154.

4 Friedrich Bergius: Background to High-Pressure Coal Hydrogenation

Bergius, born in 1884 in Goldschmieden, near Breslau, came from an old established family of scientists, theologians, civil servants, army officers, and businessmen. His grandfather had been an economics professor in Breslau and Silesia; his father, Heinrich Bergius (1848–1906), in 1884 owned a small alumina-producing plant, thereby exposing Bergius to the chemical industry at an early age. After Bergius completed his studies at the *Realgymnasium*, his father sent him to Mülheim in the Ruhr district for six months to observe the operation of a large metallurgical plant. In 1903, following his stay in Mülheim, Bergius entered the University of Breslau, where he studied with Richard Abegg (1869–1910). Later he went to Leipzig to study under Arthur Hantzsch (1857–1935), obtaining the PhD in chemistry in 1907 with a dissertation on the solvent properties of absolute sulfuric acid. Bergius then spent a year with Nernst at Berlin and six months with Haber at Karlsruhe.[16]

With Nernst and Haber, Bergius worked on the ammonia synthesis and its equilibrium at high pressures. At this time, 1909, chemists were just beginning to recognize the influence of pressure on reaction rates, in particular how pressure influenced a reaction's equilibrium point according to Le Chatelier's 1884 principle. With Nernst's heat theorem and several experimental determinations of the equilibrium point at different temperatures and pressures, they could calculate an equilibrium mixture's composition over a broad temperature and pressure range.

Describing his eighteen-month association with Nernst and Haber, at that time world leaders in high-pressure theory and application, Bergius said that he tried his hand "in these laboratories at syntheses by high-pressure techniques, with the then imperfect apparatuses, and with little success." Yet the time he spent with Nernst and Haber provided the impetus for his later investigations. By 1909 Bergius already had decided to work "in this new field which

16 The Melchett Lecture. "An Inventor's Difficulties in Building Up a Large-Scale Industry" and "The Career of Dr. Bergius" (note 2), pp. 68–79; Edgar von Schmidt-Pauli, *Friedrich Bergius* (Berlin: E. S. Mittler & Sohn, 1943), pp. 14–30; Friedrich Bergius, "Chemical Reactions under High Pressure," Nobel Lecture, 21 May 1932 (note 3); oral interview with Johannes Bergius (1916–88), Bergius's son, Bonn, Germany, 4 July 1981. Renate Bergius Burgess, PhD (1910–88) also provided biographical information on her father in my correspondence and meetings with her in the 1970s–80s in Bonn and in her London home.

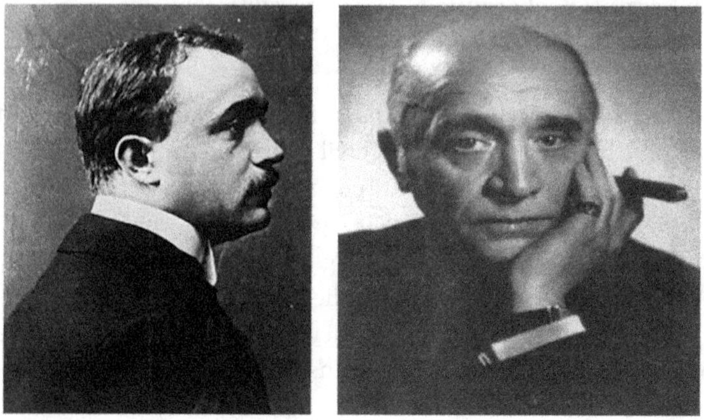

FIGURE 2.1 Friedrich Bergius
SOURCE: COURTESY OF JOHANNES BERGIUS AND RENATE BERGIUS BURGESS

appeared promising… on a wider scale." That year he left Karlsruhe to take a teaching position with Max Bodenstein (1871–1942) at Bodenstein's Institute of Physical Chemistry at the Technische Hochschule in Hanover, where he wrote his *Habilitationsschrift*, published as a monograph in 1913. The theme of Bergius's foreword was faith in new methods and instruments. Here and in other publications Bergius argued that introducing the new experimental methods of physical chemistry and new instruments to the laboratory, in this case, high-pressure apparatus, often led to progress in chemistry.[17]

Bergius's high-pressure research program at Bodenstein's institute began with a detailed study of calcium peroxide's equilibrium. He investigated the compound's dissociation pressure at high temperatures, 300–400° C, calculated its heat of formation using Nernst's heat theorem, and compared this value with his experimentally determined value. Bergius studied the reactions at high temperatures because the molecular velocities increased enough to make accurate measurement possible. At the institute Bergius also began to work on improved designs for high-pressure autoclaves, which he had first

17 Friedrich Bergius, *Die Anwendung hoher Drucke bei chemische Vorgangen und eine Nachbildung des Entstehungsprozesses der Stienkohle* (Halle: Wilhelm Knapp, 1913), p. 1; Friedrich Bergius, "Production of Hydrogen from Water and Coal from Cellulose at High Temperatures and Pressures," *J. Soc. Chemical Industry*, 32 (1913): 462–67, on p. 462. Bergius was a *Privatdozent* teaching physical and industrial chemistry; see Bergius, "Chemical Reactions," p. 244; Schmidt-Pauli, *Friedrich Bergius*, p. 31 (both note 16).

encountered in Haber's Karlsruhe laboratory, constructing apparatus for pressures up to 300 atm.[18]

Because the institute's facilities did not permit expansion of his research program, Bergius left Bodenstein's institute in 1910 and with his own funds established a well-equipped private laboratory in Hanover that gradually grew to include several workshops and plants. With the assistance of Kurt Meyer (1883–1952), Richard Tillman, John Billwiller (1886–1975), Paul Kalnin (b. 1886), Stephan Löffler (1877–1929), Arnold Debo, and Hugo Specht (1885–1964), several of whom such as Tillman and Specht remained lifelong associates and friends, Bergius continued his program of theoretical and applied high-pressure research. Their initial investigations included liquid behavior at 200 atm and temperatures well above the liquid's normal boiling point, at which temperatures the liquids normally reacted as gases; reaction of superheated water with coal to give hydrogen gas and with iron to produce a purer hydrogen gas; reaction of organic chlorides such as chlorobenzene and ethylene chloride with sodium hydroxide solution to give products like phenol or ethylene glycol;[19] and the high-pressure hydrogenation of unsaturated fats and oils.[20]

Three aspects of the early research at the Hanover laboratory contributed to the invention of high-pressure coal hydrogenation in the years 1912–13. First, Bergius gained extensive experience with high-pressure apparatus. He adopted the autoclave design that Haber and Le Rossignol first used in their ammonia synthesis, and by 1910 he was working with pressures of 300 atm and temperatures as high as 450° C. Haber and Le Rossignol had constructed airtight autoclaves by designing the autoclave cover to fit within the autoclave's cone-shaped chamber. They clamped the two parts along their line of contact and forced them together by a screw. Heating produced no gas leakage because chamber and cover, constructed of the same metal, usually steel, expanded equally. Because of the autoclave design's reliability, Bergius remarked that he

18 Bergius, *Anwendung* (note 17), pp. 5–8, 28–29. For descriptions of Haber's high-pressure autoclave, see Haber and Le Rossignol, "Bestimmung des Ammoniakgleichgewichtes" (note 9), pp. 181–96 (diagrams follow p. 184); and Haber and Le Rossignol, "Über die technische Darstellung von Ammoniak" (note 10), pp. 58–59, 61.

19 Friedrich Bergius, German Patent 254,593 of 9 December 1912; French Patent 447,000 of 9 August 1912; Friedrich Bergius, "Untersuchungen über chemische Vorgänge bei hohen Drucken," *Z. angew. Chem.*, 15 (1912): 1171–73.

20 Bergius, "Chemical Reactions" (note 3), pp. 245, 253–54.

could work with high-pressure apparatus almost as conveniently as with ordinary test tubes and flasks.[21]

Bergius's colleague Ludwig Landsberg (1858–1923), director of the Aktiengesellschaft für Petroleumindustrie (AGP) in Nuremberg, provided a second incentive. In 1910 Landsberg encouraged Bergius to study the cracking of heavy petroleum oils and oil residues into gasoline, and AGP provided financial support for two years, from 1912 to 1914.[22] Landsberg noted economic forecasts that the dramatically expanding automobile industry would ensure a considerable increase in global gasoline consumption. Henry Ford (1863–1947), founder of Ford Motor Company in Dearborn, Michigan, in 1903, had begun mass production of automobiles in the United States in 1909, and as a result automobile production rose from slightly more than a hundred thousand in 1905 to about nine million autos in 1920, only ten years after Landsberg's prediction.[23] Petroleum production rose to meet the increased demand for gasoline. The United States, then the world's leading petroleum producer, took forty-one years to extract its first billion barrels of petroleum after its first substantial strike in Titusville, Pennsylvania, in 1859 but only eight years for its next billion barrels. By the seventh billion, extraction time was only one year and seven months. Crude petroleum production increased from slightly less than two hundred million barrels (200×10^6) in 1908 to more than a billion barrels (1×10^9) in 1929.[24]

Once Bergius began to study petroleum cracking, he soon realized that the extreme inefficiency and high cost of producing gasoline resulted from large losses of hydrogen as methane gas and unsaturated hydrocarbons, and

21 Bergius, *Anwendung* (note 17), pp. 4–8, and "Unterschungen über chemische Vorgänge" (note 19); Haber and Le Rossignol, "Bestimmung des Ammoniakgleichgewichtes" (note 9), diagrams follow p. 184; and Haber and Le Rossignol, "Über die technische Darstellung von Ammoniak" (note 10), pp. 58–59, 61.

22 Friedrich Bergius, "An Historical Account of Hydrogenation," *Proceedings of the World Petroleum Congress*, 2 Vols. (London: Office of the Congress, 1934), Vol. 2, pp. 282–89, on p. 282.

23 John Cadman, "British Fuel Problem: Oil," *Chemistry and Industry*, 50 (1931): 908–12, on p. 909. The rise was more dramatic in the next ten years, increasing to 35 million automobiles at the end of 1930. *J. Soc. Chemical Industry* and *Chemistry and Industry* are the same journal. *Chemistry and Industry* used as early as 1928.

24 Arno C. Fieldner and R. L. Brown, "Future Trends in Automotive Fuels," *Industrial and Engineering Chemistry*, 18 (1926): 1009–14, on p. 1009; R. Wigginton, "Notes on Developments in Fuel Technology," *Fuel*, 4 (1925): 95; Howard W. Sheldon, "Twenty-Five Years of Progress in Petroleum Refining," in *Twenty-Five Years of Chemical Engineering Progress*, ed. Sidney K. Kirkpatrick (New York: American Institute of Chemical Engineers, 1933), pp. 52–76.

from coke deposits forming on the autoclave's walls. By May 1913 he showed that as Le Chatelier's principle predicted, he could eliminate these problems, which invariably gave a poor-quality unsaturated gasoline, by adding highly compressed hydrogen gas to replace the hydrogen removed in cracking while carefully controlling the reaction temperature. In effect, Bergius cracked the molecular structure of high-boiling heavy petroleum oils and added highly compressed hydrogen gas to form lower-boiling saturated-gasoline hydrocarbons and other light oils. The reaction also eliminated formation of hydrogen deficient products that caused coking. Because the autoclave's steel walls undoubtedly catalyzed the reaction, Bergius's process closely resembled the catalytic hydrocracking process introduced to the petroleum industry in the 1930s–40s.[25]

The third development that led to Bergius's invention of high-pressure coal hydrogenation occurred during his experiments on the water gas reaction, a well-known process used to produce hydrogen gas.

$$C(s) + H_2O(g) \rightarrow CO(g) + H_2(g)$$

Bergius believed that at high pressure (200 atm) he could run the reaction at a much lower temperature, 300–600° C rather than 900° C. This would shift the equilibrium point to favor the production of carbon dioxide and hydrogen, eliminating entirely the formation of carbon monoxide and the problem of separating it from hydrogen. In a 1911 experiment, Bergius substituted peat, a cellulose-containing plant substance, as the carbon source. When heated under water at 340° C and 100 atm the reaction liberated large amounts of carbon dioxide and left in the autoclave a dark residue that resembled a soft natural coal.[26]

In the decade before Bergius's high-pressure experiments, investigators had often run the water gas reaction at high temperatures in trying to convert

---

25 Bergius, German Patent 301,231 of 9 August 1913; W. H. Hoffert and G. Claxton, *Motor Benzole: Its Production and Use*, 2nd ed. (London: The National Benzole Association, 1938), pp. 94–96.

26 Bergius, "Production of Hydrogen from Water and Coal from Cellulose" (note 17); Bergius, "Die Anwendung von hohen Drucken bei chemischen und chemisch-technischen Vorgängen," *Z. Elektrochem.*, 18 (1912): 660–62; and Bergius, *Anwendung* (note 17), pp. 41–58. Bergius reported his later results on the cellulose to coal process in "Contributions to Knowledge of the Transformation of Cellulose and Lignin into Coal," *Proceedings of the Second International Conference on Bituminous Coal*, 1928, 2 Vols. (Pittsburgh: Carnegie Institute of Technology, 1929), Vol. 2, pp. 48–53.

plant substances such as cellulose and wood into coal. They had not succeeded because the reaction always released a large quantity of heat and left hydrogen deficient coke instead of coal.[27] By working at high pressure, Bergius at last solved the overheating problem. Water remained in the liquid state instead of vaporizing, and not only did it serve as a reactant, but because of its high heat capacity it absorbed and distributed the heat produced, thus preventing overheating. Analysis of the artificial coal showed that it had the formula $C_{21}H_{16}O_2$ and that his artificial coals contained 74.3–85.2% carbon and 4.5–5.2% hydrogen, which compared well with natural bituminous coal of 75–90% carbon and 4.5–5.0% hydrogen. The equation below represents Bergius's "coalification" or *Inkohlung* (end coal) reaction.

$$4\,C_6H_{10}O_5(s) \rightarrow C_{21}H_{16}O_2(s) + 3\,CO_2(g) + 12\,H_2O(l)\ \ \Delta H = -70\ \text{kcal/mole}\ C_6H_{10}O_5$$

Bergius's investigation showed that carbon's percentage in his artificial coal increased with temperature and time of heating but reached a maximum value of about 84%, and even after prolonged heating (about 64 hours) did not increase. A coal of 84% carbon represented the reaction's stable end product. Indeed, Bergius argued that the amount of carbon in natural coals formed from decaying vegetable matter reached 84% and rose above that value only when new external factors entered, namely extremely high temperatures and pressures.[28] Artificial or natural coals containing 84% carbon, according to his estimate, were about eight million years old, a figure in good agreement with the geologists' calculations.[29] Largely as a result of his successful coalification reaction, which converted plant substances into artificial coal, Bergius began further study of coal's chemical constitution.

---

27 George Paul Alexander Petzholdt, *Beitrag zur Kenntnis der Steinkohlenbildung nebst Kritik des Werkes von P.F. Reinsch* (Leipzig: Weigel, 1882); S. Stein, "Beitrag zur Kenntnis der bildung von fossilen Kohlen," *Chemisches Zentralblatt*, 2 (1901): 950–51; Peter Klason, Gustav von Heidenstam, and Evert Norlin, "Untersuchungen zur Holzverkohlung, I: Die trockene Destillation der Cellulose," *Z. angew. Chem.*, 22 (1909): 1205–14, and "II: Die trockene Destillation des Holzes von Keifer, Fichte, Birke and Buche," *Z. angew. Chem.*, 23 (1910): 1252–57.

28 Sources in note 26.

29 Henry Potonié, *Die Enstehung der Steinkohle* (Berlin: Gebrüder Borntraeger, 1910). In this study Bergius found that the time required to reach the same point in the reaction doubled when he reduced the temperature by 10 degrees. His results agreed with the approximate rule of chemical kinetics. Later in a paper published in 1921 Bergius changed the reaction equation to read, $2\,C_6H_{10}O_5 \rightarrow 2\,CO_2 + 5\,H_2O + C_{10}H_{10}O$. Bergius, "Neue Methode zur Verarbeitung von Mineralöl und Kohle," *Z. angew. Chem.*, 34 (1921): 341–47.

4.1 *Early Studies on the Composition of Coal*

In 1911 chemists knew very little about the composition and molecular structure of coal. Many believed that coal consisted of carbon mixed with hydrogen-containing impurities.[30] The two methods of coal analysis used at that time, destructive distillation and coal extraction (reacting the coal with different solvents), showed only that a typical ash-free dry coal contained about 85% carbon, with hydrogen, oxygen, nitrogen, and sulfur the other major constituents. Coal also contains the elements arsenic, mercury, and lead. Inorganic compounds such as aluminum and silicon oxides constituted the ash. Distillation always produced gases: hydrogen the chief component, ammonia, carbon monoxide and dioxide; water, tar, benzene, and other hydrocarbon vapors. Redistilling the tar gave a high-boiling pitch that consisted almost entirely of polynuclear aromatic compounds; a variety of lower-boiling compounds that included ethylene-like hydrocarbons, naphthenes (C_nH_{2n}), phenols, and aromatic compounds (homologues of naphthalene); and a small amount of a solid paraffin whose molecular weight indicated the formula $C_{26}H_{54}$ or $C_{27}H_{56}$.[31]

Richard V. Wheeler (1883–1939) and his associates at the Imperial College of Science and Technology in London carried out some of the earliest and most extensive investigations on coal's structure, beginning around 1910. They found that coal contained two types of compounds that differed in their ease of decomposition: a more stable type that decomposed with greater difficulty to give paraffin hydrocarbons and hydrogen gas, and an unstable type that gave paraffin hydrocarbons and no hydrogen gas on decomposition. The paraffin-yielding compounds were the degradation products of resinous substances originally present in the coal plants, while the hydrogen-yielding compounds

30 Friedrich Bergius, "The Transformation of Coal into Oil by Means of Hydrogenation," *Proceedings of the International Conference on Bituminous Coal*, 1926 (Pittsburgh: Carnegie Institute of Technology, 1927), pp. 102–31, on pp. 105–06.

31 Maurice John Burgess and Richard Vernon Wheeler, "The Volatile Constituents of Coal," *J. Chem. Soc.*, 97 (1910): 1917–35; Horace C. Porter and F. K. Ovitz, "Nature of Volatile Matter of Coal as Evolved Under Different Conditions," *Journal of Gas Lighting*, 104 (1908): 343–48; Horace C. Porter and G. B Taylor, "The Mode of Decomposition of Coal by Heat," *Proceedings of the American Gas Institute*, Part 1, 9 (1914): 234–88; and David Trevor Jones and Richard Vernon Wheeler, "The Composition of Coal," *J. Chem. Soc.*, 105 (1914): 140–51. P. Phillips Bedson at the Durham College of Science invented the pyridine extraction method in 1899; see Bedson, "Notes on the Proximate Constituents of Coal," *J. Soc. Chem. Ind.*, 27 (1908): 147–50. Wheeler and associates added the chloroform extraction; see Maurice John Burgess and Richard Vernon Wheeler, "The Volatile Constituents of Coal, Part II," *J. Chem. Soc.*, 99 (1911): 649–67.

resulted from the plant's cellulose degradation. The studies on French coals that Léo Vignon (1850–1923) at Lyon conducted in 1912 confirmed Wheeler's results, but those of other investigators differed, sometimes significantly, from Wheeler's.[32]

Bergius's researches on artificial and natural coal in 1910–13 led him to conclude that natural coals and lignite contain two kinds of chemical compounds; his findings therefore supported Wheeler's work. The first, the carbonaceous compounds, came from the decayed plant's cellulose and lignite; the second, the bituminous compounds, derived from the albuminous material, resins, waxes, fats, and other substances in the plants. Carbon, hydrogen, and oxygen were present in the carbonaceous compounds, the bituminous compounds contained the same three elements but more hydrogen and less oxygen. Bergius suggested that coal resembled a group of unsaturated hydrocarbons called terpenes with molecular formula $C_{10}H_{16}$ or some higher multiple and had either open-chained or cyclic structures with one or more benzenoid (benzene C_6H_6) groups. Coal, therefore, could add hydrogen and approach in composition the heavy petroleum oils, hydrocarbons with about twenty-three carbon atoms and a molecular weight of 300. It could then separate into medium and light hydrocarbons of fifteen or fewer carbon atoms and molecular weights of less than 200.

Bergius reported his conclusions to the International Congress of Applied Chemistry in New York in 1912.[33] But the benzenoid structure of coal remained

---

32 Burgess and Wheeler, "The Volatile Constituents of Coal, Part II" (note 31), pp. 649–67; Arthur Herbert Clark and Richard Vernon Wheeler, "The Volatile Constituents of Coal, Part III," *J. Chem. Soc.*, 103 (1913): 1704–15; Léo Vignon, "Distillation fractionée de la houille," *C. R. Acad. Sci.*, 155 (1912): 1514–17; V. B. Lewes, *The Carbonisation of Coal* (London: Allan & Co., 1912), pp. 24, 215; and Amé Pictet and Maurice Bouvier, "Sur la distillation de la houille sous pression réduite," *C. R. Acad. Sci.*, 157 (1913): 779–81. Progress on the structure of coal did not really occur until the 1920s. For discussion of later coal research see William A. Bone and Godrey W. Himus, *Coal: Its Constitution and Uses* (London: Longmans, Green & Co., 1936), Ch. 9–11, and numerous papers in *Fuel* and *J. Inst. Fuel*, 1922–32.

33 Bergius mentioned in several publications that he reported the results of his research before the International Congress of Applied Chemistry, New York, September 1912; see Bergius, "The Transformation of Coal into Oil" (note 30), p. 106, and "An Historical Account of Hydrogenation" (note 22), p. 284. I have examined the 29 volumes published by the Congress and found no paper by Bergius (see "Papers Presented," Vol. 29, pp. 19–71). He attended the congress, his name listed under "Membership List, Germany," in Vol. 29, p. 478. A brief summary of his paper entitled "Artificial Production of Coal: A Pretty Application of the Law of Reaction Velocity," appeared in *Chemical and Metallurgical Engineering*, 10 (1912): 616. This issue contains abstracts of many of the papers presented at the Congress. For the results of Bergius's research, therefore, I have relied on his 1926 and 1933 papers and his papers of 1912 and 1913 that deal with the artificial production of coal.

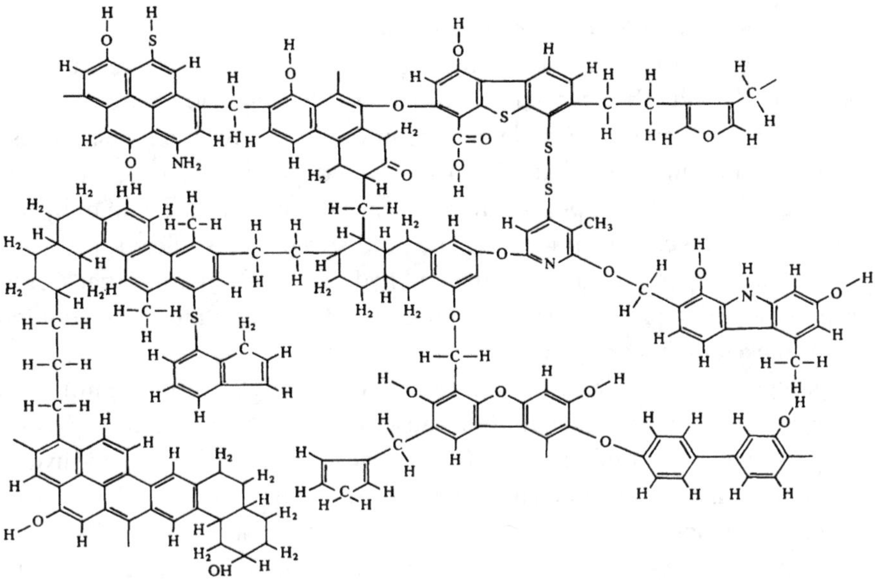

FIGURE 2.2 Schematic representation of coal structure
SOURCE: *Assessment of technology for the liquefaction of coal*, US DEPARTMENT OF ENERGY, OFFICE OF SCIENTIFIC AND TECHNICAL INFORMATION, REPORT 7049743, DECEMBER 1977, P. 14. https://doi.org/10.2172/7049743

unconfirmed until 1925, as a result of coal-extraction and oxidation studies that William Bone (1871–1938) and his research team conducted at Imperial College of Science and Technology. The benzene tri-, tetra-, and other higher carboxylic acids they obtained as oxidation products of lignite, bituminous, and anthracite coals showed that these substances had a preponderance of aromatic structures with three, four, and five fused benzene rings, as well as some structures with a single benzene ring. The simplest benzene-like structures contained eight or ten carbon atoms, while the fused-ring structures consisted of fifteen or twenty carbon atoms.[34]

34 William A. Bone et al., "Researches on the Chemistry of Coal: Part I," *Proceedings of the Royal Society*, Series A, 96 (London: The Royal Society, 1919), pp. 119–36; "Part II," 100 (1922), pp. 582–98; "Part III," 105 (1924), pp. 608–25; "Part IV," 110 (1926), pp. 537–42; "Part V," 120 (1928), pp. 523–45; "Part VI," 127 (1930), pp. 480–510; "Part VII," 147 (1934), pp. 58–67; "Part VIII," 148 (1935), pp. 492–522. Bone and Himus, *Coal: Its Constitution and Uses* (note 32), pp. 191–92, have diagrams of the benzene-like structures found in coal.

I have discussed in detail in this chapter the high-pressure hydrogenation studies of Berthelot, Sabatier, Ipatiev, Nernst, Haber and the coal composition and structure studies of Wheeler and Bone because they represented the most significant theoretical research

4.2 *New Method for Producing Hydrogen Gas*

From the very beginning of his hydrogenation experiments, Bergius, like Haber, Bosch, and others working on hydrogen-addition reactions, recognized the need for an inexpensive and reliable source of pure hydrogen gas. Because the hydrogenation of petroleum involved two independent steps, the destruction or cracking of heavy petroleum hydrocarbons and the addition of hydrogen atoms to the cracked fragments, the process required sufficient hydrogen gas to ensure equal rates of cracking and hydrogen addition. In other words, the process had to provide a sufficient number of hydrogen atoms for addition to the cracked hydrocarbon molecules.

Hydrogen obtained from the electrolysis of water was of high purity and by far the best for hydrogenation because it contained no catalyst poisons such as sulfur compounds or carbon monoxide derivatives. Electrolytic hydrogen found considerable application where the cheap cost of electricity from hydropower made its production practicable, as in the large De-No-Fa hydrogenation factory in Fredrikstad, Norway, and in northern Italian and Swiss factories. But for most of Europe, including Germany, which lacked hydropower, electrolytic hydrogen remained entirely uneconomical. The Linde-Frank-Caro process which separated the water gas reaction's products, hydrogen and carbon monoxide, by liquefying the carbon monoxide, provided an alternative hydrogen source. It yielded 97–99% pure hydrogen gas.[35]

---

done on hydrogenation and coal. The hydrogenation research (1) introduced and demonstrated the significance of high pressures in driving a chemical reaction to completion and therefore to a higher yield of products and (2) enabled chemists to determine the influence of pressure on reaction rates (kinetics). Knowing the composition and structure of coal allowed chemists to propose a mechanism showing how hydrogen molecules at a known temperature and pressure added to the coal molecules converting a solid coal to a liquid oil. They were the studies that Bergius investigated, expanded-advanced, and then applied in his own high-pressure coal hydrogenation research. Franz Fischer also recognized the importance of the earlier studies on hydrogenation and coal, and those of Bergius, in his research on the Fischer-Tropsch synthesis. With the scientific community's confirmation that high-pressure studies had opened a new field of chemical research that had large-scale application, Bergius shifted his research from theoretical-applied to applied, to establishing the best reaction temperatures and pressures for obtaining maximum coal hydrogenation yields and applying those reaction conditions in pilot plant studies.

35 Carl von Linde's description of hydrogen's production by liquefaction is in *Proceedings of the Third International Congress of Refrigeration* (Leiden, Holland: E. Ydo, 1913). Congress held in Washington-Chicago. Linde, *Lowest Temperature in Industry* (Munich: Gësellschaft für Lindes Eismaschinen, AG, 1912); also Linde, French Patent 427,983 of 31 March 1911; US Patents 1,020,102 and 1,020,103 of 12 March 1912; 1,027,862 and 1,027,863 of 28 May 1912.

Bergius recognized that water gas production followed by liquefaction of carbon monoxide, while certainly cheaper than the electrolytic process, had two very serious disadvantages. The water gas reaction produced hydrogen only at a rather high temperature of 900° C, and the liquefaction required expensive equipment to cool and compress the carbon monoxide for separation. Nevertheless, Bergius believed that he could improve the hydrogen production process by eliminating carbon monoxide formation from the water gas reaction. He ran the reaction at a much lower temperature, 300–360° C, but at a high pressure of 200 atm in a steel autoclave with an 80 L capacity, and found that this modification decreased carbon monoxide formation slightly by oxidizing it to carbon dioxide. He then removed the carbon dioxide through absorption in lime. Further, Bergius unexpectedly found that dissolved metallic salts, particularly thallium (III) chloride, catalyzed the reaction sufficiently to give pure hydrogen gas at a maximum daily production of 5 m^3. He patented his catalytic process in June 1911.[36]

A second unexpected result occurred in the course of the water gas investigation. Bergius discovered that water, which remained liquid under the reaction conditions employed, attacked the autoclave's walls, liberating hydrogen gas. The liberation resulted, he said, because liquid water at high temperature and high pressure behaved as an acid and therefore reacted with active metals. Bergius in fact demonstrated that under these conditions (1) finely divided iron, or iron filings, reacted extremely rapidly and liberated hydrogen gas quantitatively (2) he could reduce the iron oxide formed in the reaction with coal or carbon monoxide at 1000° C and re-use it for hydrogen production.

$$3\,Fe(s) + 4\,H_2O(l) \rightarrow Fe_3O_4(s) + 4\,H_2(g)$$
$$Fe_3O_4(s) + 4\,CO(g) \rightarrow 3\,Fe(s) + 4\,CO_2(g)$$

Because his hydrogen process produced a much purer gas without any additional purification cost, eliminated compression cost because the process compressed the liberated hydrogen gas, and used cheaper and more compact apparatus, Bergius believed that his hydrogen process represented a definite improvement over water gas liquefaction. His hydrogen process had a total

36 Bergius, German Patent 259,030 of 24 June 1911. Diagrams of Bergius's apparatus in *Anwendung* (note 17), pp. 6, 39; discussion of his results, *Anwendung*, pp. 33–41.

energy cost only 60% of the Linde-Frank-Caro process; the actual hydrogen cost, about 2¢/cubic meter (2¢/m^3).[37]

By 1913 Bergius had enlarged his hydrogenating operation to the point where it had outgrown his private laboratory and required the facilities of a small factory in a Hanover suburb. There he set up six hydrogen generators, each with about 40 liters (L) capacity that produced hydrogen gas at the rate of 28,000 liters per hour (L/hr). The gas passed directly into steel cylinders for storage at pressures as high as 200 atm. In the Hanover plant as in his laboratory, a thick concrete wall separated the hydrogen generators from the controls regulating temperature, pressure, and hydrogen gas removal. Bergius carried out much of the research on hydrogen gas production for the Chemische Fabrik AG (formerly Moritz Milch and Company) in Posen. By late 1913 he had several joint patents with the Milch company.[38]

4.3 *Early Experiments on Coal Hydrogenation*

During this same period Bergius began the high-pressure hydrogenation of naturally occurring heavy oils. Working with Kalnin and Meyer and with financial support from the AGP, he hydrogenated various gas oils and crude heavy oils, some of which came from oilfields in Galicia and Romania. In these early experiments Bergius, Kalnin, and Meyer used a stationary autoclave and did not stir the reactants. They achieved three important results: when cracked, heavy oils reacted with hydrogen gas; the resulting lighter oils were more saturated after hydrogenation; and no coke formed during the reaction. The success of these experiments in May 1913 led Bergius with AGP to file his first patent on the high-pressure hydrogenation of heavy oils.[39]

Every inventor, of course, believed in the importance of patents, and for Bergius patent protection ranked highest for insuring further development of an invention and especially for securing sufficient financial support in the later development stages. He believed it would take about twenty years to bring his

37 Bergius, "Production of Hydrogen from Water and Coal from Cellulose" (note 17), p. 463; German Patents 254,593 of 9 December 1912 and 277,501 of 30 November 1913; US Patent 1,059,818 of 22 April 1913. Bergius, "Eine neue Methode zur technischen Herstellung reinen, hochkomprimierten Wasserstoffs: Die Zersetzung Flüssigen Wassers durch Eisen," *Z. angew. Chem.*, 26 (1913): 517–18.

38 Bergius, "The Transformation of Coal into Oil" (note 30), pp. 102–31; Bergius, "An Historical Account of Hydrogenation" (note 22), pp. 282–89; Bergius, German Patents 202,831 of 23 July 1913; 286,691 of 21 November 1913; 277,501 of 20 November 1913.

39 Bergius, German Patent 304,348 of 17 September 1919 (filed 6 May 1913); US Patent 1,344,671 of 29 June 1920.

invention from the laboratory to the industrial stage, which required considerable financing, and he feared most that without early patent protection the possibility of parallel patents would jeopardize his invention.

Other significant improvements in the hydrogenation process followed quickly. To ensure a more thorough mixing of the heavy oil and hydrogen gas, Bergius, Kalnin, and Meyer introduced a rotating autoclave. They also discovered that adding metals and metallic oxides, such as iron (III) oxide, enhanced hydrogenation, and they patented their improvements in August 1913. Converting the hydrogenation from a batch to a continuous flow operation marked another major advance: hydrogen gas entered the autoclave at one end and low-molecular-weight hydrogenation products such as gasoline passed continuously out the other end. The conversion required a return to a stationary autoclave but one now equipped with a mechanical stirrer.[40]

By the beginning of 1914 Bergius's high-pressure hydrogenation of heavy oils had progressed sufficiently to permit them to carry out reactions in 40-L autoclaves heated externally to 430° C and 120 atm pressure. They avoided coke formation, a long-time problem of petroleum cracking, by adding and mixing sufficient hydrogen gas with the heavy oil. The hydrogen also removed any sulfur contained in the oil by reacting with it to form hydrogen sulfide gas, although adding iron (III) oxide to the autoclave worked equally well. The process gave about a 50% conversion of crude petroleum to a highly saturated gasoline.[41]

Bergius simultaneously conducted experiments with Specht and Billwiller to test whether they could hydrogenate the unsaturated artificial coal they had produced from peat. He also wanted to learn whether their coal, like the heavy petroleum oils, broke down when hydrogenated into medium- and low-molecular weight hydrocarbons.

The experiments, conducted in the summer of 1913, were a success. Reacting 2–3 g samples with hydrogen gas at about 150 atm pressure and temperatures of 400–430° C, the Bergius team converted 80% of the coal into gaseous, liquid, and solid products soluble in benzene. Encouraged by the successful hydrogenation of artificial coal, Bergius, Specht, and Billwiller proceeded to experiment with natural coal. They obtained practically the same results, showing for the first time that the hydrogenation of natural coal yielded solid,

40 Bergius, German Patent 301,231 of 9 August 1913; British Patents 4,574 of 21 February 1914 and 18,232 of 1 August 1914; French Patent 470,551 of 6 April 1914; German Patent 303,272 of 25 December 1914; Bergius, "An Historical Account of Hydrogenation" (note 22), p. 282.

41 Bergius, German Patent 290,563 of 30 November 1913.

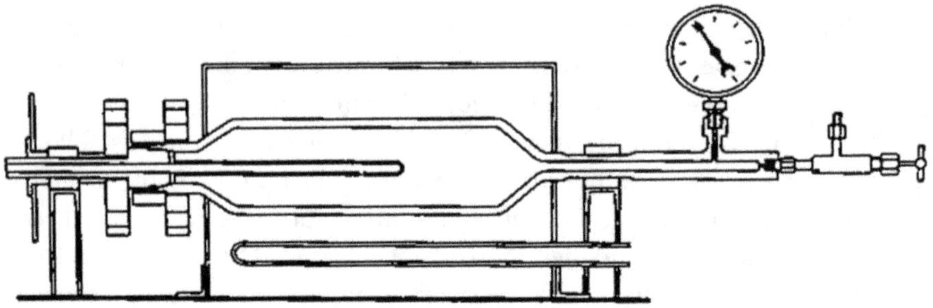

FIGURE 2.3 Five-liter rotary autoclave developed by Bergius and his associates
SOURCE: BERGIUS, "NEUE METHODE ZUR VERARBEITUNG VON MINERALÖL UND KOHLE," *Z. angew. Chem.*, 34 (1921): 341–47

liquid, and gaseous products. In the autumn of 1913 Bergius filed the first patent on the high-pressure hydrogenation of coal.[42]

Franz Fischer, director of the newly established Kaiser Wilhelm Institute for Coal Research in Mülheim, challenged Bergius's patent claim, arguing that the coal and heavy crude oil conversion reaction would not occur. Bergius repeatedly invited Fischer and the other institute members to Hanover for a demonstration, but no one ever came. As a result of Fischer's challenge, the German Patent office asked Bergius to provide a more detailed explanation of the conversion which he did to the Patent Office's satisfaction. But Fischer's challenge and the onset of World War I on 1 August 1914 delayed Bergius's receiving patents until 1919, after the war ended. Fischer remained at odds with Bergius until about 1922 when he acknowledged that high-pressure coal and crude oil hydrogenation occurred as Bergius had claimed.[43]

Bergius carried out the research on coal hydrogenation, like that on heavy oil hydrogenation, in much larger 400 L rotating and stationary steel autoclaves. All experiments were batch operations at 400° C and 200 atm hydrogen pressure. In a typical experiment Bergius mixed 150 kilograms (kg) of powdered coal with an equal weight of a heavy oil solvent to form a paste. He then added about 5 kg of hydrogen gas to the autoclave and after 12 hours stopped

42 Bergius, German Patents 304,348 of 17 September 1919 and 301,231 of 9 August 1913; British Patents 18,232 of 1 August 1914 and 5,021 of 31 March 1915; Bergius, "Chemical Reactions" (note 16). These patents came under Goldschmidt's control once Bergius entered into agreement with the Goldschmidt company.

43 Friedrich Bergius, "The Transformation of Coal into Oil by Means of Hydrogenation," *Proceedings of the International Conference on Bituminous Coal*, 1926 (Pittsburgh: Carnegie Institute of Technology, 1927), pp. 102–31; Franz Fischer, "The Synthesis of Petroleum," *Proceedings International Conference*, pp. 234–6.

the reaction. Inspection of the products showed 85% of the coal had liquefied and dissolved in a benzene solvent and only 15% remained as a dark residue that contained the ash. Bergius also discovered as a result of these experiments that the solvent, by dissolving the products, shortened the reaction time and significantly increased the yield. Distillation separated the hydrogenated products, which consisted of phenol, phenol-like derivatives, and petroleum hydrocarbons, from the solvent.

By 1914, after a large number of trials in his Hanover factory, Bergius had reached several important conclusions about high-pressure coal hydrogenation. Of all the coals tested, the younger brown coals (lignite $C_{39}H_{35}O_{10}NS$) proved most suitable for hydrogenation. They had a carbon content of less than 85% in the ash-free coal, and on liquefying, their hydrogen content rose from about 6% to almost 15%. However, Bergius could not hydrogenate anthracite ($C_{45}H_{29}O_2NS$) or any coals low in hydrogen and whose ash-free carbon content exceeded 85%. The reaction always produced methane gas, and any oxygen, nitrogen, or sulfur present in the coal reacted to form water, ammonia, and hydrogen sulfide respectively. Most of the carbon appeared in the light, medium, and heavy oils, anywhere from 1 to 20% remained in the ash residue.[44]

Bergius's trials also demonstrated that adding small amounts of metallic oxides, especially iron (III) oxide, to the reactants promoted (catalyzed) the hydrogenation of coal.[45] He attributed this effect to the oxides' ability to prevent polymerization of the hydrocarbon fragments produced when heating the coal, but Bergius had neither experts skilled in catalysis nor the funds to investigate catalytic influence on the reaction rate systematically. He attempted to perfect the hydrogenation of coal without catalysts, using high pressure to compensate for uncatalyzed hydrogenation reactions.

One serious difficulty quickly became apparent when Bergius expanded from his small laboratory operation to the bigger factory. When scaled up, the hydrogenation of coal emitted a considerable quantity of heat, producing large amounts of coke rather than liquids. Because the solids and gases present in the autoclave could not distribute the heat of the reaction effectively, the reaction temperature increased so much that the coal rapidly decomposed and then distilled to give coke. Once again it became necessary to introduce a heavy oil solvent, which absorbed and distributed the heat produced in the

44 Bergius, US Patent 1,251,954 of 1 January 1918 (filed 3 August 1914); British Patent 18,232 of 1 August 1914; German Patent 301,231 of 9 August 1913.

45 Bergius, German Patent 290,563 of 30 November 1913.

reaction, thereby preventing any overheating. The high temperature cracked the oily solvent so that it too reacted with hydrogen gas.

Bergius's aimed to transform his small-size coal hydrogenation process into an industrial-size operation. The introduction of a solvent to dissolve the hydrogenation products and to eliminate coking significantly improved hydrogenation. But the key to industrializing coal hydrogenation required its conversion from a batch process to a continuous-flow operation. By 1914 Bergius had succeeded, although it took several years to reach a sufficiently advanced stage of development.[46] Mixing the heavy oil solvent with powdered coal produced a paste that he could pump continuously into the high-pressure reactor while simultaneously removing and collecting the liquid and gaseous products in a condenser.

The hydrogenation experiments at the Hanover factory, which had given Bergius a clear grasp of the temperature and pressure conditions required for reaction and some idea of coal's composition and structure, enabled Bergius to arrive at a mechanism for coal hydrogenation. The experiments showed clearly that the process consisted of two separate reactions: conversion of the coal into heavy pitch-like petroleum hydrocarbons as the result of adding hydrogen atoms at temperatures of 300–400° C and 200 atm pressure, and cracking or splitting of the heavy pitch-like hydrocarbons into lighter liquid hydrocarbons at about 450° C. Coal hydrogenation was both an addition and a cracking reaction.

Before the outbreak of World War I on 1 August 1914, Germany had been exporting coal while importing practically all of its petroleum. In 1913 imports totaled 1.75 million metric tons of petroleum valued at 170 million Gold Marks. World demand for petroleum increased significantly at this time, rising from about 45 million tons in 1910 to 55 million tons in 1913.[47] The mass production of automobiles signaled an even more dramatic increase in petroleum consumption. Germany's only domestic source of petroleum was the destructive

46 Bergius, "Neue Methode zur Verabeitung" (note 29), pp. 341–47. Because of the outbreak of World War I and perhaps because of the delay in his receiving patents on coal hydrogenation, Bergius published nothing on his process until 1921. This is the first article, along with US Patents 1,342,790 of 8 June 1920 (filed 18 April 1916) and 1,391,664 of 27 September 1921 (filed 18 April 1916), in which Bergius discussed the continuous process and temperature control.

47 Heinrich Brückmann, "Coal Liquefaction and Its Importance in World Economy," *The Mining Journal*, 160 (28 January 1928): 69–70; 160 (4 February 1928): 94; 160 (11 February 1928): 116; Carleton Ellis, *Hydrogenation of Organic Substances*, 3rd ed. (New York: D. Van Nostrand, 1930), p. 552. Germany adopted and used the Gold Mark from 1873 to 3 August 1914.

distillation of coal, heating coal in the absence of air to produce coke, gases, and liquids. But the yield of liquid products obtained from distilling a metric ton of coal never exceeded 3%. Bergius thus saw a bright future for industrial high-pressure coal hydrogenation. It would be the way to decrease or perhaps eliminate Germany's dependence on imported petroleum.[48]

4.4 *Moves to Essen and Rheinau: World War I*

At the Bunsen Society's May 1911 meeting in Heidelberg, Bergius read a paper in which he discussed the results of his high-pressure research program in Hanover. Karl Goldschmidt (1857–1926), owner and director of Theodor Goldschmidt AG in Essen, sitting in the audience, listened attentively. Goldschmidt AG since 1847 manufactured inorganic chemicals such as tin, copper, and sodium compounds and especially thermite, an incendiary mixture of iron (III) oxide and powdered aluminum. Karl's brother Hans Goldschmidt (1861–1923) had invented thermite in 1900. When ignited with magnesium ribbon, the thermite-magnesium reaction temperature reached 2,200° C, which made it important in the welding of railroad tracks and in the chlorine de-tinning of scrap tin plate. By 1910, with the thermite ignition and de-tinning processes experiencing global growth, Goldschmidt began exploring ways to expand the company's chemical production. The company in 1911 had grown from a limited partnership to a joint stock company (AG) and had affiliated with and taken over the Gernsheim-Heubruch Works, a sulfate manufacturer in Gernsheim. On 1 January 1913 Goldschmidt acquired the old Rhenania Works in Rheinau, a producer of hydrochloric acid, sulfur, and especially sulfates.[49]

Goldschmidt had looked for several years for an eager, ambitious chemist to direct the company's inorganic chemistry research. Bergius's presentation so impressed him that he quickly visited Bergius's Hanover Laboratory and offered Bergius a generous consultant's contract and the administration of Goldschmidt's large research laboratory undergoing construction in Essen. Bergius seriously deliberated and after a short time accepted Goldschmidt's

48 Franz Fischer and Hans Tropsch in Germany did not develop their process for the production of liquid and gaseous hydrocarbons from coal until after World War I. See Fischer and Tropsch, "Die Erdölsynthese bei Gewöhnlichen Druck aus den Vergasungsprodukten der Kohlen," *Brennstoff-Chemie*, 7 (1926): 97–104; Fischer, "Liquid Fuels from Water Gas," *Industrial and Engineering Chemistry*, 17 (1925): 574–76.

49 History.Evonik.com. Excellent website, giving company history and biographies of scientists associated with the company. Theodor Goldschmidt AG in 2000 became Degussa AG in Düsseldorf, and in 2007 Evonik Industries AG in Essen.

offer. It included an annual salary of 20,000 Gold Marks, a remarkably high salary in 1911; Goldschmidt's purchase of all the apparatus Bergius had in his Hanover laboratory; and the hiring of Bergius's three assistants, John Billwiller, Hugo Specht, and Paul Kalnin. Their hiring caused a problem when World War I began because of their foreign citizenship, Billwiller a Swiss citizen; Specht and Kalnin, Russians and therefore enemies. Goldschmidt personally took responsibility for their behavior, and the Essen police department allowed Bergius's assistants to continue their work in Goldschmidt's factory.[50]

After reaching the agreement with Goldschmidt, Bergius did not immediately move to Essen. He already had planned to move from his private laboratory to a small factory in Hanover, and Goldschmidt had not completed construction of his new Essen research laboratory. By late 1913 Bergius was finding it increasingly difficult to finance his expanding program, and when his current agreement with Goldschmidt ended, he and his three assistants moved to Essen where, on 1 January 1914, he became technical director of the newly constructed research laboratory.

Bergius's arrival required the Goldschmidt company to make a significant shift from the sole production of inorganic compounds to include the production of organic compounds, particularly the synthesis of oil from coal, or coal hydrogenation. Bergius showed considerable skill in convincing the Goldschmidt company to make such a commitment although the outbreak of World War I made his task easier. Karl Goldschmidt, fortunately for Bergius, did not need much convincing because he feared that Germany, which relied almost exclusively on petroleum imports for its liquid fuel requirements, faced the certainty of an enemy naval blockade. He continued to support Bergius's coal hydrogenation experiments for two years after the war ended and discontinued financing Bergius only when the cost reached 6 million Marks (Papiermark ≡ Mark, 1.2 Gold Marks = 50 Marks) which exceeded Goldschmidt's financial capacity. His brother Hans and some of the other company administrators remained opposed to financing Bergius, and their internal dispute led to a split between Karl and Hans. Hans broke completely from the company in 1918 and moved to Berlin where he established his own research laboratory and factory.[51]

50 Goldschmidt took responsibility, Schmidt-Pauli (note 16), p. 63.

51 Bergius "Die Anwendung" (note 26), pp. 660–62; Schmidt-Pauli, *Friedrich Bergius* (note 16), p. 62; Evonik History website (note 49). Papiermark or Mark used in Germany from 4 August 1914 to 15 November 1923 when Rentenmark replaced it. Reichsmark RM replaced Rentenmark on 30 August 1924. Deutsche Mark DM replaced Reichsmark on 20 June 1948.

The start of World War I only months after Bergius moved to Essen, revealed Germany's critical need for petroleum. When hostilities began on 1 August 1914, the German general staff counted on a quick military victory but soon realized that the war would last longer than expected. Commitment to a swift victory was the basic war philosophy of Count Alfried von Schlieffen (1833–1913), chief of the general staff from 1895 to 1906. From the time of Germany's quick victory over France in 1870–71 the general staff had reworked and refined the Schlieffen Plan. In the current war against Britain, France, and Russia the plan entailed launching a tremendous assault on France and retaining only a minimal force to hold off the Russians. With the defeat of France, the German army's full force would easily prevail over Russia. Britain, now alone, would have no choice but to surrender.[52]

The Schlieffen Plan almost worked. But at the First Battle of the Marne 6–12 September 1914, an unexpected French counterattack halted the German army's rapid advance to Paris. German industry up to this time had played no part in the Schlieffen Plan. Now, with all hope for a quick victory banished, the German high command contemplated the grim prospects of fighting a war with limited supplies of such strategic materials as sodium nitrate and petroleum.

Bosch and Mittasch at BASF eliminated the nitrate shortage that occurred after the British sea blockade effectively ended the nitrate supply from Chile. Wilhelm Ostwald (1853–1932) at Leipzig in 1902 had patented a catalytic process (Ostwald process) in which he oxidized ammonia with air. By May 1915 Bosch and Mittasch successfully developed an industrial-size Ostwald process for oxidizing ammonia at the Oppau plant. Their process used a catalyst of iron (III) oxide mixed with oxides of bismuth and cerium. It had a production capacity of 6,000–7,000 metric tons of sodium nitrate per month, an 80–85% efficiency, and converted the large quantities of synthetic ammonia produced by the Haber process to nitric acid and other nitrates essential for fertilizers and explosives.[53]

---

52 Gerhard Ritter, *The Schlieffen Plan: Critique of a Myth* (New York: Praeger Press, 1958); Alfried Graf von Schlieffen, *Gesammlte Werke*, 2 Vols. (Berlin: Ernst Siegfried Mittler and Son, 1913).

53 BASF, German Patents 249,447 of 23 July 1912; 254,344 of 2 December 1912; 254,437 of 3 December 1912; 254,571 of 6 December 1912; Bosch, "The Procedure of the Synthesis of Ammonia" (note 14), pp. 451–56; Bosch, "Wo ein Wille ist" (note 14), pp. 90–94; Bosch, "The Development of the Chemical High Pressure Method," in *Nobel Lectures: Chemistry 1922–1941* (note 3), pp. 197–235; Mittasch, *Geschichte der Ammoniaksynthese* (note 11), pp. 93–116; Mittasch and Frankenburger, "Zur Geschichtlichen Entwicklung" (note 11), pp. 920–27.

$$N_2 + 3\,H_2 \rightarrow 2\,NH_3 \text{ (Haber Process)}$$
$$NH_3 + 2\,O_2 \rightarrow H_2O + HNO_3 \text{ (Ostwald Process)}$$

Bergius's effort to establish an industrial-size coal hydrogenation process for the production of synthetic petroleum was not as successful. AGP agreed to work with Bergius and Goldschmidt on the hydrogenation process, and shortly after World War I began the two companies established the Bergin-gesellschaft, each contributing 220,000 Marks in cash, the Bergius patents on the hydrogenation of oil from AGP which operated a petroleum fractionating plant in Fürstenwalde, and the Bergius patents on the hydrogenation of coal from Goldschmidt. With very little domestic production, the loss of imports and confiscation of the available supplies of benzol and gasoline inevitably following Germany's mobilization for war, Bergius and Goldschmidt saw the need to accelerate production of gasoline and other light hydrocarbons from crude oil and coal a duty and an opportunity. Goldschmidt believed that either new project could be a bigger success than the Haber-Bosch synthetic ammonia process. By September 1914 they had planned construction of a one metric ton per day coal hydrogenation plant in Rheinau near Mannheim on land at the site of the existing Goldschmidt factory.[54]

The War Ministry and especially the army, desperate for liquid fuel, already had directed and supported Franz Fischer at the KWI and Germany's alcohol distillers through an Alcohol Central Board and the Imperial Brandy Office established in April 1916 to provide more liquid fuel, but Fischer and the distillers contributed little liquid fuel. Fischer resorted to the low temperature carbonization (destructive distillation) of coal and coal gasification, which gave gasoline yields of gallons per ton of coal rather than barrels. Nor could the distillation of alcohol obtained from fermentation of beets and other sugar- and starch-rich substances yield sufficient supplies of alcohol for fuel.

The War Ministry also had an interest in Bergius's coal and oil hydrogenation. It realized that Bergius's hydrogenation process was still in its preliminary stages and therefore long term, whereas it had an immediate need. Nevertheless, the War Ministry placed Bergius in charge of the Rheinau-Mannheim plant's construction and arranged for Paul Schäffer (1891–1962), an inexperienced Regierungsbaumeister (Government Architect), to prepare plans for the hydrogenation plant. Schäffer proposed an impractical plant-converter design,

54 Schmidt-Pauli writes in his biography of Bergius that Bergius had to report for service in the German army and received a discharge two days later once the German War Ministry learned of his plan to produce petroleum from coal (note 16), p. 65.

and because Bergius lacked experience constructing industrial-size apparatus the War Ministry arranged for Alois Riedler (1850–1936), a well-known industrial apparatus designer, and Bergius's associate Stephan Löffler now at the Technical University in Berlin to design the high-pressure converters. Friedrich Krupp AG, the Essen arms manufacturer, would construct the converters.[55]

Bergius remained in Essen but frequently traveled to Berlin to confer with Riedler. By 1915 they completed the design for the first high-pressure converter and its order sent to Krupp for construction. The reactor, which had a double-wall design similar to the reactors Bosch used in the ammonia synthesis, measured 8 m in length with an 80 cm inside diameter. The plant's entire operation–feeding, mixing, reaction, and product removal–would be continuous, just as in Bergius's smaller Hanover plant. Riedler estimated quite accurately that the completed plant would cost 1 million Marks, which exceeded Bergius and Goldschmidt's calculation.[56]

In view of the projected high plant construction and operating costs, AGP, which had a capitalization of 1.150 million Gold Marks in 1913, realized it had no funds to contribute to the hydrogenation research, and after several months of negotiations, October 1917–April 1918, Goldschmidt acquired control of AGP. By 31 December 1917 Goldschmidt already had spent 3.25 million Marks on experiments and personnel, and the expanded Goldschmidt AG increased its own capitalization from 15 to 17 million Marks and to 25 million Marks on 12 June 1920. But the problems continued for Bergius and Goldschmidt. Not only did operating costs increase steadily and sometimes unpredictably, but prohibitions on the delivery of plant apparatus and problems with the release from military service of necessary personnel resulted. Additionally, the War Ministry reversed its interest in coal hydrogenation, and looked for short-term oil producing projects to support. Germany's access to Romania's oil fields in late 1916 added to Bergius's and Goldschmidt's financial worries.[57]

Despite Goldschmidt's generous support, by 1916 Bergius needed substantial additional funding to continue the coal hydrogenation research and complete construction of the Rheinau-Mannheim plant. Negotiations that started

55 Schmidt-Pauli (note 16), p. 65; "Economic Development of the Bergius Process," Konsortium für Kohlenchemie, 1 November 1921, Nr. 5, pp. 1–21, on pp. 2–4 (in German), Bergius Papers, BASF Archives, Ludwigshafen, Germany.

56 Harald Beck, *Friedrich Bergius, ein Erfinderschicksal* (Munchen: R. Oldenbourg, 1982), 61 pp. I received a pre-publication manuscript from Beck. See p. 10; "Economic Development of the Bergius Process" (note 55), pp. 2–4.

57 "Early Difficulties of Hydrogenation," *The Iron and Coal Trades Review*, 135 (3 December 1937): 937; Bergius, US Patent 1,391,664 of 27 September 1921; Bergius, "Transformation of Coal into Oil" (note 30), p. 118.

on 18 July 1916 with the Silesian coal industrialist Fritz von Friedländer-Fuld (1858–1917) and ended with his death on 16 July 1917, resumed with the banker Robert Friedländer on 22 December 1917. The negotiations led to the establishment on 10 May 1918 of the Konsortium für Kohlenchemie which in July 1918 came under control of the holding company Erdöl- und Kohleverwertung AG, or EVAG established in Berlin in 1912. The transfer of AGP to the Konsortium für Kohlenchemie in 1918 gave EVAG a capitalization of 30 million Marks. Bergius contributed his expertise and his patents on the hydrogenation of coal and crude oil and the wood conversion-to-sugar patents on which he was also working. The Bergius patents had an estimated value of about 5 million Marks. Goldschmidt contributed 3 million Marks and the Rheinau-Mannheim plant for which he had paid most of the operating expenses up to that time, Friedländer provided 2 million Marks, and the Fürstlich-Henckel Donnersmarksche Güterverwaltung (Administration) added another 25 million Marks for a total of 30 million Marks for plant construction.[58]

EVAG (Konsortium) distributed all profits exceeding 5% accordingly: 25% Donnersmarksche, 45% Goldschmidt, and 30% Friedländer. Under the general management of Bergius its technical director and Heinrich Brückmann its financial director, EVAG's (Konsortium) research expenses totaled 6.5 million Marks by 31 December 1918. As a result of the rapidly escalating inflation that existed in post-World War I Germany, research expenses increased to 18 million Marks including operating expenses for 300 workers that averaged about 50,000 Marks per month, and to 50 million Marks (1.2 million Gold Marks) in the beginning of September 1920 when EVAG transferred plant ownership to the International Bergin Company, a new holding company. Bergius did not solve the major operating problems until long after the war ended, however. In fact, Bergius did not resume active research at Rheinau-Mannheim until 1919, first with crude oil hydrogenation and later with coal. EVAG completed the plant's construction only in 1924.[59]

Bergius faced two major technical problems at Rheinau-Mannheim: how to make the industrial-size hydrogenation process a continuous-flow operation, and how to design heating apparatus that permitted temperature control to within 5 or 10 degrees of the hydrogenation temperature. After many trials Bergius, Löffler, Tillman, and Debo managed to run the plant for 26 hours in

58 "Economic Development of the Bergius Process" (note 55), pp. 4–5; Brückmann, "Coal Liquefaction" (note 47), p. 69. The plant's estimated consumption increased to 1 metric ton of crude oil per day in 1923, 18,000 metric tons per year of solid material in 1923, and 80 metric tons of coal per day in 1925; Schmidt-Pauli (note 16), p. 64.

59 Brückmann, "Coal Liquefaction" (note 47).

August 1919, and in 1921 they finally achieved a continuous flow operation. As in the smaller Hanover plant, they first ground the coal into pieces of about 2 millimeters (mm) in diameter, then mixed it with one-half its weight of a heavy petroleum oil to form a thick paste. They obtained the oil from a previous hydrogenation or from crude petroleum. Specially designed hydraulically operated pistons pushed the paste, after preheating, into the first of two horizontally arranged reactors, each 8 m in length and 1 m in diameter. Finally, they pumped the preheated hydrogen gas into the reactor, where it mixed with the paste.[60]

Pitch formed in the first reactor, and from there the entire mass passed to the second reactor for the addition of more hydrogen gas. An outlet at the bottom of the second reactor permitted removal of the products, which consisted of gas, liquefied coal, untransformed coal, and ash, at the reaction's completion. After the mixture cooled and the pressure reduced to one atmosphere through release of the gaseous products through valves, the gases, mainly hydrogen, separated from the liquid and solid products. Bergius and his assistants operated the coal hydrogenation process at 400–500° C and 150–200 atm. They used a heat engine to recover the heat energy of the compressed gas.[61]

The second technical problem Bergius faced, to design heating apparatus that permitted accurate temperature control, required immediate attention because of coal hydrogenation's sensitivity to temperature change. This made it significantly different from and more difficult than the slightly endothermic crude oil hydrogenation. At very high temperatures coke formed, whereas low temperatures led to a slow reaction rate. The new temperature-control apparatus, in operation at the Rheinau-Mannheim plant by 1921, used the heat-exchange principle, with relatively unreactive nitrogen or carbon dioxide gas as the heat-exchange medium. The apparatus required enclosing each reactor in a steel container or jacket, heating and compressing the heat-exchange gas to the pressure inside the reactor, and then circulating the gas in the space between the reactor and jacket. Equalizing the two pressures prevented the strength of the reactor's wrought-iron or pressed-steel walls from decreasing considerably at the working temperature of 400–500° C.

60 Bergius, "Neue Methode zur Verarbeitung" (note 29), pp. 341–47; Friedrich Bergius, "The Transformation of Coal into Oil by Means of Hydrogenation," *Canadian Chemistry and Metallurgy*, 10 (1926): 275–79; Richard Tillman, German Patent 307,853 of 29 February 1916; and Bergius, US Patents 1,342,790 of 8 June 1920 and 1,391,664 of 27 September 1921.

61 Bergius, US Patent 1,391,664 of 27 September 1921; Bergius, "Transformation of Coal into Oil" (note 30), p. 118.

Applying exterior pressure permitted construction of a reactor with relatively thin walls, because only the outer jacket's walls had to withstand the high pressure. A water column, one end connected by a pipe to the reactor, the other end to the jacket space, indicated any pressure difference. This indirect heating arrangement, according to Bergius, permitted very accurate temperature control and was highly economic, for it recovered all the heat exchanged in the process.[62]

Over a three-year period at Rheinau-Mannheim, 1922–25, Bergius's research group tested crude oil and more than two hundred different kinds of coal. One of the crude oils hydrogenated, a Mexican crude oil, contained only 6% oil that fell in the gasoline boiling point range, boiling below 210° C. Hydrogenating it for one hour in one of the reactors increased its hydrogen content to 42% oil in the gasoline boiling range. In general hydrogenating crude oil gave about 45% gasoline, 25% kerosene, and 27% fuel oil. They hydrogenated crude oil at the rate of one metric ton per day and then treated it in a 50 metric ton/d industrial plant.

Starting on a relatively small scale for the more difficult coal hydrogenation, Bergius's research group eventually hydrogenated quantities as large as 1000 kg (one metric ton). A typical reaction run contained 100 kg of powdered coal mixed with 40 kg of a heavy oil, 5 kg of hydrogen gas, and 5 kg of iron (III) oxide to remove any sulfur present in the coal. The reaction yielded 20 kg of gas, consisting of excess hydrogen and such low molecular-weight hydrocarbons as methane and ethane, and about 128 kg of oil and solids. Distilling the oil produced 20 kg of gasoline with boiling point to 230° C, 10 kg of liquid with a boiling-point range 230–330° C, and 51 kg of heavy oils boiling above 330° C. The heavy oils contained all other constituents present in the coal and amounted to 10% of the coal consumed.[63]

$$\text{Powdered Coal} + \text{Heavy Oil} + H_2(g) \rightarrow \text{Oils-Solids mainly} + \text{Gases}$$
$$(\text{Gases} = H_2 \text{ excess}, CH_4 \text{ methane}, C_2H_6 \text{ ethane})$$
$$\text{Oils (heated)} \rightarrow \text{Gasoline} + \text{Liquids} + \text{Heavy Oils}$$

62 Bergius, US Patent 1,592,772 of 13 July 1926 (filed 30 August 1921); Bergius and Stephan Löffler, British Patents 192,849 and 192,850 of 8 March 1923 (filed 30 November 1921).

63 Bergius, "Transformation of Coal into Oil" (note 30), p. 123; Bergius, "An Historical Account of Hydrogenation" (note 22), p. 287. Bergius, "Neue Methode zur Verarbeitung" (note 29), pp. 341–47; Bergius, "Transformation of Coal" (note 60), pp. 275–79; Richard Tillman, German Patent 307,853 of 29 February 1916; and Bergius, US Patents 1,342,790 of 8 June 1920 and 1,391,664 of 27 September 1921.

FIGURE 2.4 Bergius high-pressure reactor
SOURCE: COURTESY OF JOHANNES AND RENATE BERGIUS BURGESS

Expanding the coal hydrogenation program forced Bergius to seek a larger and more economical supply of hydrogen gas than what he obtained from the reaction of iron with superheated water.

$$Fe(s) + H_2O(l) \rightarrow H_2(g) + FeO(s)$$

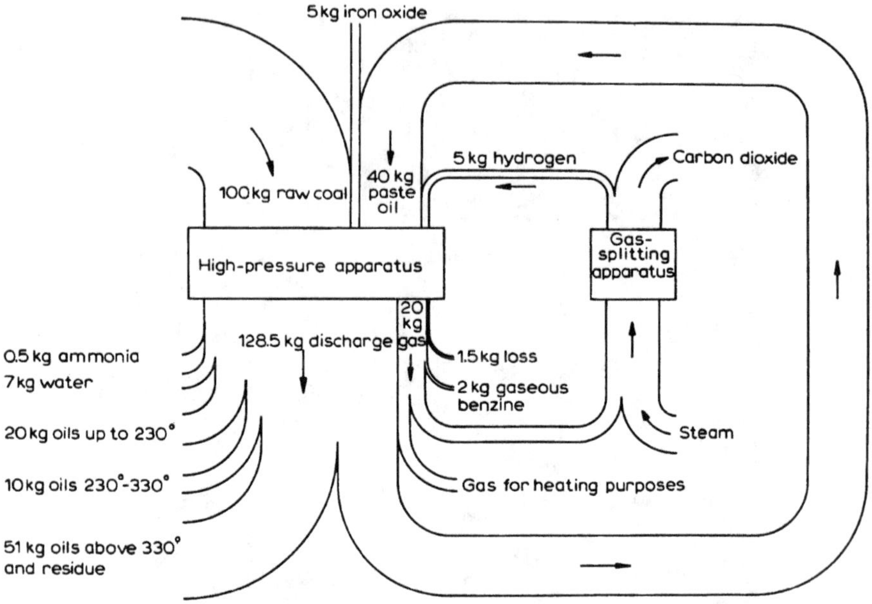

FIGURE 2.5 Flow diagram of continuous method of operation
SOURCE: FRIEDRICH BERGIUS, "THE TRANSFORMATION OF COAL INTO OIL BY MEANS OF HYDROGENATION," *Proceedings of the International Conference on Bituminous Coal*, 1926 (PITTSBURGH: CARNEGIE INSTITUTE OF TECHNOLOGY, 1927), PP. 102–31

Bergius recognized that the economic success of coal hydrogenation depended on a cheap and convenient supply of hydrogen gas. At the Rheinau-Mannheim plant he therefore decided to experiment once more with the well-known water gas reaction, hoping to improve its efficiency and yield by making two modifications: using the methane and ethane gases present in high concentrations in the gaseous products as starting materials and carrying out the reaction in two steps. Bergius first heated methane or ethane and steam at 1100° C to give carbon monoxide and hydrogen. He then treated the gaseous mixture with additional steam at 700° C and a catalyst, so that the reaction yielded carbon dioxide and more hydrogen.

$$\begin{array}{c} CH_4(g) + H_2O(g) \rightarrow CO(g) + 3\,H_2(g) \\ \underline{CO(g) + H_2O(g) \rightarrow CO_2(g) + H_2(g)} \\ CH_4(g) + 2\,H_2O(g) \rightarrow CO_2(g) + 4\,H_2(g) \end{array}$$

With these modifications one volume of methane gas supplied four volumes of hydrogen gas, whereas in the standard water gas reaction the coke reactant

FIGURE 2.6 Rheinau-Mannheim plant
SOURCE: COURTESY OF JOHANNES BERGIUS AND RENATE BERGIUS BURGESS

yielded only one volume of hydrogen. Compressing the two gases and then dissolving the carbon dioxide in water under pressure permitted their easy separation.[64]

At the Rheinau-Mannheim plant, one metric ton of coal produced 490–650 kg of liquids at a cost of 71 Marks. These products (gasoline, diesel oil, and lubricating oil) had a combined value of 141 Marks, giving a profit of 70 Marks/metric ton of hydrogenated coal. As EVAG's financial director Brückmann pointed out in a 1928 article, because Germany imported annually 2.5 million tons of petroleum costing RM 500 million, development of a synthetic petroleum industry would benefit enormously Germany's trade balance. Their projected plant's gasoline production of 100,000 metric tons/yr, based on an average price of RM 6/L of gasoline and RM 5/L of illuminating oil, could provide a total revenue of RM 42 million.[65]

4.5 *International Development of the Bergius Process*

Not only technical obstacles stood in the way of Bergius's promising research on coal hydrogenation. Obtaining sufficient research funds proved such a serious problem in postwar, inflation-plagued Germany that by 1920 EVAG had exhausted its entire capital. Bergius's reputation also appeared to fall within the scientific and professional community who regarded him as a poor young optimist who had invested his and his colleagues' brains, as well as his and his supporters' money, in hopeless and costly experiments. He was a chemical adventurer who ran enormous risks endangering himself, his family, and his collaborators and supporters. To secure the necessary operating funds, Bergius and Brückmann decided to license the foreign rights to the coal hydrogenation process. They established on 27 September 1920 a new company, the Deutsche Bergin Aktiengesellschaft für Kohle und Erdöl-Chemie AG, or Deberg, in the Hague, to control all coal hydrogenation rights outside of Germany. Deberg had a 1 million Mark capitalization.[66]

---

64 Bergius, "Production of Hydrogen from Water and Coal from Cellulose" (note 17), p. 463; Bergius, German Patents 254,593 of 9 December 1912; 277,501 of 30 November 1913; US Patent 1,059,818 of 22 April 1913; Bergius, "Transformation of Coal into Oil" (note 30), pp. 123–24; Bergius "An Historical Account of Hydrogenation" (note 22), p. 289.

65 Brückmann, "Coal Liquefaction" (note 47), p. 116.

66 *Bergius-Verträge*, Summary Survey, pp. 1–8, Bergius Papers, BASF Archives, Ludwigshafen, Germany; "Economic Development of the Bergius Process" (note 55), pp. 1–22. Schmidt-Pauli (note 16), pp. 64–66; *Bergin Protection Abroad*, 12 May 1926, contains a list of the expiration date of all the Bergius patents protected abroad, Bergius Papers, BASF Archives, Ludwigshafen, Germany.

Extremely complicated licensing arrangements resulted from lengthy negotiations between J. Duiker, EVAG's representative in Holland, and Royal Dutch-Shell group's general managing director Sir Hendrik Deterding (1866–1939) and director Hendrikus Colijn (1869–1944). The negotiations led to Royal Dutch-Shell's entering into a one-year option agreement on 28 October 1920 that allowed it to evaluate the Bergius process by placing a Royal Dutch chemist in the Rheinau-Mannheim plant, and required Royal Dutch to share the operational costs involved. The agreement gave Royal Dutch-Shell 50% ownership of Deberg through one of its divisions, Bataafsche Petroleum Maatschappij Company in the Hague. Bataafsche agreed to pay up to 2 million Marks in costs during the first six months of the option period and 200,000 Marks monthly from the seventh through the twelfth month for the delivery of 11,000 metric tons of crude oil and 600 metric tons of gas oil. Bataafsche envisioned the Bergius process as a way to upgrade its Mexican high asphalt content (50%) crude oil sources.

The complicated 1920 licensing arrangement brought two new corporations into existence in 1921: (1) Maatschappij voor Kolen en Olietschniek (Makot), a holding company that Bergius and Brückmann owned outright and (apparently with Deberg) managed Bergius process foreign rights. German rights remained with EVAG and the Rheinau plant which had a 100,000 metric ton/yr hydrogenation license. (2) the International Bergin Company (IBC) in the Hague. Makot contributed the Rheinau plant, the Bergius patents and his expertise; Bataafsche provided the required capital.[67] Bataafsche and Makot divided equally membership on IBC's governing board. Bataafsche had responsibility for all matters relating to hydrogenation of crude oil whereas Makot handled coal hydrogenation. By 1923 Royal Dutch-Shell's general managing director, Deterding, reported that Bataafsche had contributed 9.7 million Marks (450,000 Gulden) toward developing high-pressure coal hydrogenation at Rheinau.[68]

The licensing arrangement led to a new financial issue that same year, 1923, and resulted in the breakup of the Konsortium, established only five years earlier in May 1918. Karl Goldschmidt who owned shares in EVAG, the Konsortium's holding company, accused Friedländer and Brückmann, the Konsortium's financial director, of blackmail, threats, and behavior contrary to

67 "Zür Geschichte des Bergin-Verfahrens," *Erdöl und Teer*, 40 (7 October 1926): 630–32. "Economic Development of the Bergius Process" (note 55), pp. 1–22. The Weimar government reconstituted and issued the new Reichsmark (RM) in 30 August 1924. It set the Reichsmark, RM 1 = $0.23 US.

68 "Zür Geschichte des Bergin-Verfahrens" (note 67); "Economic Development of the Bergius Process" (note 55), pp. 1–22; Schmidt-Pauli (note 16), pp. 64–66.

their arrangement. Goldschmidt specifically denounced the unwillingness of the original members, Friedländer and the Donnersmarksche, to pay 500,000 Gulden to IBC as part of the arrangement establishing IBC and to support further research on the Bergius process. Germany's high rate of inflation aggravated an already bad financial situation for the Konsortium.

Bergius and Brückmann in 1920 had welcomed Bataafsche's investment in IBC, the arrangement pleased them and provided them with considerable revenue. They feared that after Goldschmidt and their other German partners ended their participation, foreign control of the process was inevitable without renewed German investment. Their fears never materialized. Instead, from 1923 to 1925 several Ruhr coal companies, namely the Gesellschaft für Teerverwertung, Harpener Bergbau, Hibernia Bergwerkgesellschaft, and Essener Steinkohlenwerke AG, became the major shareholders in EVAG and contributed RM 2.5 million to the research.

The German military also expressed interest in coal hydrogenation, and in 1925 Lieutenant-General Ludwig Wurtzbacher (1870–1926), head of the Army Weapons Agency (Heereswaffenamt), discussed constructing a coal hydrogenation plant in eastern Germany's lignite fields, in the Leipzig or Braunschweig (Saxony) region. That same year Wurtzbacher and EVAG's directors spoke with government ministers of defense, economics, labor, commerce, transportation, and others about constructing a plant in the Silesian coal area. The plants would provide Germany with a source of liquid fuel, reducing the cost of oil imports, and increase employment in the economically depressed Silesian coal mines.

While all the ministers supported plant construction, the financing never materialized for any of these ventures mainly because of the high costs involved and the German banks' and coal companies', some of whom were EVAG shareholders, reluctance to provide funds for what they considered a high-cost risk venture in troubled times. Rheinische Stahlwerke in 1925 estimated a cost of at least RM 2.5 million, possibly RM 5.0 million in case of unexpected difficulties, for a plant hydrogenating 20,000 metric tons of coal per year. They estimated a profit of RM 30/ton, about one-half that obtained in the gas manufacturing industry. Discussions with governments in Baden, Russia, and Bavaria in 1925 also failed to generate any financial support.

The only other coal hydrogenation operation underway in Germany at this time was the Gesellschaft für Teerverwertung in Duisberg. Adolf Spilker (1863–1954) directed the Gesellschaft, an association of German mining companies whose members included most of the Ruhr's coal and steel firms. Spilker obtained the rights to bituminous coal hydrogenation in 1922, intending to hydrogenate the by-product pitch from the Ruhr coke ovens in an exper-

TABLE 2.1 Bergius process industrialization

1912	Friedrich Bergius and Heinrich Brückmann established Erdöl- und Kohl-Verwertung (EVAG 1912–41) in Berlin
August 1914	Aktiengesellschaft für Petroleumindustrie (AGP) and Theodor Goldschmidt AG established Bergingelleschaft
10 May 1918	Konsortium für Köhlenchemie (KfK) established, capitalized at RM 30 million
July 1918	EVAG acquired Konsortium für Köhlenchemie
1918	AGP transferred to Konsortium für Köhlenchemie
27 September 1920	Deberg established to control coal hydrogenation internationally
	EVAG owned Deberg 100%
28 October 1920	Royal Dutch Shell (Bataafsche) acquired 50% Deberg
1921	Bergius and Brückmann established Makot for international coal hydrogenation
10 September 1921	Makot sold 50% of its stock to Bataafsche
1921	Makot (50%) and Bataafsche (50%) established International Bergin Corporation (IBC).
	Konsortium für Köhlenchemie transferred Rheinau-Mannheim plant ownership to IBC
	Makot and Bataafsche now owned 100% Deberg
	EVAG (Konsortium für Köhlenchemie and AGP) transferred to Makot
1924	British Bergius Syndicate acquired 20% of Makot
26 May 1925	BASF acquired controlling interest in IBC
1925	IBC consisted of Bataafsche 20%
	BASF 30%
	Makot 50%
1925	Makot consisted of BASF 40%
	Bergius 40%
	British Bergius Syndicate 20%
February 1927	ICI bought control of British Bergius Syndicate

SOURCE: COURTESY OF AUTHOR

imental plant having 15,000 metric tons/yr coal capacity. The Gesellschaft became the AG für Steinkohleverflussigung und Steinkohleveredelung (Corporation for Bituminous Coal Liquefaction) in 1927 and the following year began construction on a much larger 30,000 metric ton/yr capacity plant in Meiderich near Duisberg. Bergius had established a high-pressure construction division separate from Deberg, called the Gesellschaft für Hochdrucktechnik, that he integrated with the Berlin Anhaltische Maschinenbau Gesellschaft (Bamag), and it constructed Spilker's plant.

Initial tests were successful, and by 1930 Spilker's plant was producing about 75% light, middle, and heavy oils and about 25% solids that it used to manufacture graphite electrodes. The German Benzol Association, an organization of benzol producers that controlled the benzol market, purchased the light oil for automobile fuel, and the navy the remaining oil for its fuel. The collapse of global oil prices brought on by oil discoveries in East Texas in 1931, the increased flow of oil from the Soviet Union which reduced the price of graphite from 9.5 pfennig/L in 1927 to 4–5 pfennig/L in 1931–33, forced Spilker's Meiderich plant to cease production in 1931.[69]

Bergius's and Brückmann's need for research and development funding continued into the mid-1920s, but beginning in 1925 they reduced considerably their financial dependence on Bataafsche when BASF invested in IBC.[70] In the 1920s BASF had expanded its high-pressure synthesis of ammonia and methanol to include brown coal hydrogenation. The success of its brown coal program led BASF to believe in the readiness of high-pressure coal hydrogenation for industrial development, and on 26 May 1925 BASF acquired for RM 2.85 million the Bergius patents and controlling interest in IBC.[71]

Only one foreign subsidiary resulted from the establishment of IBC and the licensing of Bergius hydrogenation rights in 1920. A group of British investors in London in 1924 organized the British Bergius Syndicate to experiment with the hydrogenation of British coals to produce liquid fuel for Britain's oil-burning navy and if successful to develop the process within the British Empire. The British Syndicate, largely as a consequence of Brückmann's marketing skill, also acquired a 20% interest in Makot. Once BASF and the British Bergius Syndicate became shareholders in IBC, IBC's organization consisted of

69 Note 68. Also at this time the Gesellschaft für Teerwerkung, an association of German mining companies, established Steinkohlen-Bergin, or Kobergin, a much smaller operation than BASF's program, to investigate the Bergius process.

70 Schmidt-Pauli (note 16), p. 70.

71 Company Report of 6 June 1941 to Maria Höring, Höring Papers, BASF Archives, Ludwigshafen. Also on 15 October 1925 Bergius signed a secret contract with BASF that would pay him RM 75,000 annually (in quarters) from 31 December 1925. The contract began on 1 October 1925 and was valid for 10 years. Contract between Dr. Friedrich Bergius, Heidelberg and Badischen Anilin- & Soda-Fabrik in Ludwigshafen a/Rh. Bergius Papers, BASF Archives, Ludwigshafen, Germany.

Bataafsche 20%; BASF 30%; and Makot 50%. Makot's ownership consisted of BASF 40%; Bergius-Brückmann 40%; and the British Bergius Syndicate 20%.[72]

Sir James Calder (1869–1962) at Distillers Company Limited in London and an IBC board member, headed (chairman) the British Bergius Syndicate. The Syndicate comprised fifteen leading British industries, among them Brunner Mond and Company, South Metropolitan Gas Company, United Steel Companies, and International Sugar and Alcohol Company. The Syndicate worked closely with the British government's Department of Scientific and Industrial Research (Greenwich Fuel Research Station), which in 1922 had invited Bergius to present his results to a group of British fuel researchers that included Sir John Cadman (1877–1941), a mining engineer and petroleum technologist with the Anglo-Persian Oil Company in London; and Cecil Howard Lander (1881–1949) then at the Greenwich Fuel Research Station Research.

Bergius's presentation favorably impressed Cadman and Lander, and that same year the Department of Scientific and Industrial Research conducted studies on the hydrogenation of British coals at its Greenwich Fuel Research Station. The success of these experiments resulted in the government's providing financial support to the British Bergius Syndicate in 1926. William R. Ormandy (1870–1941), who beginning in 1924 would work closely with Bergius for ten years and whom Bergius described as a good friend; Lander, and other members of the Greenwich Fuel Research Station actively participated in the research at Bergius's Rheinau-Mannheim plant.

In February 1927 the British government completed construction on a similar plant that processed one metric ton of coal per day.[73] Cadman had an

---

72 "British Interests in Bergius Coal-Oil," *Chemical Age*, 15 (7 August 1926): 150; *Bergius-Verträge* (note 66), p. 2. Bergius and Brückmann owned EVAG. EVAG owned 100% of Deberg and its debt of 18 million Marks. Bergius and Brückmann transferred EVAG and therefore Deberg to Makot, a corporation capitalized at 25,000 Gulden that they also owned outright and had organized for this purpose. Makot then sold for 242 million Gulden, 50% of its stock (500,000 Mark shares) to Bataafsche on 10 September 1921. Makot and Bataasche then established IBC in 1921, with a capitalization of 200,000 Gulden (50% Bataafsche and 50% Makot) and now owned 100% of Deberg. In 1942 IBC consisted of IG Farben 50%, Shell 30%, English investors 10%, Bergius 10%. Conversion rate is below.

1914 rate $ US 1 = 2.46 Gulden
1938 rate $ US 1 = 1.82 Gulden
1940 rate RM 1.5 = 1 Gulden

73 "Early Difficulties of Hydrogenation" (note 57), p. 937; "Report of the Fuel Research Board," *Chemical Age*, 19 (6 October 1928): 318–19. For complete description of the coal hydrogenation research done at Fuel Research Station, see Fuel Research Board, Depart-

interest in Bergius's hydrogenation process because Anglo-Persian saw hydrogenation as a way of upgrading the heavy oil remainders from Persian oil it distilled at its large Swansea Plant. During the early 1920s Anglo-Persian tested several hundred metric tons of remainders and as a result of these tests considered obtaining a license for the Bergius process.[74] Royal Dutch-Shell however, intervened as it could according to its agreement with Makot, and the negotiations with Anglo-Persian ended. Royal Dutch-Shell saw Anglo-Persian's plan for heavy oil hydrogenation in competition with its own plan.

By late 1926 another British company, Imperial Chemical Industries (ICI), considered hydrogenation ready for industrial development, and in February 1927 it bought control of the British Bergius Syndicate. ICI also entered into an agreement with the Department of Scientific and Industrial Research and IBC that called for the Fuel Research Station to do basic research on coal hydrogenation and for ICI to proceed with the industrial development. Consequently, in October 1935 ICI completed construction of a plant at Billingham in northeast England that employed 12,000 workers and produced metric 100,000 tons of gasoline per year. Billingham provided high-quality synthetic gasoline for the British Air Force in World War II, but economic conditions forced ICI to end its coal hydrogenation program in 1958.[75]

4.6 *Bergius's Related Scientific Projects and Later Activities*

In addition to his Nobel Prize winning invention of high-pressure coal hydrogenation, Bergius helped to develop several other industrial processes. During his World War I years with the Goldschmidt Company, Bergius developed a high-temperature hydrolysis of chlorobenzene (C_6H_5Cl) to give phenol (C_6H_5OH) and a process for the conversion of ethylene (C_2H_4) to ethylene glycol (CH_2OHCH_2OH), both of which were potential starting materials for the preparation of explosives. Goldschmidt had a government contract for a glycol production of 120 metric tons/month, but the plant they constructed produced only 60 metric tons/month in 1918. Later, during the 1930s and World War II years, Bergius attempted to industrialize the acidic (hydrochloric acid)

ment of Scientific and Industrial Research, *Annual Reports and Technical Papers* for 1930s. Anglo-Persian Oil Company Ltd. registered on 14 April 1909; renamed Anglo-Iranian Company Ltd. in 1935; renamed British Petroleum Company Ltd in 1954, and current name BP in 1982.

74 Schmidt-Pauli (note 16).

75 William J. Reader, *Imperial Chemical Industries: A History*, 2 Vols. (London: Oxford University Press, 1970 and 1975), Vol. 2, pp. 164, 180–81, 265; on the agreement see also "The Hydrogenation of Coal," *Chemical Age*, 19 (6 October 1928): 310; D. J. Payton-Smith, *Oil* (London: Her Majesty's Stationery Office, 1971), p. 20.

hydrolysis of wood cellulose to the sugar dextrose (glucose, $C_6H_{12}O_6$) and to ethanol (C_2H_5OH), or to a nutrient yeast containing 50% albumin. In his hydrolysis reaction the addition of a single water molecule converted non-edible cellulose to edible products.

Bergius had collaborated earlier, in 1917, with Richard Willstätter (1872–1942), the 1915 Nobel Prize recipient in chemistry, and Erik Hägglund (1887–1959) at the Goldschmidt plant in Essen on the wood-to-sugar hydrolysis. In the 1930s–40s he received financial support from the Nazi Government to expand on the original work and to construct a plant in Rheinau-Mannheim with an annual sugar production of 6,000 metric tons. The wood-to-sugar hydrolysis never reached industrial development because of poor economics but found limited use in producing food supplies for animals during World War II. After World War II Bergius failed to secure suitable employment, and he left Germany for Lenzig in upper Austria where he continued his research on the wood-to-sugar process and began research on a new method of synthesizing penicillin. Both programs had brief lifetimes, and in 1946–47 Bergius went by invitation of the Spanish government to Madrid where he served as a consultant and resumed research on his wood-to-sugar process. The Argentine government invited Bergius to Buenos Aires, and he left Madrid to accept appointment as a scientific adviser. He died in Buenos Aires of complications from kidney problems and diabetes on 30 March 1949.[76]

Bergius came from a family of industrialists and intellectuals and from a democratic tradition that he held to throughout his life. He grew up in an affluent environment and lived a lifestyle that took him to the limits of his financial means and in the opinion of some beyond his means. His spacious home in Heidelberg where he moved in 1921 from Berlin, and which is today part of the University of Heidelberg, during the 1920s and early 1930s was a gathering place for many of Germany's leading intellectuals and political figures. They included the liberal anti-Nazi novelist and 1929 Nobel Prize winner Thomas Mann (1875–1955), 1912 Nobel Prize winner dramatist Gerhart Hauptmann (1862–1946), and Weimar chancellors Heinrich Brünning (1885–1970) and Gustav Stresemann (1876–1929), who shared the 1926 Nobel Peace Prize with French foreign minister Aristide Briand (1862–1932). Bergius joined Stresemann's People's Party, which consisted of right-liberal members of the former National Liberal Party, in the 1920s. For this reason as well as for his democratic beliefs, the Nazi government never trusted Bergius, despite its supporting his wood-to-sugar process, support which Bergius always believed extremely inadequate.

76 See article on Bergius at Wikipedia.

FIGURE 2.7 Bergius home in Heidelberg
SOURCE: EDGAR VON SCHMIDT-PAULI, *Friedrich Bergius* (BERLIN: E. S. MITTLER & SOHN, 1943), P. 65

Nevertheless, Bergius acknowledged that he had to survive in Nazi Germany. He never joined the Nazi Party but possessed a strong nationalism, and in early 1933 and 1935 he gave two addresses that indicated his nationalism and which he probably aimed at securing more financial support for his perpetually financially plagued projects. In an address on 3 April 1933 delivered in Berlin under the auspices of the International Radio Forum and which NBC broadcast in the United States, Bergius discussed the effect of technical progress on world economic conditions. Referring to the Depression's widespread unemployment, Bergius believed that the global nations' policy of self-sufficiency and not the introduction of new machines caused the widespread unemployment. The scarcity of raw materials and finished products that resulted from the dislocation of world commerce during World War I led to a turning away from old economic methods and to national self-sufficiency. This policy, he said, determined each nation's economic policies, and it meant holding in check the rest of the world.[77]

77 "Germany Orderly, Bergius Declares," *The New York Times*, 3 April 1933, p. 9.

In a 13 March 1935 address before the American Chamber of Commerce in Berlin, Bergius advanced a new concept of national political economy that he called potential autarchy. This policy enabled a nation to make itself self-sufficient if it had no access to the world's raw materials either because of financial difficulties, as Germany experienced in 1933, or because of a blockade as Germany endured during World War I when potential autarchy guided Germany's *ersatz*, or new working materials policy. Bergius acknowledged that any newer synthetic products resulting from his potential autarchy policy initially would cost more than natural products, but their costs would fall with technological improvements leading to new production.[78]

Bergius's acceptance of the British Institute of Fuel's Melchett Medal in November 1934 for his research and development of coal hydrogenation appears to be the last time he made a statement on Germany's dire political-social situation. In his address Bergius spoke of Britain's using its total moral, physical, and scientific possibilities to improve its conditions of life and to lead the country to prosperity. Germany, he said, had done the same. Despite "the enormous difficulties surrounding international trade and debt payments" Germany demonstrated "a similar hopeful effort with the aid of all moral and physical force to improve the conditions of the life of the nation and generally to contribute as in the past, to the advancement of civilization throughout the world."[79]

When Bergius concluded his research on high-pressure coal hydrogenation in the mid-1920s he had demonstrated the value of applying scientific principles to a new technology. His use of chemical thermodynamics enabled him to determine the temperature and pressure conditions that gave the greatest yield of liquid products. His study of coal's composition and structure showed the probable arrangement of its atoms, and therefore the chemical bonds between the atoms most likely to break during hydrogenation. Bergius then could propose a possible mechanism for the reaction and begin a search for substances to catalyze it. His research marks the beginning of coal hydrogenation's transition from an empirical technology to a more exact science and provides an excellent example illustrating how chemical technology became increasingly scientific.

Bergius continued a movement that began in the late nineteenth century with the work of J. Willard Gibbs (1839–1903) at Yale, Wilhelm Ostwald (1853–1932) in Leipzig, Jacobus van't Hoff (1852–1911) in Amsterdam,

78 "Reich Warned Self-Sufficiency is Visionary; Synthetic Materials Termed a Makeshift," *The New York Times*, 13 March 1935, p. 10.

79 Melchett Lecture (note 2).

and Svante Arrhenius (1859–1927) in Uppsala. Up to that time, empiricism dominated the sciences, including chemistry. Scientists had only a few scientific principles or theories to guide them in their experimental work. Technology was even more of an empirical experience. When Gibbs published his important phase rule in 1875–76 he provided the theoretical foundation that transformed mineralogy, metallurgy, alloying, steel making, and chemical technologies into more exact sciences. Within a few years, in 1887, Ostwald, van't Hoff, and Arrhenius established the *Zeitschrift für physikalische Chemie*, the first journal devoted to theoretical research in chemistry. Six years later in 1903 Nernst published the first edition of his famous *Theoretische Chemie* and then his heat theorem in 1906. By 1909 Haber had applied thermodynamic principles in synthesizing ammonia from its elements. His investigations led to the industrial production of ammonia and the emergence of one of the first science-based technologies. Other applications of theoretical principles to chemical technology, among them Bergius's coal hydrogenation process, rapidly followed.

5 IG Farben Process and Its Industrialization in Germany

In 1927 Bergius brought his coal hydrogenation program to a successful conclusion. In the Rheinau-Mannheim plant, where he spent millions of Reichmarks in research and employed 150 men, Bergius had demonstrated the industrial potential of coal hydrogenation. But two problems remained. First, Bergius had not studied the influence of different catalysts on the reaction. Second, in his process the hydrogenation and decomposition of coal into petroleum took place in one step. As a consequence he obtained a smaller than expected yield and the gasoline had a low quality. Unless refined it could not compete with gasoline obtained from natural petroleum.[80]

BASF in Leuna solved these problems. In 1925 BASF had reached an agreement with Bergius and Brückmann, to purchase their 40% of Makot, and Bergius withdrew from further experimental work on coal hydrogenation. Bosch, who had successfully developed BASF's industrial ammonia operations during 1909–13, now made the industrialization of coal hydrogenation his personal crusade. Despite considerable opposition from his company's executive

80 Bergius, "An Historical Account of Hydrogenation" (note 22), p. 289; Bergius, "What Impact can an Artificial Production of Oil have on the Crude Oil Business," *Petroleum*, 34 (22 December 1926): 1275–85, lecture of 26 May 1926, Vienna Industry Building.

directors, because of cost, Bosch succeeded.[81] His success ensured Germany's developing Bergius's high-pressure coal hydrogenation rather than a second coal-conversion process, Franz Fischer and Hans Tropsch's indirect synthesis (F-T) of petroleum from carbon monoxide and hydrogen.[82]

On 9 December 1925 BASF merged with five other German chemical companies to form the industrial giant IG Farben. By 1926, Matthias Pier, now IG Farben's director of research, succeeded in separating Bergius's one-stage operation into two stages: hydrogenation of coal to a heavy oil in the first or liquid phase, decomposition of the heavy oil to gasoline-size molecules in the second or vapor phase. Pier also directed the successful search for catalysts that accelerated each phase of the process. He discovered the insensitivity of metallic sulfides to sulfur, one of the worst catalyst poisons and almost always present in coal.[83]

From the first coal-hydrogenation plant constructed in April 1927 at Leuna, which by 1931 had the capacity to produce annually 300,000 metric tons (2.16 million barrels) of synthetic petroleum, the German coal hydrogenation industry grew to twelve large plants in 1944: Böhlen, Scholven, Magdeburg, Welheim, Lützkendorf, Gelsenberg, Wesseling, Leuna, Zeitz, Brüx, Pölitz, and Blechhammer. At their peak in 1944, Germany's twelve plants produced over three million metric tons (21.6 million barrels) of synthetic petroleum. In World War II these plants provided most of the German military's fuel.[84] For

81 Joseph Borkin, *The Crime and Punishment of IG Farben* (New York: The Free Press, 1978), pp. 48–50; Thomas Parke Hughes, "Technological Momentum in History: Hydrogenation in Germany 1898–1923," *Past and Present*, 44 (1969): 106–32.

82 Franz Fischer and Hans Tropsch, "Über die Reduktion des Kohlenoxyds zu Methan am Eisenkontakt unter Druck," *Brennstoff-Chemie*, 4 (1923): 193–97; Fischer and Tropsch, "Über die Herstellung synthetischer Ölgemische (Synthol) durch Aufbau aus Kohlenoxyd und Wasserstoff," *Brennstoff-Chemie*, pp. 276–85; Fischer and Tropsch, "Methanol und Synthol aus Kohlenoxyds als Motorbetreibstoff," *Brennstoff-Chemie*, 6 (1925): 233–34; and Fischer, "Liquid Fuels from Water Gas," pp. 374–76; Fischer and Tropsch, "Die Erdölsynthese bei Gewöhnlichem Druck," pp. 97–104 (both note 48).

83 *High-Pressure Hydrogenation at Ludwigshafen-Heidelberg*, FIAT Final Report No. 1317, 9 Vols. (Dayton, Ohio; published for Joint Intelligence Objectives Agency by Central Air Documents Office, 1951), Vol. 1, pp. 24–26, 32; Matthias Pier, "The Hydrogenation of Bituminous Coal," *Fuel*, 14 (1935): 136–46; Alfred von Nagel, *Methanol Treibstoffe* (Ludwigshafen: Schriftenreihe des Firmenarchivs der Badischen Anilin- und Soda-Fabrik AG, 5, 1970), pp. 41–47; Matthias Pier, US Patent 1,955,829 of 24 April 1934 (filed 27 August 1930); Matthias Pier and Karl Winkler, US Patent 1,946,108 of 6 February 1934 (filed 3 January 1929); E. E. Donath and Maria Höring, "Early Coal Hydrogenation Catalysis," *Ullmann's Enzyklopädie der technischen Chemie*, 3rd ed. (Munich: Urban & Schwarzenberg, 1958), Vol. 10, pp. 483–569, on pp. 486–535.

84 *Report on the Petroleum and Synthetic Oil Industry of Germany*, Ministry of Fuel and Power (London: His Majesty's Stationery Office, 1947), pp. 46–69.

wartime Germany, with only a limited supply of natural petroleum, no alternative to the construction of coal hydrogenation plants existed. In this way Germany utilized its abundant supplies of bituminous and brown coal.[85]

6 Fischer-Tropsch Synthesis

The history of the synthetic fuel industry's fits and starts remains only partially written, with much of the historical interest having focused on Germany's coal hydrogenation process because of its advanced development and significantly greater contribution to Germany's liquid fuel supply than the F-T synthesis. Coal hydrogenation produced high quality aviation and motor gasoline, whereas the F-T synthesis gave high quality diesel and lubricating oil, waxes and some lower quality motor gasoline. The two processes actually were complementary rather than competitive, but because only coal hydrogenation produced high quality gasoline it experienced much greater expansion in the late 1930s and war years than the F-T synthesis, which hardly grew at all. F-T products mainly provided the raw materials for further chemical syntheses with little upgrading of its low-quality gasoline by cracking because of unfavorable economics. Hydrogenation also experienced greater development because brown coal (lignite), the only coal available in many parts of Germany, underwent hydrogenation more readily than a F-T synthesis. In addition, the more mature and better developed hydrogenation process had the support of IG Farben, Germany's chemical leader, which successfully industrialized coal hydrogenation beginning in 1927.[86]

Despite its smaller size and lower production, the nine F-T plants contributed 455,000–576,000 metric tons (3.28–4.15 million barrels) of coal-derived oil per year during the war years, 12–15% of Germany's total liquid fuel requirement. The historical analysis that follows examines the F-T synthesis' invention and industrial development during several decades of German social, political, and economic unrest and complements the historical literature on Germany's coal hydrogenation process. Examination of the two

85 *Report* (note 84), pp. 1–2, 46–48, 50.

86 A third process, the distillation or carbonization of coal at either a high temperature (HTC) of 700–1000° C or a low temperature (LTC) of 500–700° C produces petroleum. The process is not a synthesis but a decomposition and gives small yields of only gallons per metric ton of coal rather than barrels. It is a derived process and was never a major contributor to Germany's liquid fuel requirements. Its simplicity, distilling the petroleum from coal, not its yield resulted in its use.

processes provides a more complete history of Germany's synthetic fuel industry.

6.1 *Early Development of the F-T Synthesis: Catalysts, Reaction Conditions, and Converters*

In 1926, thirteen years after Bergius patented his coal hydrogenation process, Fischer and Tropsch at the Kaiser-Wilhelm Institute invented a second process for the synthesis of liquid fuel from coal. Fischer and Tropsch reacted coal with steam to give a gaseous mixture of carbon monoxide and hydrogen and then converted the gaseous mixture at low pressure P = 1–10 atm and temperature T = 180–200° C to petroleum-like liquids. Fischer and his co-workers in the 1920s–30s developed the cobalt catalysts that made the F-T synthesis a success, and on 27 October 1934 Ruhrchemie AG, a company Ruhr coal industrialists founded in 1927 (AG in 1928), acquired the patent rights to the synthesis.

Fischer received his PhD at Giessen in 1899, where he studied under Karl Elbs (1858–1933). His research focused on the electrochemistry of the lead storage battery. He continued his electrochemical studies spending a semester with Henri Moissan (1852–1907) in Paris, 1901–2 in Freiburg's chemical industry, and 1902–4 at the University of Freiburg's physiochemical institute. Upon leaving Freiburg, Fischer conducted additional research from 1904 to 1911 in the institutes of Wilhelm Ostwald (1853–1932) in Leipzig and Emil Fischer in Berlin and from 1911 to 1914 at the Technische Hochschule in Berlin-Charlottenburg.

Emil Fischer (1852–1919) had an interest in Franz Fischer's electrochemical work, and as a leading figure in establishing the KWIs beginning in 1912, he invited Fischer to direct the new institute for coal research planned for Mülheim in the Ruhr valley. The institute, the first KWI located outside of Berlin-Dahlem, opened on 27 July 1914, and like the others, the Imperial Ministry of Education provided funding for the operating and administration costs whereas private industrial firms paid for the building and equipment. The Ruhr industries, particularly Hugo Stinnes, supported the Mülheim institute.

Fischer had planned to study a coal-to-electricity direct path conversion, but with the institute's opening four days before World War I began and Germany's lack of petroleum quickly becoming apparent, the institute's program shifted from basic research on coal to methods of converting coal to petroleum. This wartime work was the institute's first comprehensive research program. It involved the decomposition of coal and the production of tar from the low-temperature carbonization (LTC) of different coals, giving yields of 1–25%, and the extraction (solution) of a coal with different organic solvents such as alcohols, pyridine, benzene, and petroleum ether at various temperatures and pressures. The extraction studies showed that decreasing the coal's particle

FIGURE 2.8 Franz Fischer (left), Hans Tropsch (right)
SOURCE: MAX-PLANCK-INSTITUT FÜR KOHLENFORSCHUNG, MÜLHEIM AN DER RUHR

FIGURE 2.9 Kaiser Wilhelm Institute for Coal Research, Mülheim
SOURCE: COURTESY OF CHRISTOPH KIEßLING, MAX-PLANCK-INSTITUT FÜR KOHLENFORSCHUNG, MÜLHEIM AN DER RUHR. PHOTOGRAPHER: ISE KÖNIG, 1939

size by grinding increased tar yields. With benzene as the solvent at 270° C and 55 atm, Fischer and W. Gluud in 1916 obtained tar yields many times the low yields obtained at atmospheric pressure. These early studies on coal also led Fischer and Hans Schroder in 1919 to propose their controversial lignin theory of coal's origin that during the peat-bog stage of coal's formation the cellulose material in the original plant material decomposed leaving only the more resistant lignin that then changed into humus coal.

With the wartime coal investigations well underway, Fischer's interest shifted to a different hydrocarbon reaction. In 1913 BASF in Ludwigshafen patented a process for the catalytic hydrogenation (reduction) of carbon monoxide to give hydrocarbons other than methane, alcohols, ketones, and acids. According to the patent, hydrocarbon synthesis occurred best with a 2:1 carbon monoxide: hydrogen volume mixture, at 300–400° C, 120 atm, and the metals cerium, cobalt, or molybdenum, or their alkali-containing (sodium hydroxide) metallic oxides as catalysts. Because of World War I and priority given to industrializing the ammonia and methanol syntheses, BASF never continued its hydrocarbon synthesis.[87] Upon learning of BASF's patent Fischer decided to test its claims. Working with Tropsch he began investigating the catalytic hydrogenation (reduction) of carbon monoxide at various temperatures and pressures but using excess hydrogen gas, a 2:1 hydrogen: carbon monoxide volume mixture they called synthesis gas. This avoided carbon monoxide decomposition which deposited carbon (soot) on the catalyst and rendered it ineffective.

$$2\,CO \rightarrow C + CO_2$$

The experiments with synthesis gas continued into the 1920s, and in 1923 Fischer and Tropsch showed that reacting synthesis gas in a tubular, electrically heated converter at high temperature and pressure, 400–450° C and 100–150 atm, and with alkali-iron instead of metallic oxide catalysts, gave a mixture of oxygen-containing organic compounds, such as higher alcohols, aldehydes, ketones, and fatty acids, that they called synthol. The reaction produced no hydrocarbons.[88] Additional studies in 1925–26 at different reaction conditions, showed that using small hard glass combustion tubes 4–5 mm

87 BASF, German Patent 293,787 of 8 March 1913; BASF, British Patent 20,488 of 10 September 1913; BASF, French Patent 468,427 of 13 February 1914; BASF (Alwin Mittasch and Christian Schneider), US Patent, 1,201,850 of 17 October 1916.

88 Franz Fischer and Hans Tropsch, "Über die Reduktion des kohlenoxyds zu Methan am Eisenkontakt under Druck," *Brennstoff-Chemie*, 4 (1923): 193–97; Fisher and Tropsch, "Über die Herstellung synthetischer Ölgemische (Synthol) durch Aufbau aus Kohlenoxyd und Wasserstoff," *Brennstoff-Chemie*, 4 (1923): 276–85; Fischer and Tropsch, "Methanol und Synthol aus Kohlenoxyds als Motorbetreibstoff," *Brennstoff-Chemie*, 6 (1925): 233–34; Fischer, "Liquid Fuels from Water Gas," *Industrial and Engineering Chemistry*, 179 (1925): 574–76; Fischer and Tropsch, German Patent 484,337 of 22 July 1925; Fischer and Tropsch "Die Erodölsynthese bei gewöhnlichem Druck aus den Vergangsprodukten der Kohlen," *Brennstoff-Chemie*, 7 (1926): 97–104; Fischer and Tropsch, German Patent 524,468 of 2 November 1926; Fischer and Tropsch, "Über Reduktion und Hydrierung des Kohlenoxyds," *Brennstoff-Chemie*, 7 (1926): 299–300; Franz Fischer, "Über die synthese der Petroleum

inner diameter, placed in a 495 mm long, gas-heated horizontal aluminum block furnace, and cobalt-iron catalysts at 250–300° C and 1 atm eliminated completely the oxygenated compounds. The product contained only hydrocarbon gases (ethane, propane, butane) and liquids (octane, nonane, isononene) with a boiling point range of 60–185° C. By 1928 Fischer and Tropsch had modified reaction conditions once again. Using an iron-copper catalyst, a much lower temperature of 190° C, and 1 atm, they eliminated completely the oxygenated compounds and produced a mixture containing only hydrocarbon gases (ethane, propane, butane) and liquids (octane, nonane, isononene) with a boiling point range of 60–185° C.[89]

Fischer continued his investigations into the 1930s, constructing a small pilot plant in Mülheim in 1932. The plant contained a series of converters 5 m high, 1.2 m wide, 12 mm thick walls, immersed in an oil bath for cooling, and operated at the same conditions he had used earlier (2:1 hydrogen: carbon monoxide volume mixture, 190–210° C, 1 atm) but with a catalyst having the weight ratio 100 nickel-25 manganese oxide-10 aluminum oxide-100 kieselguhr. The catalyst, containing previously untested nickel (Z = 28), which differed in atomic number Z from iron (Z = 26) and cobalt (Z = 27) catalysts only by one and two protons, had a short four-six-week lifetime because of

---

Kohlenwasserstoffe," *Brennstoff-Chemie*, 8 (1927): 1–5 and *Berichte*, 60 (1927), 1330–34; Fischer and Tropsch, "Über das Auftreten von Synthol bei der Durchführung der Erdölsynthese under druck und über die Synthese hochmolekular Paraffin Kohlenwasserstoffe aus Wassergas," *Brennstoff Chemie*, 8 (1927): 165–67; BASF, British Patents 227,147, 228,959, 229,714, 229,715 all 28 August 1923; Fischer, "Zwölf Jahre Kohlenforschung," *Z. angew. Chem.*, 40 (1927): 161–65; "Zur Geschichte der Methanolsynthese," *Z. angew. Chem.*, 40 (1927): 166; BASF (Alwin Mittasch, Matthias Pier, and Karl Winkler), German Patent 415,686 of 27 July 1925 (application 24 July 1923); BASF, US Patent 1,558,559 of 27 October 1925; BASF (Mittasch and Pier), US Patent 1,569,775 of 12 January 1926; BASF, French Patents 571,285 of 29 September 1923; 571,354, 571,355, and 571,356 all 1 October 1923; 575,913 of 17 January 1924; 580,914 of 30 April 1924; 580,949 of 1 May 1924; 581,816 of 19 May 1924; 585,169 of 2 September 1924; Fischer and Tropsch, "Über die direketer synthese von Erdöl-Kohlenwasserstoffen bei gewöhnlichem Druck," (Erste Mittelung), *Berichte*, 59 (1926): 830–31 and 832–36; Fischer and Tropsch, "Uber Einige Eigenschaften der aus Kohlenoxyd bei gewohnlichem Druck Hergestellten Synthetischen Erdöl-Kohlenwasserstoffe," *Berichte*, 59 (1926): 923–25. Henry H. Storch, Norma Golumbic, and Robert B. Anderson, *The Fischer-Tropsch and Related Syntheses* (New York: John Wiley & Sons, Inc., 1951), p. 115.

89 Franz Fischer and Hans Tropsch, German Patent 484,337 of 22 July 1925; Fischer and Tropsch publications (note 93); Fischer, "The Synthesis of Petroleum," *Proceedings of the International Conference on Bituminous Coal*, 1926 (Pittsburgh: Carnegie Institute of Technology, 1927), pp. 234–46; Storch, Golumbic, and Anderson, *Fischer-Tropsch* (note 88), pp. 116–17.

FIGURE 2.10 F-T apparatus, aluminum block furnace, at the Kaiser Wilhelm Gesellschaft 1930
SOURCE: F. FISCHER AND H. TROPSCH, "THE SYNTHESIS OF PETROLEUM," *Ges. Abhandl. Kenntnis Kohle*, 10 (1932): 313–501; REPRINTED IN HENRY H. STORCH, NORMA GOLUMBIC, AND ROBERT B. ANDERSON, *The Fischer-Tropsch and Related Syntheses* (NEW YORK: JOHN WILEY & SONS, INC., 1951), P. 116

sulfur poisoning. For every cubic meter of synthesis gas consumed the process gave a total yield per of only 70 g of a 58-octane number gasoline and a diesel oil boiling above 220° C.[90]

Two years later Fischer's decade-long research moved to the next level with the construction in 1934 of the first large pilot plant in which he planned to solve the synthesis' three main problems and produce hydrocarbons from carbon monoxide and hydrogen. Ruhrchemie envisioned the F-T synthesis as an outlet for its surplus coke, and upon acquiring the patent rights to the synthesis in 1934, constructed the pilot plant in Oberhausen-Holten (Sterkrade-Holten), near Essen. The plant operated at the conditions used in Fischer's

90 Storch, Golumbic, and Anderson, *Fischer-Tropsch* (note 88), p. 135; Franz Fischer, Helmet Pichler, and Rolf Reder, "Überblick über die Möglichkeiten der Beshaffung geeigneter Köhlenoxyd-Wasserstoff-Gemische für die Benzinsynthese auf Grund des heutigen Standes von Wissenschaft und Technik," *Brennstoff Chemie*, 13 (1932): 421–28; Franz Fischer, Otto Roelen, and Walter Feisst, "Über die nunmehr erreichten technischen Stand der Benzinsynthese," *Brennstoff Chemie*, 13 (1932): 461–68; Franz Fischer and Herbert Koch, Über den Chemismus der Benzinsynthese und über die motorischen und senstigen Eigenschaften der dabei auftretenden Produkte (Gasol, Benzin, Dieselöl, Hartparaffin), *Brennstoff Chemie*, 13 (1932): 428–34; Herbert Koch and Otto Horn, "Vergleichende Untersuchung über das motorische Verhalten eines synthetischen Benzins nach Franz Fischer (Kogasin I) und eines Erdöl-Benzins, *Brennstoff Chemie*, 13 (1932): 164–67.

small pilot plant and had an annual capacity of 1,000 metric tons (7,200 barrels) of motor gasoline, diesel oil, and lubricating oil.

Although the larger pilot plant demonstrated the overall success of the F-T synthesis, its three main problems: removing the large amount of heat released in the gas stream during the reaction, the nickel catalyst's short lifetime, and the significant loss of catalytic metals (nickel, manganese, aluminum) during their recovery (regeneration) for reuse, persisted throughout the operation. The nickel catalyst's poor performance forced Fischer and Ruhrchemie to abandon its use for industrial development. At this time research resumed with the more active but expensive cobalt catalysts. Oberhausen-Holten subsequently became the production center for a standardized cobalt catalyst used in all the F-T plants constructed later in the 1930s, for all the development work on synthetic motor fuel and lubricating oil, and for the oxo process.[91]

The successful pilot plant research and development at Oberhausen-Holten was the major turning point in the F-T synthesis. By November 1935, less than three years after Germany's Nazi government came to power and initiated the push for petroleum independence, construction had begun on four industrial-size Ruhrchemie licensed F-T plants. They had a total annual capacity of 100,000–120,000 metric tons (724,000–868,000 barrels) of motor gasoline, diesel oil, lubricating oil, and other petroleum chemicals. All the plants operated at atmospheric pressure (1 atm) or medium pressure (5–15 atm) syntheses and 180–200° C, and used a cobalt catalyst (100 Co-5 ThO_2-8 MgO-200 kieselguhr) that Ruhrchemie's chief chemist Otto Roelen at Ruhrchemie's F-T plant in Oberhausen-Holten (Sterkrade-Holten) developed from 1933 to 1938.[92]

By 1936 the four Ruhrchemie licensed F-T plants constructed or under construction had a total annual capacity of 145,000 metric tons (1.05 million barrels) of gasoline, diesel oil, and other petroleum hydrocarbons.

---

91 Storch, Golumbic, and Anderson, *Fischer-Tropsch* (note 88), p. 337; "Synthetic Petrol by the Fischer Process," *Gas World*, 105 (1936): 362–63; H. H. Storch, "Synthesis of Hydrocarbons from Water Gas," in *Chemistry of Coal Utilization*, ed. H. H. Lowry, 3 Vols. (New York: John Wiley & Sons, Inc., 1945 and 1963), Vol. 2, pp. 1797–1845, on p. 1800. The motor fuel's estimated average cost was 22 pfennig/kg. According to Arno Fieldner and American engineers the wartime cost of either the F-T or the hydrogenation process was 20–30¢/US gallon. Arno C. Fieldner, "Frontiers of Fuel Technology," *Chemical and Engineering News*, 26, no. 23 (7 June 1948): 1700–01. For the conversion of metric tons to barrels, 1 metric ton = 7.2 barrels.

92 *Report* (note 84), pp. 82–90; "Germany's Home Production of Motor Fuels," *Gas World*, 104 (9 May 1936): 421; Franz Fischer, "The Conversion of Coal into Liquid Fuels," *Chemical Age*, 35 (24 October 1936): 353–55; "The Fischer Process," *Chemical Age*, 35 (31 October 1936): 367.

The four plants were (1) Ruhrchemie AG (Ruhrbenzin AG) in Oberhausen-Holten (Sterkrade-Holten), Ruhr; (2) Rheinpreussen in Moers-Meerbeck (Homberg), Niederrhein; (3) Gewerkschaft Viktor AG in Castrop-Rauxel, Ruhr; and (4) Braunkohle Benzin AG (Brabag), the largest plant, completed 1937 in Ruhland-Schwarzheide, Dresden area. Total annual capacity increased to 300,000 metric tons (2.17 million barrels) by 1937–38, and with the completion of five additional plants, (5) Mitteldeutsche Triebstoff und Öl Werke, Wintershall AG, completed 1937 in Lützkendorf-Mücheln, Leipzig area; (6) Krupp Treibstoffwerk GmbH in Wanne-Eickel, Ruhr; (7) Essener Steinkohle AG in Kamen-Dortmund, (Bergkamen), Ruhr; (8) Schaffgotsch Benzin GmbH in Deschowitz-Beuthen (Upper Silesia); and (9) Hoesch-Benzin GmbH in Dortmund, Ruhr, giving a total of nine, plant capacity rose to 525,000 metric tons by the end of 1938 and to 740,000 metric tons (5.4 million barrels) at the outbreak of World War II in September 1939. Actual peak production reached 570,000 metric tons (4.15 million barrels) in 1944. The older F-T plants operated at 1 atm and 180–200° C, whereas newer medium pressure plants operated at 5–15 atm and 180–200° C.

7 Conclusion

Several factors contributed to Bergius's invention of coal hydrogenation in 1913 just prior to World War I, to his technological successes and setbacks, and to his constant struggle to secure funds to continue his research. Financial problems led Bergius to establish a complex network of companies and to enter into licensing agreements with several oil and chemical companies including IG Farben which purchased his patents in 1925 and began industrialization of the Bergius process in 1927. For his important contributions to high-pressure coal hydrogenation, Bergius shared the 1931 Nobel Prize in chemistry with IG Farben's Carl Bosch. Fischer's synthetic fuel research paralleled that of Bergius, starting with Fischer and Tropsch's invention of the gas synthesis or F-T synthesis in 1926, their development of suitable cobalt catalysts, and the licensing agreement concluded with Ruhrchemie on 27 October 1934.

Bergius's and Fischer and Tropsch's pioneering inventions made Germany the first industrialized nation to succeed in synthesizing petroleum from coal. Germany developed a technologically successful synthetic fuel industry that grew from a single coal hydrogenation plant constructed in Leuna in 1927 to twelve coal hydrogenation and nine F-T plants that in 1944, a year before World War II ended, reached a peak production of 26 million barrels (3.6 million metric tons) of synthetic fuel.

CHAPTER 3

Germany's Synthetic Fuel Industry 1927–45

1 Introductory

Petroleum was clearly the fuel of the future, and to insure that Germany would never lack a plentiful supply, German scientists and engineers invented and developed two processes that enabled them to synthesize petroleum from their country's abundant coal supplies and to establish the world's first technologically successful synthetic liquid fuel industry. Friedrich Bergius in Rheinau-Mannheim began the German drive for energy independence with his invention and early development of high-pressure coal hydrogenation or liquefaction in the years 1910–25. A decade after Bergius began his work Franz Fischer and Hans Tropsch at the Kaiser-Wilhelm Institute for Coal Research (KWI) in Mülheim, Ruhr, invented a second process for the synthesis of liquid fuel from coal. By the mid-1930s IG Farben, Ruhrchemie, and other chemical companies had started to industrialize synthetic liquid fuel production, resulting in the construction of twelve coal hydrogenation and nine Fischer-Tropsch (F-T) plants by the time World War II ended in 1945. Several breakthroughs contributed to the success of coal hydrogenation the most significant of which were the sulfur resistant catalysts and the two-stage liquid-vapor phase hydrogenation that Matthias Pier at BASF (IG Farben) developed in the late 1920s. For the F-T synthesis the cobalt catalysts that Fischer and his co-workers and Otto Roelen at Ruhrchemie prepared in the 1920s–30s were crucial to its success.

Because of synthetic liquid fuel's high production cost the industry benefited from the financial incentives Germany's Nazi government offered beginning in December 1933, and because liquid fuel was crucial to Germany's war effort the synthetic fuel industry became a major part of Adolf Hitler's Four Year Plan (Vierjahresplan) of 9 September 1936. As the war dragged on, the synthetic fuel industry, like many German industries, experienced serious labor shortages, and to avoid any loss of production and slowdown of the war effort some of the plants used forced labor that the German government provided.

Of the two processes hydrogenation was the more advanced, 20% less expensive, and contributed much more significantly to Germany's liquid fuel supply than the F-T synthesis. Coal hydrogenation produced high quality aviation and motor gasoline, whereas the F-T synthesis gave aliphatic compounds (carbon-hydrogen chain compounds), high quality diesel and lubricating oil,

 | DOI:10.1163/9789004690912_004

TABLE 3.1 Synthetic fuels in Germany 1911–45

Period	Activity
High-pressure coal hydrogenation: Research and development falls into four divisions	
1911–20	Bergius, laboratory research and pilot plant development BASF-IG Farben laboratory research and pilot plant development
1920–33	BASF and IG Farben research and development at Ludwigshafen and Leuna; most important breakthroughs occur such as introduction of two-phase conversion and sulfur resistant catalysts
1933–39	IG Farben use of 700 atm pressure (1936) First hydrogenation of bituminous coal (1936) New hydrogenation companies established seven plants constructed New steel alloys introduced Research and development continue
1939–45	Five plants constructed Mainly a period of development, not much research other than catalysis
Fischer-Tropsch: Research and development falls into two divisions	
1922–35	Laboratory research and pilot plant development
1936–45	Industrial-size development, most development occurring prior to World War II

SOURCE: COURTESY OF AUTHOR

waxes, and some lower quality motor gasoline. The two processes were complementary rather than competitive, but because only coal hydrogenation produced high quality gasoline it experienced much greater expansion in the late 1930s and war years than the F-T synthesis, which hardly grew at all. F-T products were mainly the raw materials for further chemical syntheses with little upgrading of its low-quality gasoline by cracking because of unfavorable economics. Hydrogenation also had experienced greater development because brown coal (lignite), the only coal available in many parts of Germany, underwent hydrogenation more readily than a F-T synthesis. In addition, the more mature and better developed hydrogenation process had the support of IG Farben, Germany's chemical leader, which had successfully industrialized coal hydrogenation beginning in 1927. A historical analysis of the industrial development of the two synthetic fuel processes during several decades of German social, political, and economic unrest follows.

2 Germany's Energy Plan

The growth of the German synthetic fuel industry remains inseparably linked to events taking place there in the 1930s and 1940s. A special relation existed between the industry and Nazi government, and without it Germany's emerging synthetic fuel industry might have collapsed. The small German oil industry had remained reasonably free from government interference and benefited from an April tariff increase in 1930 that raised the duty on imported oil from RM 77.40/metric ton (8.5¢/US gallon) to RM 129/metric ton (14.3¢/US gallon). But government policy began to change with the German banking crisis that followed the failure of the Kredit Anstalt on 3 May 1931 and the Darmstaedter National Bank on 15 July 1931 and led the Weimar government to impose a number of controls and regulations which the Nazi government expanded and intensified beginning in 1933. The Weimar government established Supervisory Boards to allocate raw materials and placed these boards under control of the Reichswirtschaftsministerium (Ministry of Economics) which in 1939 renamed them Reichsstellen (Reich Offices). The Reichsstelle für Mineröl (Office of Mineral Oil) regulated the oil industry; additional regulations came from the Reichsstelle für Rohstoffamt (Office of Raw Materials) and its subdivision Wirtschaftsgruppe Kraftstoffindustrie (Economic Group for Liquid Fuels). All liquid fuel producers reported their production and import figures and any new construction of plants and refineries to the oil regulatory boards. In addition to regulatory boards, the government established four industry associations that had responsibility for the production and allocation of the fuels under their control.

The government-industry relation also resulted in risk-free partnership agreements between the government and those industries (coal, chemical) involved in synthetic fuel production. The earliest of these was the Fuel Agreement (Benzinvertrag) that IG Farben, the only company then producing synthetic fuel, and the Reichswirtschaftsministerium signed on 14

December 1933. It required IG Farben to produce at least 300,000–350,000 metric tons (2,160,000–2,520,000 barrels) of synthetic gasoline per year by the end of 1935 and to maintain this production rate until 1944. The agreement set the production cost, which included depreciation, 5% interest on IG Farben's investment, and a small profit, at 18.5 pfennig/L (1.1 quarts). The government

TABLE 3.2 Government-industry associations

Association for Crude Oil Production and Refining (REV)
Association for Hydrogenation, Synthesis, and Low Temperature Carbonization (ARSYN)
Association of German Benzol Producers (ARBO)
Association for Allocation of German Bituminous Coal Tar Products (AVS)

SOURCE: COURTESY OF AUTHOR

not only guaranteed the production cost but agreed to pay IG Farben the difference between that cost and any lower market price, and to buy the gasoline if no other market emerged. Alternatively, IG Farben had to pay the government the difference between the production cost of 18.5 pfennig/L, which was at that time more than three times the world market price, and any higher price obtained on the market. Because of increasing petroleum costs, as well as improvements in the hydrogenation process, IG Farben paid RM 85 million to the government by 1944.[1]

Eight months after signing the fuel agreement with IG Farben the government took two additional steps to assist the synthetic fuel industry. The first was the establishment on 24 August 1934 of the Wirtschaftliche Forschungsgesellschaft (WIFO, Economic Research Company), a completely government-owned company capitalized at RM 20,000 and charged with the construction and operation of liquid fuel (natural and synthetic) storage depots. German fuel producers sent WIFO their lubricating oil, and their aviation grade products for blending and leading, which WIFO stored and eventually distributed mainly to the air force and minimally to the army.

In 1938 WIFO had a storage capacity of 630,000 metric tons of motor and aviation gasoline and 84,000 metric tons of lubricating oil. It actually stored 500,000 metric tons of aviation gasoline, most of it in bombproof underground locations within Germany. The government in 1938 planned to increase Germany's total storage to 6,000,000 metric tons of liquid fuel and lubricating oil by 1943 and projected the following contributions: WIFO 2,900,000

1 *Trials of War Criminals before the Nürnberg Military Tribunals* (Washington, DC: Government Printing Office, 1953), Vol. 7, the IG Farben Case (Case 6), Testimony of Defendant (Carl) Krauch. The Fuel Agreement is *Nürnberg Industrialists Document* NI-881, reel 9, 14 December 1933. See also Wolfgang Birkenfeld, "Leuna, 1933," *Tradition*, 8 (1963): 107–08; Wolfgang Birkenfeld, *Der synthetische Treibstoff 1933–1945* (Göttingen: Musterschmidt Verlag, 1964), pp. 23–34; Peter Hayes, *Industry and Ideology* (Cambridge: Cambridge University Press, 1987), pp. 115–20.

metric tons; German industrialists 1,250,000 metric tons; and the navy, which had underground storage tanks and a smaller number of surface tanks in the North Sea and Baltic Sea areas and in the German interior, the remaining 1,800,000 metric tons. But the government's lofty projection fell short. Germany's total storage reached 2,400,000 metric tons of liquid fuel on 21 June 1941. WIFO's contribution, 500,000 metric tons of aviation gasoline, was a significant amount that represented about one-third the total 1940 US production of aviation gasoline (40,000 bbl/d ≡ 5,555 metric tons/d) and almost equal to Germany's refining capacity of 420,000 metric tons/month or 5 million metric tons/yr, about half of it refined in the Hamburg and Hannover areas.

Two months after establishing WIFO the German government took the second step when it forced the establishment of Braunkohlen Benzin AG (Brabag) to promote and carry out industrial-size synthesis of synthetic liquid fuel and lubricating oil from coal and tar. Brabag was an association of IG Farben and nine central German brown coal producers (Compulsory Union of German Lignite Producers) that accounted for 90% of Germany's brown coal. At the time of its formation on 26 October 1934, Brabag had a capitalization of RM 100 million financed entirely with a fifteen-year loan that the German brown coal producers guaranteed. Gesellschaft für Mineralölbau GmbH, a division of Brabag established two years later in November 1936 by the ten brown coal producers, carried out the design and engineering of the Brabag plants, using technical information that the government required IG Farben, Ruhrchemie, and other synthetic fuel producers to provide upon their entering into government licensing agreements. Brabag and Mineralölbau built and operated three coal hydrogenation and one F-T plant during the 1930s–40s.

Additional government commitment to the synthetic fuel industry, and indicative of the supportive government-industry relation, emerged at a Nazi party rally in Nuremberg on 9 September 1936. At that time Adolf Hitler announced his Four Year Plan to make the German military ready for war in four years and the economy independent and strong enough to maintain a major war effort. Hitler put Hermann Göring (1893–1946) in charge of the plan in October 1936, gave him the title Commissioner General for the Four Year Plan, and had Göring officially approve the plan in May 1937. With Hitler's war strategy requiring large supplies of petroleum, a petroleum-independent Germany became the Four Year Plan's major thrust. Of the 289 projects scheduled for the period 23 October 1936–20 May 1937 at a cost of RM 1,369 million, 42% costing RM 570 million were synthetic fuel projects. In fact in 1936 Hitler urged the petroleum industry, including synthetic fuel produced by both coal and tar hydrogenation and F-T synthesis to become independent of foreign production in eighteen months and called for synthetic fuel production to increase from 630,000 metric tons in 1936 to 3,425,000 metric tons in 1940. As an incentive toward the synthesis of petroleum from coal the German government in December 1936 raised the tariff on imported petroleum from a 1931 levy of RM 219.30/metric ton (24.4¢/US gallon) to RM 270.90/metric ton (30.1¢/US gallon). By this time only four coal hydrogenation and two F-T plants were operating with a combined production far less than required for petroleum independence. The high tariff enabled the synthetic fuel plants to show a profit despite their high inefficiency and production costs that greatly exceeded the cost of natural petroleum.

In the beginning, private capital coming from bank loans and from the synthetic fuel companies' own funds, stock, and bond issues provided practically all of the plants' financing. But by 1939, as the cost of the program increased significantly and private capital dried up, the German government provided more and more of the funding. A report of 21 March 1939 showed that of the RM 132 million spent on synthetic fuel in 1939, the government provided RM 70 million to Minerölbau GmbH for the purchase of plant equipment. Additional government support came in the form of guaranteed purchases of synthetic fuel at prices high enough to allow for short term amortization of plant costs. Total synthetic fuel production from the seven coal hydrogenation and seven F-T plants operating in September 1939 was 1,280,00 metric tons increasing to almost 1,900,000 metric tons in May 1940. It exceeded Germany's refining of crude oil from natural sources (1,256,000 metric tons) and imports mainly from Romania (1,085,000 metric tons) in 1939.

The Reichsamt für Wirtschaftsausbau (Office of Economic Development) constantly revised the Four Year Plan. Its general concerns were the raw material and manpower requirements and the never-ending iron and steel shortages, and in particular for the synthetic fuel industry the anticipated shortages of aviation gasoline and fuel oil. The first revision at Karinhall, Göring's vast country palace in the Schorfheide (Berlin-Postdam), on 12 July 1938 gave priority to hydrogenation plants for the production of aviation gasoline and to bituminous coal distillation for the production of fuel oil. The Welheim hydrogenation plant, which had begun production of aviation gasoline and fuel oil for the navy in 1937–38, already had received priority; and the Brüx (Most) hydrogenation plant, which produced diesel oil, later benefited from the revised plan. The 1938 revision also dealt with the number of workers required for the construction, operation, and maintenance of the synthetic fuel plants. It called for 30,000 construction workers in 1938, increasing to 57,600 workers on 1 July 1939, and expanding from a projected 70,000 workers on 1 October 1939 to 135,000 workers during the last quarter of 1941. The actual construction force, however, numbered about 35,000 at the outbreak of war, about 70,000 by mid-1941, and peaked at 85,000 in spring 1943.

Steel production also failed to meet the Four Year Plan's requirements. A second Karinhall revision of 1 January 1939 called for the production of 4.5 million metric tons of steel by the end of 1943. With this amount of steel Germany expected to expand existing plants and construct new plants to increase synthetic fuel production from 3.7 million metric tons in 1938 to 11 million metric tons per year by 1944. According to the US Strategic Bombing Survey's postwar report the required 4.5 million metric tons of steel equaled the amount necessary to build a fleet 3.5 times the size of the British navy that existed on 1 January 1940.[2]

3 BASF-IG Farben's Coal Hydrogenation Program

3.1 *Period of Development: Synthesis of Ammonia and Methanol*

Friedrich Engelhorn (1821–1902) and eight others that included a banker, industrialists, and chemists, established BASF in April 1865 in Ludwigshafen, a small Bavarian town on the Rhine River's left bank, near Mannheim. The chemical industry was experiencing a period of growth especially in the development of synthetic organic products and catalyzed reactions. BASF shared in the growth, and within a few years it had become the world's leader in developing new industrial techniques and chemical products. The synthetic dye, "indigo pure BASF," was one of its early successes. Adolf von Baeyer (1835–1917) at the University of Munich had determined indigo's structure in 1883, after a fourteen-year investigation, but yields remained low until the early 1890s when Karl Heumann (1851–94) at the University of Zurich improved the synthesis. Heumann's improved synthesis enabled BASF's Heinrich von Brunck (1847–1911) to begin indigo's industrial production in 1897.

Another notable breakthrough occurred in 1888 when BASF's Rudolph Knietsch (1854–1906) invented the Knietsch compressor and made possible the industrial production of liquid chlorine. Two years later, in 1890, Knietsch industrialized the platinum-catalyzed sulfur dioxide-trioxide oxidation that Clemens Winkler (1838–1904) at the University of Freiberg first demonstrated

2 The United States Strategic Bombing Survey, The German Oil Industry Ministerial Report Team 78, Oil Division, 1st ed. 5 September 1945, 2nd ed. January 1947, pp. 19–20. The United States Strategic Bombing Survey, Summary Report (European War), 30 September 1945, 18 pages. The word benzin (benzene) refers to petroleum benzin, a petroleum fraction.

in 1875. Knietsch's contact process for producing sulfuric acid from sulfur trioxide replaced the lead chamber process that chemical manufacturers had been using since John Roebuck (1718–94) in England invented the chamber process in 1746. The contact process was the first of BASF's catalyzed gaseous synthesis reactions and a progenitor of the catalytic high-pressure syntheses that BASF developed later, specifically the 1909–10 Haber-Bosch ammonia synthesis for the production of fertilizers and explosives; the 1913 Alwin Mittasch-Christian Schneider (1887–1972) water gas synthesis of a petroleum-like mix of hydrocarbons, alcohols, ketones, and acids; and the 1923 Matthias Pier water gas synthesis of methanol.[3] Ammonia's synthesis from nitrogen and hydrogen led naturally to the others because water gas, a carbon monoxide-hydrogen mixture that provided the hydrogen used to synthesize ammonia, was the reactant for the other syntheses.

Three years after Bosch's 1910 high-pressure laboratory synthesis of ammonia, BASF inaugurated the high-pressure chemical industry. On 9 September 1913 in the nearby village of Oppau it completed construction of the world's first plant for the high-pressure synthesis of ammonia. With the outbreak of World War I on 1 August 1914 and Germany's increased need for fertilizers and explosives, BASF constructed a second larger synthetic ammonia plant in Leuna, near Merseberg and the Saale River, twenty miles west of Leipzig. Fearing possible war damage BASF did not want to concentrate its ammonia production at one location, and in the summer of 1915, while searching for a suitable site, Bosch selected Leuna and had the plant built according to his design. Production began on 29 April 1917, and at the end of the war Oppau and Leuna had annual capacities of 50,000 and 130,000 metric tons of ammonia, respectively, although their actual production totaled 95,000 metric tons of ammonia or about 50% of Germany's production of nitrogen compounds.[4] Their production enabled Germany to prolong the war beyond the early years despite having neither stockpiles of fertilizers and explosives nor imported nitrates from Chile because Britain controlled the seas. Synthetic ammonia production failed to prevent Germany's defeat, however.

After opening its Leuna plant in 1917, BASF continued high-pressure research on the water gas reaction but now with an expanded program that resulted in the addition of Matthias Pier. Pier had studied under Wilhelm

3 Wilhelm Roggersdorf, *In the Realm of Chemistry* (Düsseldorf: Econ-Verlag, 1965), pp. 8–15; *High-Pressure Hydrogenation at Ludwigshafen-Heidelberg*, FIAT Final Report No. 1317, 9 Vols. (Dayton, Ohio: Joint Intelligence Objectives Agency, 1951), Vol. 1, p. 28. Not all of the CIOS, BIOS, and FIAT reports gave construction costs of the German synthetic fuel plants.

4 L. F. Haber, *The Chemical Industry: 1900–1930* (New York: Oxford University Press, 1971), p. 203.

Röntgen (1845–1923) in Munich and Nernst at the Kaiser Wilhelm Institute for Physical Chemistry in Berlin where he received his PhD in 1908. Nernst had pioneered in high-pressure reactions, and in his laboratory Pier investigated the high-pressure water gas synthesis, establishing the most favorable conditions of temperature, pressure, and concentration for reaction. Pier remained as Nernst's assistant until 1910 when he accepted a position with the Zentralstelle für wissenschaftlich-technische Untersuchungen GmbH (Center for Scientific-Technical Research) in Neubabelsberg, the research home of the German explosives industry. There he investigated the catalytic high-pressure ammonia synthesis. World War I interrupted the research, and Pier served as a captain in the German army on the Western front before his serious wounding near Winterberg at Chemin des Dames during Germany's successful campaign of 27 May-6 June 1918. After his recovery Pier returned to Neubabelsberg in 1919, but in June 1920, with the Zentralstelle's closing under the terms of the Versailles Treaty, he accepted a position with BASF in Ludwigshafen.[5]

In the Ludwigshafen Ammonia Laboratory located in Building 35 Pier resumed his high-pressure research, working for a time with Mittasch on the synthesis of ammonia and of methanol. The ammonia experiments aimed to test the effectiveness of different iron catalysts at extremely high pressures, up to 1,000 atm as in the Claude process. Upon completing these experiments, Mittasch and Pier turned to the water gas synthesis of methanol. Mittasch and Schneider earlier, in 1913, had produced a petroleum-like mix of hydrocarbons, alcohols, ketones, and acids, but because they used an excess of carbon monoxide, considerable carbon (soot) deposited on the metallic or alkali-containing metallic oxide catalysts rendering them ineffective.

Pier continued the catalytic studies, and by the end of 1922 he showed that, in addition to the problematic but preventable soot deposits, iron (iron

5 Maria Höring and Ludwig Raichle, "Zur Technologie der Kohle und Ölhydrierung, Matthias Pier zum Gedachtnis," *Chemie Ingenieur Technik*, 38, 3 (March 1966): 205–08. American military authorities arrested Pier on 2 January 1946, interrogated him and other IG Farben officials at the Nuremberg hearings (14 November 1946–24 June 1947) but did not charge him personally. From June-October 1947 he was at Lager Neuengamme (concentration camp) near Hamburg, then at Entlassungslager Darnmstadt (8 November 1947) before returning to Heidelberg. Two of Pier's close associates Ernst Donath and Hans Schappert at BASF (IG Farben) insisted that Pier had no Nazi sympathies. Mrs. Schappert told me in a telephone interview (17 July 1978) that Pier had defended Schappert who openly expressed his anti-Nazi beliefs and that German newspapers had attacked Schappert for his defense of Jewish people. Donath, who had a Jewish-Catholic parenthood, was not as outspoken as Schappert but confirmed everything Schappert said about Pier. (Anthony N. Stranges, Donath interview 9–12 August 1982).

FIGURE 3.1
Matthias Pier
SOURCE: WERNER LANG, "MATTHIAS PIER, EHRENBÜRGER VON NACKENHEIM," *Nackenheimer Heimatkundliche Schriftenreihe*, HEFT 16 (HERAUSGEGEBEN VOM VERKEHRSVEREIN E.V., NACKENHEIM AM RHEIN, 1982). PIER COLLECTION, BASF ARCHIVES, LUDWIGSHAFEN

carbonyl) interfered with the synthesis. This was highly significant because once Pier recognized that iron, a commonly used catalyst and familiar to him from the ammonia research, was the remaining problem, advances in the methanol synthesis came rapidly. On 16 January 1923, he tested a zinc oxide catalyst and obtained pure methanol from water gas at 400° C and 1,000 atm. Additional testing showed that 1,000 atm pressure was unnecessary, and on 15 May Pier synthesized methanol in pilot plant experiments at 200 atm, the same pressure BASF used in the ammonia synthesis. Industrial-size methanol production began at BASF's Leuna plant on 1 July using spare equipment from the ammonia synthesis and experienced no problems from the start. Three weeks later, on 24 July, BASF submitted a patent for the methanol synthesis using metallic oxide catalysts, particularly those of zinc and chromium which Pier found quite resistant to sulfur poisoning.[6]

3.2 *Hydrogenation at Ludwigshafen 1924–27*

By summer 1924, BASF had succeeded in industrializing the ammonia and methanol syntheses, both of which were high-pressure hydrogenations of gases. It also followed with considerable interest Bergius's high-pressure hydrogenation of coal, coal tar, and crude oil to give gasoline and other light petroleum distillates. Bergius had recognized that petroleum was clearly the fuel of the future, and because Germany lacked domestic petroleum sources he committed himself to using high-pressure hydrogenation to convert Ger-

6 BASF (Alwin Mittasch, Matthias Pier, and Karl Winkler), German Patent 415,686 (application 24 July 1923, awarded 27 July 1925).

many's vast coal supplies to petroleum. Pier was ready to expand his high-pressure program to include coal and tar, and in 1924 he received a directive from Carl Krauch (1887–1968), head of BASF's Nitrogen Division in Oppau and a strong supporter of high-pressure research, to begin experiments in the Nitrogen Division's High-Pressure Section.

The hydrogenation of coal and tar, as Bergius's ten years of research had demonstrated, was by far a much more difficult synthesis than either ammonia or methanol. First, the reaction involved working with heavy solids and dense viscous liquids rather than easily flowing and relatively pure gases. Second, coal's complex ring structures permitted a large number of side reactions in addition to the desired hydrogenation. Third, because of side reactions the catalysts had to increase the hydrogenation reaction rate and favor hydrogenation over the other reactions. The presence of catalyst-poisoning sulfur in the reactants also required catalysts to have a higher resistance to attack than the metallic and metallic oxide catalysts developed for the ammonia and methanol syntheses.

Initial laboratory high-pressure experiments involved catalytic tests with inexpensive iron (III) oxide (Fe_2O_3) and other metallic oxides. Pier selected phenol and phenol derivatives obtained from coal tar as the reactants because they gave easily identifiable products, mainly neutral hydrocarbon oils and water, and permitted a large number of tests in a short time. As expected, the iron (III) oxide catalyst performed very well until sulfur or sulfide poisoning occurred. Sulfur poisoning remained the most outstanding and enduring difficulty catalytic chemistry had experienced, and in the 1920s many chemists began to doubt they could solve the sulfur problem. Free sulfur or sulfides present in coal was the main reason Bergius had not attempted to find an effective hydrogenating catalyst and had added iron (III) oxide to react with and eliminate their disastrous effects. Catalyst development also had become an expensive industrial undertaking, often with high risks and low returns. Therefore in December 1924 when Pier decided to counteract "like with like" and prepared an iron (III) sulfide catalyst (Fe_2S_3) by treating a previously used iron oxide catalyst with hydrogen sulfide and found it resistant to sulfur poisoning he made one of the most important breakthroughs in the study of catalysis and especially in high-pressure coal hydrogenation. Pier later showed cobalt sulfide's superiority as a hydrogenating catalyst and that the oxides of cobalt, tungsten, and molybdenum effectively resisted sulfur poisoning; molybdenum oxide performed better than all previously tested metallic oxide catalysts.[7]

7 *High-Pressure* (note 3), Vol. 1, p. 32; *Interrogation of Dr. Pier and Staff IG Farbenindustrie AG Ludwigshafen-Oppau*, FIAT Final Report No. 426 (Washington, DC: Joint Intelligence Agency, 14 November 1945).

By the end of 1924 Pier's work had reached the point where BASF would have to decide the future course of its high-pressure coal hydrogenation program. Pier's sulfide compounds solved the catalyst poisoning problem, but he had not yet hydrogenated coal and coal tar. Nor had he established the synthesized liquids' quality and production cost, although from Bergius's published accounts he knew that the liquids needed upgrading and had high production costs. Bergius had spent RM 48 million on the construction and operation of his Rheinau-Mannheim plant in two years, from 1918 to 1920. He had obtained 490–650 kg of liquids from each metric ton of coal hydrogenated at a cost of RM 71. According to Brückmann, EVAG's co-director, the liquids, consisting of gasoline, diesel oil, and lubricating oil, had a market value of RM 141, giving a profit of RM 70/metric ton of hydrogenated coal.[8] Yet, despite such optimistic claims Bergius and EVAG remained hard-pressed for funds to keep the Rheinau-Mannheim plant operating and in the early 1920s had entered into a financial agreement with Royal Dutch-Shell and organized a British Bergius syndicate in England, in their never-ending search for operating funds.[9]

Bosch, BASF's managing director (1919), was the driving force behind the directors' decision to continue BASF's coal hydrogenation program, looking towards possible industrialization. Bosch won the Board's support by pointing out that coal hydrogenation already was a technological accomplishment and that an expanded program would enable BASF to use its excess high-pressure ammonia equipment that remained idle because of a serious decline in fertilizer consumption. Not only the fertilizer market, but most of the chemical industry recently had experienced an eighteen-month depression that lasted from mid-1920 to the end of 1921. The depression made evident the chemical industry's excess capacity, which although disappearing briefly after World War I, had reappeared in 1920. Despite the downturn, BASF had averaged a 20% annual profit for 1920–22 (based on its net assets), but because of Europe's unsettled political and economic situation and particularly the German monetary crisis, the directors initially were quite content to continue with synthetic ammonia and methanol production rather than venture into another expensive high-pressure program. Kaiser Wilhelm II's abdication on 9 November 1917 and the establishment of the first of seventeen Social Democratic govern-

---

8 Heinrich Brückmann, "Coal Liquefaction and its Importance in World Economy," *The Mining Journal*, 160 (28 January 1928): 69–70; 160 (4 February 1928), p. 94; and 160 (11 February 1928), p. 116.

9 Anthony N. Stranges, "Friedrich Bergius and the Rise of the German Synthetic Fuel Industry," *Isis*, 75 (1984): 643–67, on 663–65. Average exchange rate for 1919: RM 1 = 3¢; for 1920: RM 1 = 1.75¢; for 1934–41: RM 1 = 40¢. No figure is available for 1918, *Banking and Monetary Statistics, 1914–1941*, Federal Reserve System (Washington, DC: Government Printing Office, 1947), p. 671.

ments in Weimar Germany had put the country and the economy in a state of flux. The directors very much feared that a bad decision at this time could lead to disaster in the future.[10]

Germany's loss of important industrial areas, with its obvious link to postwar reparations payments and government debt, also dealt a serious blow to the country's economic recovery. France and Belgium occupied the Ruhr valley, Germany's industrial center, in January 1923 (and remained until the summer of 1925) to ensure Germany's payment of reparations, but the Allies' action reduced Germany's production of iron and coal by 80%. Inflation made matters worse. In July 1914, on the eve of World War I, the German-US currency exchange rate was RM 4 per US dollar, in November 1923 the rate increased to an unbelievable RM 4.2 trillion per US dollar. During this period the wholesale price index rose from a scale of one in 1913 to 750 trillion in 1923. A pound of butter cost RM 6 billion, a pair of shoes RM 32 billion.[11]

In view of these economic hardships Bosch believed that the establishment of a large German synthetic fuel industry would stimulate postwar recovery in two ways. First, it would reduce the government's debt and aid the balance of trade, and at the same time reverse Germany's shortage of foreign currency by eliminating the huge amounts spent on petroleum imports. Germany depended almost totally on imported petroleum and in the early 1920s imported annually 2.5 million metric tons of petroleum costing RM 500 million. In addition to the expenditure problem, a continuous supply of imports remained uncertain because of the widespread belief in the 1920s that global petroleum supplies faced rapid exhaustion. Second, the new industry clearly would reduce Germany's unemployment.

Having decided to proceed with the industrial production of synthetic fuel, BASF had to reach an agreement with Bergius. Bergius and EVAG held the German rights to high-pressure coal, tar, and crude oil hydrogenation, and to avoid any patent infringement BASF needed an agreement with Bergius. Bergius's poor financial situation worked to BASF's advantage. The high cost of operating his Rheinau-Mannheim plant, where he employed 150 men and by the mid-1920s had spent more than RM 18 million on research, had drained Bergius

10 Thomas Parke Hughes, "Technological Momentum in History: Hydrogenation in Germany 1898–1923," *Past and Present*, 44 (1969): 106–32; E. E. Donath and Maria Höring, "Early Coal Hydrogenation," *Fuel Processing Technology*, 1 (1977): 3–20.

11 Tony Edwards, *The New Republic, History Broadsheet, No. 4*, (London: Heinemann Educational Books, Ltd., 1972). Germany abandoned its old currency in November 1923. One new Gold Mark now equaled one trillion old Marks, and $1.00 = 4.28 Gold Marks.

financially, and on 26 May 1925 he gladly sold his patents to BASF for RM 2.850 million and withdrew from further experimental work.[12]

BASF's decision to continue the coal hydrogenation program led in 1924 to the establishment of a new high-pressure experiment group that consisted of Pier, its leader; Fritz Winkler (1888–1950), an organic chemist and in 1921 inventor of the Winkler generator for the production of water gas, fuel and power gases from brown coal; and Walther Rumpf, a mechanical engineer, both of whom had worked with Pier on the methanol synthesis. Membership quickly increased, and at the end of 1924 the group included two chemists, two mechanical engineers, and forty laboratory workers. Two more chemists, Ernst Donath (1902–83) and Maria Höring (1899–1984), who became Pier's long-time associates, joined the group in 1926, and by 1927 membership had expanded to seventy graduate chemists and engineers and one thousand workers. During its twenty-year existence the high-pressure experiment group tested the laboratory-size hydrogenating apparatus, developed the industrial-size coal hydrogenation equipment used later in IG Farben's Leuna plant and in other synthetic fuel plants, and trained the plants' operators.[13]

The first reactions Pier and his high-pressure experiment group investigated after completing the phenol hydrogenation early in 1925, were the hydrogenation of brown coal and the coal tar obtained from the low-temperature carbonization or distillation (LTC) of brown coal. Coal tar and coal were considerably more difficult to hydrogenate because they required the hydrogenation and removal of ever-present oxygen, sulfur, and nitrogen by converting them to water, hydrogen sulfide, and ammonia, before any splitting and hydrogenating of the complex ring structures in coal and tar occurred. Pier carried out the coal tar hydrogenation at 450° C and 200–300 atm in small, continuous-flow, externally heated, high-pressure, carbon steel converters of 25 mm inner diameter containing pellets of molybdenum oxide catalyst (MoO_3) arranged in a fixed bed that filled a space of 50 ml. This arrangement gave maximum concentration of the catalyst in the converter's reaction space. The experiments were a success, the first conversion occurring on 14 January 1925 when Pier hydrogenated Oppau brown coal tar to give an almost 100% yield of a water-clear, oxygen-free gasoline. The reaction produced no coke or other residue,

12 Stranges, "Bergius" (note 9), pp. 663–65; *High-Pressure* (note 3), Vol. 1, p. 41. BASF concluded only a patent agreement not an exchange of information with Bergius. On 9 December 1925 BASF merged with five other German chemical companies to form the IG Farben cartel.

13 Donath and Höring, "Early Coal" (note 10), p. 5.

and the product after slightly refining was equivalent to good commercial grade gasoline.[14]

Brown coal hydrogenation was identical to tar hydrogenation except for the grinding of coal to a powder before reacting with hydrogen gas. The reaction initially produced 55% tar and other oils which when hydrogenated under the same conditions gave gasoline. Pier's experiments clearly revealed the superiority of hydrogenation to LTC in producing coal tar and oil; yields averaged about 55% compared to 15% from LTC.

Shortly after beginning the small converter hydrogenations, Pier began the design, construction, and operation of more efficient and considerably larger converters that measured 12 m in length and 300 mm, 500 mm, and 800 mm in diameter. Unlike the externally heated 25 mm converters, ring-shaped electric heaters that minimized radiation loss, internally heated the large converters. Testing began in February 1925 with the hydrogenation of brown coal tar at 450° C and 200 atm over a fixed bed molybdenum oxide-aluminum oxide catalyst in a 300 mm converter (100 L volume). By November-December 1925 Pier had hydrogenated coal-oil paste and crude oil (petroleum residues) from Elwerath in a 500 mm converter (166 L volume). Three more 500 mm converters were operating by January-March 1926, and two weeks later hydrogenation began with the 800 mm converters (266 L volume). Because of their increased reaction volume the large converters allowed sufficient contact between reactants without having to increase the velocity of the incoming hydrogen gas. This eliminated the problem of having a too-rapid hydrogen flow through the converter which reduced the reactants' contact time and consequently the yield of liquid products. The much greater throughput and lower hydrogen recycle rate, because of greater conversion to liquid products, clearly improved the economics of coal and tar hydrogenation, and significantly increased the possibility of industrial success.

A second important advance resulted from the large converter hydrogenations. Pier had observed that as the coal-oil paste, tar, or crude oil throughput increased, catalytic activity decreased. Follow-up investigations to determine why the decrease occurred showed that instead of vaporizing, the high-boiling, asphalt-like hydrocarbons present in the reactants had deposited as a film on the catalyst and greatly inhibited its effectiveness. What unexpectedly emerged as a serious obstacle to industrializing the hydrogenation process, Pier eliminated in 1925 with the simple yet brilliant innovation of dividing the reaction into two stages, both taking place at the same temperature and

14 German Patents 608,466 and 633,185 (both 14 February 1925).

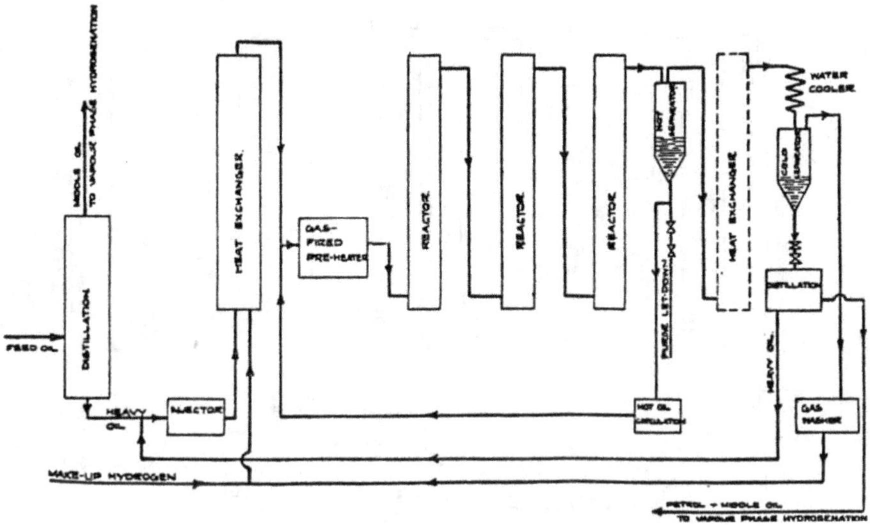

FIGURE 3.2 Liquid phase hydrogenation converters (reactors)
SOURCE: *Report on the Petroleum and Synthetic Oil Industry of Germany* (LONDON: MINISTRY OF FUEL AND POWER, 1947), P. 54. SYNTHETIC FUELS COLLECTION, TEXAS A&M ARCHIVES

pressure, 400–500° C and 200–300 atm: (1) a liquid phase with a suspended (dispersed) catalyst for converting coal, tar, and crude oil into middle oil with a boiling point range of 165–230° C and (2) a vapor phase for converting middle oil into gasoline with a boiling point range of 25–165° C by passing the vapors over a fixed bed catalyst. Pier at first had arranged the liquid phase catalyst in a fixed bed, but he soon realized that the permeation of inorganic coal ash throughout the catalyst bed and the high absorption of asphalt-containing substances seriously reduced catalytic action. To eliminate the problem Pier added the catalyst in a finely dispersed state and at a higher concentration (25–30% of the reactants). He also found that sharply topping (distilling) the middle oil at 230° C extended catalyst life because the topping removed the heavier oil fractions and prevented any condensation on the catalyst. These two changes significantly improved the liquid phase hydrogenation, but the complexity of the liquid phase reactants still made the hydrogenation difficult to carry out. Coal ash and asphalt deposits presented no problem for the vapor phase hydrogenation, and the fixed bed arrangement remained the accepted design.

Extensive research on catalysts continued throughout 1925–26. Pier intended to find or prepare catalysts that most easily split the coal and tar ring structures into smaller molecules in the liquid phase and other catalysts that

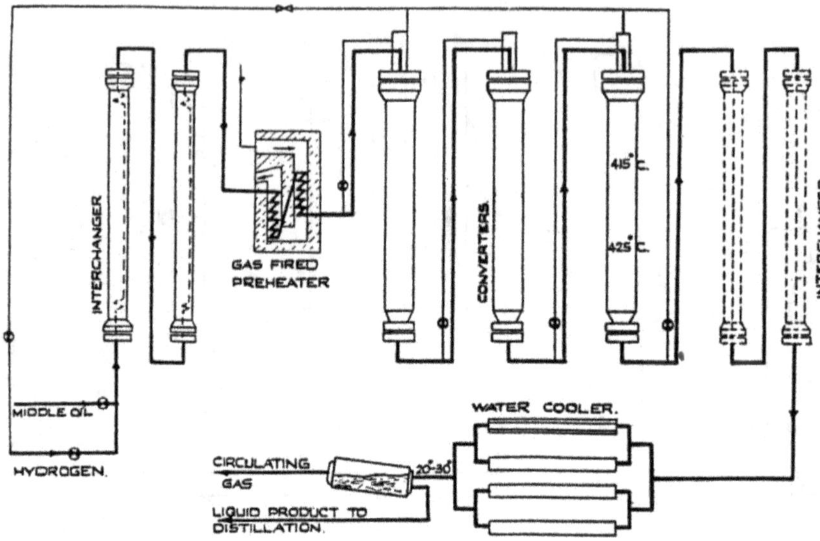

FIGURE 3.3 Vapor phase hydrogenation converters
SOURCE: *Report on the Petroleum and Synthetic Oil Industry of Germany* (LONDON: MINISTRY OF FUEL AND POWER, 1947), P. 51. SYNTHETIC FUELS COLLECTION, TEXAS A&M ARCHIVES

facilitated the hydrogenation of the smaller molecules in the vapor phase. By mid-1926 IG Farben had forty 25 mm diameter converters in operation, testing almost every chemical element in both its free and combined state. Molybdenum and tungsten compounds remained the best splitting and hydrogenating catalysts, especially molybdic acid-zinc oxide-magnesia (catalyst 2365), which Pier prepared in April 1926, and molybdic acid-chromic acid (catalyst 3884) which he prepared in May. IG Farben used both catalysts in its first industrial-size plant. Pier also found that cobalt and Group V and VI elements were good splitting and hydrogenating catalysts followed in order of decreasing activity by Group III and IV elements, silver, copper and zinc compounds, and finally nonmetallic compounds.[15]

By mid-1926 Pier's high-pressure experiment group had established the major chemical and engineering principles of coal and tar hydrogenation and was ready to begin industrial production once IG Farben selected a suitable plant location. IG Farben owned large brown coal deposits near Nachterstedt

15 *High-Pressure* (note 3), Vol. 1, pp. 34–38; Donath and Höring, "Early Coal" (note 10); French Patent 612,504, English Patents 249,156, and 274,404 (all 14 March 1925). The patents were for the two-phase process.

in central Germany, in Geiseltal, and near Ammendorf. The last two locations already supplied coal for the ammonia plant in Leuna, making Nachterstedt and Leuna the likely choices. Leuna's existing hydrogen plant, power supply, and other supporting facilities gave it a clear advantage over Nachterstedt, and on 18 June 1926 IG Farben announced plans to build in Leuna a large pilot plant employing 15,000 workers and having an annual production of 100,000 metric tons of gasoline. Construction began on 1 November 1926 and upon completion consisted of 18 converters, 18 m long, 800 mm in diameter, arranged in three groups, with each group containing three double (pairs of) converter stalls (enclosures). Four Winkler generators installed in 1926 had a water gas output (hydrogen source) of 60,000 m^3/hr/generator. Plant operation started on 1 April 1927, producing gasoline from crude oil, brown coal, and coal tar obtained from LTC of coal.[16]

3.3 *Hydrogenation 1927–32: Beginning of Leuna*

With construction underway in Leuna, research continued in IG Farben's high-pressure laboratory in Ludwigshafen. By mid-1927 Pier's high-pressure experiment group tested stalls, containing four 300 mm diameter, five 500 mm diameter, and five 800 mm diameter converters, each 18 m long. The tests provided Pier with data for the Leuna plant and showed, as their earlier tests showed, that increasing converter size, lowered operating costs. Because Pier lacked suitable hydrogen-resistant steel alloys for converter construction until 1929, he had to heat separately the hydrogen gas and the reactants before reaction. An electric heater inside the liquid and vapor phase converters heated the hydrogen; fusion baths and later hot water coils heated the brown coal-oil paste, brown coal tar, or crude oil. The converters also had an internal heat exchanger to control the reaction's temperature. Unlike the methanol synthesis, an almost trouble-free reaction from the start, synthetic liquid fuel production had to overcome a number of technical and economic problems such as the choice of brown coal, brown coal tar, or crude oil as the reactant; the corrosion and failure of equipment; and the preparation of suitable liquid and vapor phase catalysts. Pier's group experimented with about 1,000 possible catalysts each year, while continuing to give special emphasis to molybdenum catalysts.[17]

16 *High-Pressure* (note 3), Vol. 1, pp. 39–41.

17 *High-Pressure* (note 3), Vol. 1, pp. 53–54; H. H. Storch, "Hydrogenation of Coal and Tar," in *Chemistry of Coal Utilization*, ed. H. H. Lowry, 3 Vols. (New York: John Wiley & Sons, 1945 and 1963), Vol. 2, pp. 1750–96, on p. 1768.

Pier tested the first of the new molybdenum catalysts on 20 August 1927. He found that substituting molybdic anhydride for molybdic acid in catalyst 2365 increased 2365's activity giving the new catalyst 3510 with the percentage composition 53.5 MoO3-30 ZnO-16.5 MgO. In comparison with other catalysts, 3510 proved excellent for splitting, hydrogenation, aromatization (hydroforming or isomerization: the formation of branched chain aromatic, cyclic unsaturated, hydrocarbons from straight run gasoline), and refining. Catalyst 3510 had superb mechanical strength, and for three years it remained the best liquid and vapor phase hydrogenation catalyst. It had reduced activity in the liquid phase where the amount of ash and unused coal increased rapidly and resisted separation from the catalyst. Saturating the coal-oil paste with a solution of 3510 (0.02% by weight of coal) improved its effectiveness, giving yields of 70% and above. Pier tested a number of other active catalysts such as 3884 (90 MoO3-10 CrO3-20 kaolin-0.5 powdered aluminum), ammonium phosphomolybdate, ammonium silicomolybdate, 2500 (tungstic acid-magnesia), and 2700 (molybdic acid-chromic acid-magnesium carbonate), but they lacked the mechanical strength required for industrial use.

Continued experimentation with 3510 in 1928–29 showed that during the hydrogenation reaction molybdenum oxide changed to the sulfide, and in 1930 this led Pier to a renewed study of sulfide catalysts and to the discovery that sulfide catalysts gave highest yields when their crystalline structure contained no oxygen. Tungsten sulfide (WS_2, catalyst 5058), which Pier prepared on 27 November 1930 from decomposing sulfotungstate and by the reaction of tungstic acid with hydrogen sulfide under pressure, gave the best results. He prepared molybdenum sulfide in a similar way.

Tungsten sulfide represented the first of a new class of sulfide catalysts, and it had several advantages over the previously tested oxide and sulfide catalysts. It worked well with a higher coal or coal tar throughput at a 30–50° C lower temperature, resulting in greater conversion to liquid fuel and less production of the hydrocarbon gases methane and ethane. It had better splitting and aromatization properties giving better quality gasoline. In tests with aromatic hydrocarbons and middle oil obtained from bituminous coal, catalyst 5058 converted them to gasoline at 380–400° C versus 470° C for catalyst 3510, and it converted crude oil and brown coal middle oil to gasoline at 420° C. Its introduction at Leuna in April 1931 permitted gasoline production from brown coal and tar, bituminous coal and tar, and cracking oils.[18]

Production at Leuna also led the Ludwigshafen and Leuna Materials Testing Laboratories to the introduction and development of high-pressure steel

18 *High-Pressure* (note 3), Vol. 1, pp. 42–47.

alloys. For long-term operation the hot sections of the hydrogenation plant such as the heat exchangers, converters, preheaters, electric heaters, and their connecting lines, required steel alloys that retained their strength at high temperatures and resisted attack from ash and other solids, and from the gases hydrogen and hydrogen sulfide. These two gases proved especially troublesome because the brown coal hydrogenated at Leuna had a high sulfur content.

Bosch and Mittasch had observed hydrogen's corrosive effect, its embrittlement and decarbonization of plant equipment, during the ammonia synthesis and had eliminated the problem with a double-tubed converter that had an inner tube of low carbon steel enclosed in a tightly fitting outer tube of medium carbon steel with a large number of small holes. They also had experienced the troublesome reaction between elemental iron and carbon monoxide during the methanol synthesis and had solved that problem initially with the use of copper and copper alloys, and finally in 1929 with V2A steel, a good hydrogen resistant steel containing 96.8–97.4% iron, 1.1–1.4% manganese, 1.1–1.4% silicon, and 0.4% carbon. For coal and tar hydrogenation copper's reactivity with sulfur made it unsuitable. Aluminum resisted attack from sulfur, but because it reacted with oxygen (even when combined in the water molecule) and lost its mechanical strength at the high reaction temperature, it too was unsuitable. Galvanized steel tubes held up well in the heat exchangers but not in the other hot sections. Only the V2A steel alloy withstood corrosive attacks by hydrogen and the other substances present in either the reactants or products, and it remained the main structural material until the introduction of chromium steels in the early 1930s. Low chromium steels containing 2.5% chromium and 0.5% molybdenum provided good resistance at 350–400° C and 300–400 atm. Those with 6% chromium remained fairly resistant to 400–600° C and 300 atm, whereas a 5% chromium and 0.5% molybdenum steel provided resistance up to 600° C and 250 atm. Corrosion became negligible only when the chromium content reached 12%. By 1938 the German coal and tar hydrogenation plants were using cheaper high-temperature chromium steels such as N6, N8, and the best of them N10 (3.0–3.6% chromium, 0.5% molybdenum, 0.3% tungsten, 0.75% vanadium, and 0.18–0.22% carbon).[19]

19 Stranges, "Bergius" (note 12); *High-Pressure* (note 3), Vol. 1, pp. 55, 129; Storch, "Hydrogenation" (note 17), p. 1793; Ronald Holroyd, *IG Farbenindustrie AG Works, Ludwigshafen and Leuna. XIII. Hydrogen Resistant Steels, etc.*, CIOS XXX-103, Item No. 30 (1945); Holroyd, *IG Farbenindustrie AG Works*, US Bureau of Mines, Information Circular 7375 (Washington, DC: Bureau of Mines, August 1946), pp. 66–67; J. H. Howell and R. M. Crawford,

FIGURE 3.4 Leuna plant 1927
SOURCE: DIE BASF, AUS DER ARBEIT DER BADISCHEN ANILIN-& SODA-FABRIK AG HELFT 1/1965.15 JAHRGANG, P. 34. BASF ARCHIVES, LUDWIGSHAFEN

3.4 *Engineering and the Economics of Hydrogenation*

Because of its location in the heart of Germany's brown coal district the Leuna plant from its opening in April 1927 until July 1929 had hydrogenated brown coal along with brown coal tar and small amounts of crude oil. During Leuna's development period from 1924 to 1927, the western world's liquid fuel market faced a severe price war, preceding by only a few years the US stock market crash of 1929, the collapse of the German banking establishment in 1931, and a global depression that lasted for most of the 1930s. For the Leuna plant with a projected annual gasoline production of 100,000 metric tons, a higher or lower cost of one pfennig/kg represented a difference of RM 1 million. Because of the unfavorable economic climate and no immediate possibility of lowering dramatically its production costs, IG Farben decided in July 1929 to discontinue its less profitable brown coal hydrogenation and to hydrogenate only brown coal tar. IG Farben produced the tar in its Leuna plant from LTC of brown coal

---

The Production of Synthetic Fuels by the Hydrogenation of Solid and Liquid Carbonaceous Materials. XII. Materials of Construction, US Naval Technical Mission in Europe, Technical Report 217–45 (New York, New York, August 1945), pp. 99–117; R. Taylor, "High-Pressure Technique," *Chemistry and Industry*, 56 (1937): 41–50T.

TABLE 3.3 IG Farben coal and tar hydrogenation

Year	Production (metric tons)	Production cost (RM/metric ton)
1930	82,000 (80,000)	471 (450)
1931	108,000	272
1932	93,000	254
1939–45	65,000 (1944 capacity)	190

SOURCE: COURTESY OF AUTHOR

or obtained it from Riebeck-Montan mining company and other central German and Hessian LTC plants. It added a fifth Winkler generator with an output of 30,000 m^3/hr of water gas in 1930 and planned to hydrogenate enough tar to produce annually 300,000 metric tons of gasoline at a cost of RM 220/metric ton, but Germany's poor economy limited Leuna's production in 1930 to 80,000 metric tons of a high-quality gasoline that cost RM 450/metric ton.

The switch from brown coal to brown coal tar failed to lower the production cost, but in 1930–31 Pier introduced several engineering changes that significantly simplified and improved hydrogenation. These changes, especially in the simpler vapor phase hydrogenation, enabled IG Farben to reduce considerably the cost of gasoline produced from brown coal tar and beginning in 1931 from Volkenroda crude oil. The first improvement resulted from mixing the tar and hydrogen gas in the liquid phase converters to produce a uniform temperature throughout the system and then repumping the hot tar and hydrogen gas. This provided a high linear flow velocity and simplified converter construction by eliminating the converter insets required for heat exchange and stirring. By 1930 Pier had connected two liquid phase converters in series and three converters the following year. The series arrangement produced stalls with enough throughput to make hot repumping unnecessary because the higher linear flow velocity adequately mixed the tar and hydrogen.

Pier also connected the vapor phase converters in series in stalls, and after the introduction of chromium steels in the early 1930s heated them with an outside electrical or gas-heated tubular preheater while adding sufficient cold hydrogen gas to control the reaction temperature. As it had done since 1927 IG Farben continued to sell its synthetic gasoline through Deutsche Gasoline AG; Deutsche-Amerikanische Petroleum AG, Standard Oil's German subsidiary; and Rhenania. IG Farben delivered its first sample of aviation gasoline to Lufthansa in 1930.

IG Farben's economic problems did not disappear despite hydrogenation's improved engineering. Its synthetic ammonia production collapsed within this period and gasoline's market price fell because of the Depression and because of the gasoline wars and underselling by the Russian and Romanian oil companies supplying the German market. Gasoline sold for about the price IG Farben paid to buy tar for hydrogenation. IG Farben had no choice but to reverse its position, and in September 1931 it resumed brown coal hydrogenation in pilot plant tests, showing that if Leuna produced annually 300,000 metric tons of gasoline from brown coal, the cost fell to RM 190/metric ton. As a result, in September 1932 IG Farben abandoned its plan for gasoline production from brown coal tar and Volkenroda crude oil and continued the hydrogenation with only brown coal. The resumption of brown coal hydrogenation in Leuna effectively concluded IG Farben's brown coal research program and marked the start of its pilot plant research on bituminous coal hydrogenation.[20]

3.5 *International Developments: IG Farben-Standard Oil Connection*

During the Leuna period of coal and tar hydrogenation, IG Farben pursued several agreements for the international development of coal, tar, and crude oil hydrogenation. It already had acquired the Bergius patents for RM 2.850 million in May 1925, and from 1927 to 1929 representatives from Belgium, France, England, United States, Holland, Luxembourg, and South Africa visited Ludwigshafen and Leuna to inspect the plants and to consider the benefits of developing their own synthetic fuel industries. IG Farben concluded its most important agreements with Standard Oil (New Jersey), Imperial Chemical Industries (ICI) in Billingham, England, and Royal Dutch-Shell in Amsterdam, during the period from September 1927 to April 1931.[21]

Standard Oil first became interested in coal hydrogenation in 1922 when Frank Howard (1889–1964) engineer, patent attorney, and general manager of Standard Oil Development, sent Ross H. Dickson, a Standard engineer, to investigate the progress Bergius had made on hydrogenation during World War I. Dickson's visit became the first of several visits by Standard's officials to Bergius and later to BASF. His visit occurred at the time Bergius had just resumed postwar work at Rheinau-Mannheim, and he correctly reported to Howard that high-pressure hydrogenation of coal, tar, and crude

20 *High-Pressure* (note 3), Vol. 1, pp. 54–59.

21 Stranges, "Bergius" (note 12); Anthony N. Stranges, "From Birmingham to Billingham: High-Pressure Coal Hydrogenation in Great Britain," *Technology and Culture*, 26 (1985): 726–57.

oil appeared nowhere near industrial development. Howard remained interested in Bergius's work, nevertheless, and relied on Heinrich Riedemann (1871–1959), head of Deutsche-Amerikanische Petroleum Gesellschaft (DAPG), Standard's German sales organization in Hamburg, to keep him up-to-date on hydrogenation financing and development and financial state.[22]

Three years later, in spring 1925, when BASF's managing director Wilhelm Gaus (1876–1953) and other BASF directors arrived in New York City for patent negotiations with the DuPont Company, Howard arranged for them to tour Standard's New Jersey refineries. The BASF tour prompted Gaus to invite Howard to BASF's research center in Ludwigshafen. Standard had just begun to establish an industrial research organization in late 1919–20, and a visit to Ludwigshafen would enable Howard to see firsthand one of the world's finest research centers. Howard went to Ludwigshafen in late March 1926 which at that time still remained under French military occupancy (until June 1930). A few days later Walter Teagle (1878–1962), Standard's president 1917–37, and James A. Moffett Jr. (1886–1953), a Standard director, joined him for an inspection of the Ludwigshafen's plant and laboratories. That summer Bosch, now IG Farben's principal, visited Howard and other Standard officials in New York City.

The result of these visits occurred on 27 September 1927 when IG Farben and Standard Oil signed a twenty-five year contract to begin a joint program for hydrogenation research in the United States. At IG Farben's request, the two companies supplemented the contract with an exchange of letters indicating the good intentions of each should unforeseen developments arise. Under the contract's terms, Standard paid IG Farben no cash but committed itself to spending millions of dollars on high-pressure research and gained the right to use the hydrogenation process in the United States upon paying a fair royalty to IG Farben. Standard also retained patent ownership of improvements it made on hydrogenation and would pay IG Farben one-half of any royalties received in the United States. IG Farben retained all its US patents and controlled their licensing to other companies, but to compensate for Standard's expenditures, IG Farben agreed to pay Standard one-half of any royalties.[23]

22 George S. Gibb and Evelyn H. Knowlton, *The Resurgence Years, 1911–1927: History of Standard Oil Company (New Jersey)*, 4 Vols. (New York: Harper and Brothers, 1956), Vol. 2, p. 545. Four volumes written by different authors and published at different times (1956, 1958, 1971, 1988) give the history of Standard Oil.

23 *Ibid.*, 546; Henrietta M. Larson, Evelyn H. Knowlton and Charles S. Popple, *New Horizons, 1927–1950: History of Standard Oil Company (New Jersey)*, 4 Vols. (New York: Harper and Row, 1971), Vol. 3, p. 154. Gauss was IG Farben's chairman 1931–37.

Both firms viewed the arrangement favorably. IG Farben would acquire Standard's experience in heating, distilling, and refining petroleum and in the examination and evaluation of petroleum products. Alternatively, Standard expected to benefit from IG Farben's research on de-asphaltizing, viscosity lowering, desulfurizing hard-to-crack oils, upgrading illuminating and lubricating oils, and to use IG Farben's experience with catalytic high-pressure to maximize the yield of gasoline and other high-grade petroleum products. Standard clearly was thinking about the recently released 1926 Federal Oil Conservation Board's report that had mentioned the possible exhaustion of US petroleum reserves in six years.[24] Gasoline demand continued rising at the time, and even with the latest cracking methods two barrels of crude oil yielded only one barrel of gasoline. Standard already had looked into the possibility of extracting oil from shale deposits it owned in Colorado, Wyoming, and Utah, but it found IG Farben's process of converting crude oil and tar into gasoline much more attractive.[25]

Even before signing the 1927 contract, Standard had started construction on a large new development laboratory at its Baton Rouge (Louisiana) refinery. To head the new laboratory Howard convinced Robert J. Haslam (1888–1961), professor of chemical engineering at MIT and director of MIT's School of Chemical Engineering Practice and its Research Laboratory of Applied Research, to take a leave of absence in the spring of 1927. The leave became permanent. Robert P. Russell, Haslam's assistant director at MIT who headed a research staff of twelve chemical engineers, among them W. C. Asbury, M. W. Boyer, P. J. Byrne, Jr., G. H. B. Davis, W. V. Hanks, and J. M. Jennings, joined Haslam at Standard.[25]

Haslam and several of his staff went to Ludwigshafen early that summer to inspect IG Farben's facilities. Upon their return they constructed and tested a small, 0.5 bbl/d experimental plant with four converters. A few months later, in November-December 1927 Pier made his first visit to the United States, and at that time the two companies made plans to construct industrial-size 5,000 bbl/d hydrogenation plants at Baton Rouge, Bayway (New Jersey), and Baytown (Texas). Within a year, in August 1928, Standard began pilot plant operation, hydrogenating crude oil in a small 100 bbl/d pilot plant at Baton Rouge.

---

24 *High-Pressure* (note 3), Vol. 1, pp. 60–61; Federal Oil Conservation Board, *Public Hearing*, 27 May 1926 (Washington, DC: Government Printing Office, 1926); Federal Oil Conservation Board, *Complete Record of Public Hearings*, 10 and 11 February 1926 (Washington, DC: Government Printing Office, 1926); Anthony N. Stranges, "The US Bureau of Mines' Synthetic Fuel Program, 1920–1950s: German Connections and American Advances," *Annals of Science*, 54 (1997): 29–68, on pp. 32–34.

25 Robert P. Russell and Robert J. Haslam, "Hydrogenation of Petroleum," *Industrial and Engineering Chemistry*, 22 (1930): 1030–37.

It cooperated closely with IG Farben on hydrogenation research, and during this initial period Ludwigshafen and Baton Rouge frequently exchanged information and scientists.[26] By the end of 1928, the success of its hydrogenation research convinced Standard "that hydrogenation was the most important scientific development that had ever occurred in the oil industry." From one barrel of crude oil its refineries could produce more than one barrel of high-grade gasoline. Hydrogenation made better lubricants possible now, and unwanted by-products would become something in the past.[27]

Nevertheless, despite the cooperation with IG Farben, Standard had reservations about the 1927 contract. It believed that hydrogenation had a profitable future and wanted to broaden the contract to permit its foreign affiliates to obtain licenses for the process. IG Farben, on its part, recognized that Standard's rapid progress in hydrogenation might enable it to develop new applications for hydrogenation without adequately compensating IG Farben. Being primarily a chemical company, IG Farben also was quite willing to sell its hydrogenation process to an oil company rather than do the licensing itself. Consequently, in March 1929 officials from the two companies met in New York City to negotiate a new contract, with the result that IG Farben sold Standard not only its hydrogenation patents but all other oil-refining patents including any it might obtain in the next seventeen years. The patent rights applied globally except for Germany. In return, IG Farben received $35 million which it requested in Standard's stock (546,011 shares) rather than cash, and a 20% interest in royalties collected on all hydrogenation and related patents.

The contract established a new holding company, the Standard-IG Farben Company in which Standard held 800 shares and IG Farben 200 shares, and called on them to continue exchanging relevant technical knowledge on hydrogenation. Standard thought it had made a good bargain because it had purchased a relatively new process still beyond the capabilities of the oil industry. Despite the progress it had made, in 1929 Standard remained without a first-rate catalytic high-pressure research program.

The 1929 contract eliminated the concerns of the earlier contract, but within a year it, too, appeared inadequate mainly because of the difficulty in defining borderline processes and products as either chemical or petroleum. In October 1930, Standard and IG Farben, therefore, organized another company, the Joint American Study Company or JASCO Incorporated, to which

26 *High-Pressure* (note 3), Vol. 1, p. 61.

27 Larson, Knowlton, and Popple, *New Horizons* (note 23), p. 155; R. T. Haslam and J. T. Ward, "Progress in Fuel Utilization in 1927," *Mechanical Engineering*, 50 (1928): 26.

each assigned its borderline processes for examination on a case by case basis. IG Farben assigned JASCO its world rights, outside of Germany, to its paraffin oxidation process, acetylene arc process, and Buna process. Standard had none to contribute in 1930.[28]

During the period of negotiations, Standard completed construction of the 5,000 bbl/d industrial-size plants at Baton Rouge and Bayway. Each plant, similar in design to IG Farben's plants, and built at a cost of about $5 million had two stalls, each stall holding three or four converters arranged vertically in series. The Standard design converter had a lining of noncorrosive steel alloy and measured 12.2 m (40 ft) in length, 0.91 m (3 ft) in diameter. Both plants began operation in August 1930 using their first stall to hydrogenate and upgrade crude (illuminating and lubricating) oil. Bayway's second stall, an aromatization unit for the manufacture of anti-knock gasoline, started operation in October 1930, making Bayway the first industrial-size hydrogenation plant in the United States. At Baton Rouge the second stall aromatization unit began operation on 23 May 1931, giving the plant an actual capacity of 4,000–8,000 bbl/d. The two plants ran continuously for periods of up to one year, operating at 370° C and 200 atm, with the sulfur resistant catalyst 3510 arranged in a fixed bed inside each converter, over which passed crude oil and hydrogen gas. They produced high quality motor oil, fuel oil, and gasoline, comparable to those obtained from Pennsylvania crude oil.[29] Work at this time began on Standard's third hydrogenation plant at Baytown.[30]

Standard's research on the high-pressure, catalytic hydrogenation of petroleum aroused considerable interest among several American oil companies. Their interest provided Standard with an excellent opportunity to acquire a quick return on its $35 million investment and to maximize its royalty income. In 1930 Standard therefore organized Hydro Patents Company (HP) to license the hydrogenation process in the United States. It invited companies large enough to consider constructing hydrogenation plants of 5,000 bbl/d, minimum size, to purchase stock in the new organization. Seventeen companies participated, among them Texas Corporation, Pure Oil Company (Ohio), Standard Oil Company (Indiana), Skelly Oil Company, Gulf Oil Corporation, and Shell Oil Company. Standard retained 10,000 shares in HP and sold shares to the other companies: thirteen large oil companies purchased 1,000 shares

28 Larson, Knowlton, and Popple, *New Horizons* (note 23), pp. 155–56; John L. Enos, *Petroleum Progress and Profits: A History of Process Innovation* (Cambridge: MIT Press, 1962), p. 190.

29 Russell and Haslam, "Hydrogenation" (note 26).

30 *High-Pressure* (note 3), Vol. 1, p. 62.

each, four small companies acquired 500 shares each. The following year, in 1931, Standard organized its fully owned Hydro Engineering and Chemical Company (HEC), with Russell as general manager, to provide its HP partners with patent information on hydrogenation.[31]

A similar pattern of development took place on the international scene in the early 1930s when IG Farben and Standard discussed hydrogenation agreements with Sir Henri Deterding (1866–1939), general manager and chairman of Royal Dutch Petroleum Company 1900–36 (Royal Dutch-Shell in 1907), and Sir Harry McGowan (1874–1961) in 1930 general manager and chairman of ICI. Royal Dutch-Shell had an interest in hydrogenation even earlier than Standard, and by 1923 it had invested RM 9.7 million in Bergius's hydrogenation plant in Rheinau-Mannheim.[32] With Bergius out of the picture after selling his patents to IG Farben in May 1925, Deterding entered into negotiations with IG Farben and Standard for hydrogenation rights outside of Germany.[33]

ICI, under the direction of Kenneth Gordon (1897–1955), had begun laboratory tests on the hydrogenation of bituminous coal early in 1927, after it bought control in February 1927 of the British Bergius Syndicate, the company that Bergius had set up for the hydrogenation of British coal and tar. By late 1929 ICI had constructed a 10-metric ton/d coal hydrogenation pilot plant in Billingham, and in pilot plant tests its research team led by Gordon and Ronald Holroyd (1904–73) showed that tin (II) oxalate and ammonium chloride were good catalysts for bituminous coal hydrogenation. They were the equal of IG Farben's molybdic acid catalyst (2365) used to hydrogenate brown coal and had the advantage of being cheaper. The two catalysts were also the first bituminous coal catalysts, preceding IG Farben's preparation of catalyst 5058 for bituminous coal hydrogenation on 27 November 1930 and its first use in April 1931 at Leuna. When ICI approached IG Farben to discuss the development of coal hydrogenation it came to the acknowledged expert with its own considerable high-pressure pilot plant experience and with first-rate knowledge of catalysts. The combination of ICI's and IG Farben's expertise contributed to the successful series of pilot plant experiments with German and British bituminous coals that began in Ludwigshafen in September 1931. The success of these experiments justified Leuna's resumption of brown coal hydrogenation

31 Larson, Knowlton, and Popple, *New Horizons* (note 23), p. 157; Enos, *Petroleum* (note 29), pp. 190–91.

32 Stranges, "Bergius" (note 12), p. 663. The US dollar and RM exchange rate dropped rapidly during this period from RM 1 = 17¢ in 1917, to RM 1 = 1.2¢ in 1921, to RM 1 = 0.002¢ in 1923, before leveling off at RM 1 = 24¢ from 1925 to 1932, *Banking* (note 9).

33 Stranges, "From Birmingham" (note 21).

in September 1932 and played an important part in the research that resulted in the construction in 1935 of ICI's Billingham plant that produced annually 100,000 metric tons of gasoline from hydrogenating British bituminous coal.[34]

The culmination of the discussions and negotiations on coal hydrogenation led in April 1931 to IG Farben's, Standard's, ICI's, and Royal Dutch-Shell's organizing another holding company that they named International Hydrogenation Patents Company Limited (IHP). IHP incorporated in Vaduz, Liechtenstein, owned the world rights to coal, tar, and oil hydrogenation in all countries except Germany and the United States. That same year IHP incorporated in the Hague, the International Hydrogenation Engineering and Chemical Company (IHEC) to handle the exchange of research and patent information on coal, tar, and oil hydrogenation. IHP held all the stock in IHEC, of which ICI had only 10%. IHEC drew its staff from the four participating companies. It kept a permanent member in Pier's high-pressure experiment group in Ludwigshafen and performed hydrogenation experiments in Amsterdam, in the laboratories of Royal Dutch-Shell's subsidiary, Bataafsche Petroleum Maatschappij. IHP organized a final holding company, Hydrogenation Construction and Service Company, in 1935 for consulting and for the design and construction of hydrogenation plants. IHP and IG Farben held equal shares in the company.[35]

By 1931 the oil industry's interest in and enthusiasm for catalytic high-pressure hydrogenation had become global. But both had short lives because of the late 1930 discovery of the huge East Texas oil field. One thousand oil wells were operating within eight months after the East Texas oil field's discovery, forcing the price of oil to drop from $1.30/bbl in 1930 to 10¢/bbl and then 5¢/bbl.[36] The number of producing oil wells almost tripled, from 3,612 in 1932 to 9,372 in 1933 with production increasing from 120.4 million barrels to 171.8 million barrels.[37] With the present flood of cheap oil, the assurance of more to come, and the onset of the Depression, Standard could not justify economically hydrogenating crude oil to synthesize fuels and lubricants that refining natural petroleum made available at a cheaper cost. Gasoline obtained from hydrogenation cost about $275/bbl/d more than gasoline obtained by refining petroleum, and it became attractive to refiners only when the cost of

34 Stranges, "Bergius" (note 12); *High-Pressure* (note 3), Vol. 1.

35 Stranges, "From Birmingham" (note 21); *High-Pressure* (note 3), Vol. 1, pp. 62–63; Larson, Knowlton, and Popple, *New Horizons* (note 23), p. 157.

36 Richard O'Connor, *The Oil Barons* (Boston: Little, Brown and Company, 1971), p. 304.

37 Harold F. Williamson, *The American Petroleum Industry*, 2 Vols. (Evanston: Northwestern University Press, 1959), Vol. 2, p. 545.

petroleum gasoline exceeded 7¢/gal. In 1931 Standard therefore ceased the production of gasoline from hydrogenation and restricted hydrogenation at the Baton Rouge and Bayway plants to the production of lubricants. The Pennsylvania oil fields supplied a plentiful source of good lubricants at low cost, but a two-year testing of "Essolube," Standard's hydrogenation product, showed that some of Essolube's properties surpassed those of petroleum-refined lubricants. Produced in high yields from suitable crude oil fractions, Essolube had as high a viscosity index and resistance to oxidation, as long a lifetime, a lower pour point, and a lower carbon residue when compared to Pennsylvania oil lubricants.[38]

By 1932, Standard had spent $30 million in research and plant construction and transferred another $35 million in stock to IG Farben. Despite the technological success of its hydrogenation program, the high cost forced Standard to suspend completion of its Baytown plant and the following year to reduce lubricant production at its industrial-size Baton Rouge and Bayway plants.[39] The small 100 bbl/d Baton Rouge pilot plant continued with the aromatization of petroleum to produce high-grade specialty products such as high flash point safety fuel for motor boat and aviation motors, and special solvents for the dye, lacquer, varnish, soap, and textile industries.[40] The hydrogenation program, while not as vital as originally anticipated, nevertheless stimulated research in catalysis and led to the establishment of a large research laboratory that provided Standard with the expertise it needed to improve old products and develop new ones.[41]

In July 1933 Britain's Institution of Petroleum Technologists organized the first World Petroleum Congress. Delegates from thirty-six countries attended the week-long meeting held at London's Imperial College of Science and Technology.[42] One of the twenty-five symposia, the symposium on hydrogenation, included its foremost investigators: from Germany, Bergius, the inventor of coal, tar, and oil hydrogenation, and Pier of IG Farben; from England, Gordon of ICI; from the United States, Haslam, Russell, and Asbury of Standard. All of them spoke of the great technical advances made in catalytic high-pressure

38 R. T. Haslam, R. P. Russell, and W. C. Asbury, "Comparison of Cracking and Hydrogenation and Methods of Producing Gasoline," *Proceedings of the World Petroleum Congress*, 2 Vols. (London: Office of the Congress, 1934), Vol. 2, pp. 309–16.

39 *High-Pressure* (note 3), Vol. 1, p. 63; Larson, Knowlton, and Popple, *New Horizons* (note 23), p. 158.

40 *High-Pressure* (note 3), Vol. 1, p. 64.

41 Larson, Knowlton, and Popple, *New Horizons* (note 23), pp. 158–59.

42 See the papers published in *Proceedings of the World Petroleum Congress* (note 38).

hydrogenation in the last twenty years, but no one mentioned a bright future for hydrogenation. They must have felt a sense of gloom. Because of the global Depression, petroleum consumption had decreased while the potential to produce more petroleum had increased significantly with the discovery of the East Texas oilfield in October 1930. A steady supply of low-cost petroleum from Iraq, Kuwait, and Iran in the mid-east reinforced the feeling that supplies were going to outrun consumption for a long time.[43] Why else would Jean Baptiste August Kessler Jr. (1884–1972), Henri Deterding's right-hand man, lead a campaign in the early 1930s to organize the major oil producers into one large world trust.[44] Yet, at the same time, Germany, England, and Japan, which lacked large deposits of natural petroleum and sought a way to reduce or even eliminate their dependence of petroleum imports, became the leaders in establishing a global coal-based synthetic fuel industry.

3.6 *Research and Development at IG Farben in the 1930s–40s: Technical Advances*

The ongoing research of Pier and his high-pressure experiment group resulted in a number of significant improvements that benefited the high-pressure synthetic fuel industry's science-engineering and economics. Pier's earlier catalytic studies, had shown the catalytic effectiveness of ammonium molybdate, aluminum chloride, and tin (II) chloride or tin (II) oxalate, about 1% by weight of the coal-oil paste, in the liquid phase at 250–300 atm. New studies enabled Pier in 1934 to replace those scarce compounds with cheaper yet effective iron oxides and iron (II) sulfate catalysts for brown coal hydrogenation at 300 atm and in 1936 for bituminous coal hydrogenation at 700 atm. Pier's high-pressure group had begun small-size tests at 700 atm in 1933, and by late 1936 it had shown that the higher pressure broke down the asphalt present in bituminous coals and high hydrogen content brown coals, such as Rhine coal, to give an asphalt-free heavy oil. This increased the coal-oil paste throughput and improved coal utilization.

Pier also knew that dividing the vapor phase hydrogenation into successive saturation and splitting reactions that took place in separate stalls at different temperatures and pressures, and with different catalysts, gave better product control. Tungsten sulfide (catalyst 5058) was an excellent saturation catalyst and a good splitting catalyst but produced a relatively saturated mediocre

43 John M. Blair, *The Control of Oil* (New York: Vintage Books, 1978), pp. 33, 42–43.

44 Glyn Roberts, *The Most Powerful Man in the World, the Life of Sir Henri Deterding* (Westpoint, CT: Hyperion Press, 1976), p. 290.

gasoline. Pier eliminated its splitting deficiencies by developing catalyst 6434 in November 1936. This was an activated terrana earth catalyst (Fuller's or active earth, aluminum silicates impregnated with 10% tungsten sulfide) that gave a better quality aromatic gasoline in the splitting reaction. It increased octane numbers from 64–74 to 87 and met all the requirements for aviation gasoline. Tungsten, chromium, and vanadium oxides, and activated carbon impregnated with iron were other active splitting catalysts that Pier developed.

The introduction of 700 atm in 1936 also improved vapor phase hydrogenation. Higher pressure eliminated the need for oil scrubbing (washing or purifying with oil from the process) of the recycled hydrogen gas to remove the other gases, mainly oxygen, nitrogen, methane, and ethane that accumulated in the circuit and permitted a combined saturation-splitting hydrogenation of oils unsuitable for a combined vapor phase hydrogenation at 300 atm. By this time Pier had prepared better splitting catalysts of zinc, chromium, and molybdenum on terrana earth for hydrogenating at 700 atm. Six of the German plants constructed from 1937 to 1944 at Welheim, Gelsenberg, Pölitz, Lützkendorf, Wesseling, and Blechhammer used these catalysts for bituminous and brown coal and tar hydrogenation at 700 atm in either the liquid or vapor phase, or in both hydrogenation phases.

Studies in the 1930s by Pier and particularly by the British chemists J. Ivan Graham and H. G. Shatwell at the University of Birmingham and James G. King, L. Horton, and F. A. Williams at the Fuel Research Station in East Greenwich led to a good understanding of the hydrogenation reaction's mechanism. These studies suggested that initially coal broke down into smaller molecules (depolymerization), followed by dispersal of these molecules in the solvent (pasting) oil. Some coals dissolved completely in anthracene oil, and all coals ranking below anthracite in carbon content (bituminous and brown) dispersed considerably even in the absence of hydrogen gas. Easily dehydrogenated oils (hydroaromatics of condensed ring structures containing phenolic groups) were the most effective solvents and served as hydrogen donors to the depolymerized coal molecules. The addition of high-pressure hydrogen gas permitted a higher solution temperature resulting in greater solubility to give a tar-like product. Hydrogenation, therefore, required a continuous breakdown and hydrogenation of the coal molecules while opposed by a repolymerization reaction. Temperature, pressure, reaction time, and catalyst determined the direction and extent of reaction. Too high a reaction temperature resulted in excessive gaseous hydrocarbon formation whereas increasing the amount of catalyst and pressure at a higher temperature accelerated the reaction and reduced repolymerization.

The industrial-size coal hydrogenation process that IG Farben developed at Ludwigshafen and Leuna, and which remained unchanged, occurred in two stages. In the first stage or liquid phase hydrogenation, moisture- and ash-free coal (5% ash, 8–10% water), after crushing in hammer mills, drying, and adding a catalyst, entered a pasting mill for mixing with an equal weight of heavy oil recycled from the process. Next, the coal-oil paste passed into tanks equipped with stirrers and heated to 100° C before entering hydraulically operated paste (piston) pumps for compression to the reaction pressure of 250–300 or 700 atm. From there the coal-oil-catalyst paste traveled to a heat exchanger(s) for heating to 300–350° C by heat exchange with hot steam and mixing with a large excess of hydrogen gas, also at 250–300 or 700 atm. The reactants then passed into a gas-heated preheater(s) for additional heating to 400–420° C before flowing through 3–4 converters connected in series where liquid phase hydrogenation occurred. Cold hydrogen gas added to the converters controlled the reaction temperature.

The converters, each 18 m long, 800–1,200 mm in diameter and having 8 m^3 of reaction volume, comprised a coal hydrogenation stall. The 18 m × 800–1,200 mm converter became the standard size installed in the coal hydrogenation plants. They converted about 95% of the coal to middle oil vapors and heavy oil liquids, the proportions depending on the reaction time and temperature; the oil products also contained excess hydrogen and the gases nitrogen, methane, and ethane. Upon completion of the liquid phase hydrogenation the products left the last converter and entered a hot catchpot (separator) to reduce coking. There the heavy oils, mixed with coal ash, unconverted coal, and catalyst were let down, or removed continuously, from the bottom of the catchpot at a temperature selected to give a liquid (heavy oil let down or HOLD) containing 25–30% solid residue. Diluting the HOLD with additional heavy oil and centrifuging, separated the solid residue which after carbonizing (distilled in the absence of air) to recover more oil, was burned as fuel. The separated heavy oil, after mixing with the oil obtained from carbonization, was recycled to the pasting mill and used to paste more coal for reaction.

The middle oil vapors, which contained most of the oil synthesized in the reaction, and gases including excess hydrogen, passed from the top of the hot catchpot into a heat exchanger where they gave their heat to the incoming coal-oil-catalyst paste. From there the oil vapors traveled to a water cooler(s) where they condensed to oil liquids and then entered a cold catchpot for a two-step release of the pressure from 250–300 atm or 700 atm to atmospheric pressure. The oil liquids flowed to a distillation column for separation into light oil (gasoline), middle oil, and heavy oil; the heavy oil, after mixing with heavy oil from the hot catchpot, returned to the pasting mill; the light oil and middle oil passed to the second stage vapor phase for saturation and splitting hydrogenation.

Excess hydrogen from the cold catchpots, after oil scrubbing, flowed back to the liquid phase heat exchangers for reaction with more coal-oil-catalyst paste. The rate of hydrogenation depended directly on the hydrogen pressure and required oil scrubbing to purify the recycled hydrogen and maintain a maximum hydrogen pressure. The products contained oxygen, nitrogen, and sulfur compounds, and as a result, the liquid phase hydrogenation had very little effect on the quality of the products.

Second stage or vapor phase hydrogenation was similar but much simpler than liquid phase hydrogenation because the light and middle oil reactants had much simpler compositions. IG Farben also simplified and significantly improved the vapor phase hydrogenation by omitting the hot catchpot, so that oil and hydrogen leaving the last converter in both the saturation and the splitting stall directly entered the first of two heat exchangers and a water cooler for a two-step pressure release from 250–300 or 700 atm to atmospheric pressure before passing to the distillation column. In the saturation reaction, an injector forced liquid phase light and middle oil, mixed with an excess of hydrogen gas (2,000 m^3 hydrogen per metric ton of oil) and heated in heat exchangers and a preheater to the reaction (vaporizing) temperature, into the saturation stall's converters at 300 atm or 700 atm. The stall, as in the liquid phase hydrogenation, had 3–4 converters in series, each having a reaction volume of 8 m^3 and a catalyst of tungsten sulfide pellets (catalyst 5058) arranged in a fixed bed. Cold hydrogen gas admitted at various levels of the catalyst beds kept the reaction temperature at about 400° C.

The catalyst helped remove oxygen, nitrogen, and sulfur as their hydrogenated compounds, water (H_2O), ammonia (NH_3), and hydrogen sulfide (H_2S), thereby preventing them from suppressing the catalyst's activity in the splitting reaction. The hot vapors, which contained some light oil produced by hydrogenating the phenol present in the middle oil vapors, left the saturation stall, entered the heat exchangers and water cooler where they condensed and passed into the cold catchpot for the two-step pressure release, and then traveled to the distillation column for separation into light and middle oil. Excess hydrogen gas, which for the 700 atm hydrogenation required no oil scrubbing and only a water washing to remove ammonia after it left the cold catchpot, returned to the vapor phase heat exchangers to hydrogenate more oil. Washing the light oil gave a high-quality gasoline. Distilled middle oil entered the splitting stall, which also consisted of 3–4 converters in series, each containing a fixed-bed catalyst of tungsten sulfide on terrana (catalyst 6434), for hydro-

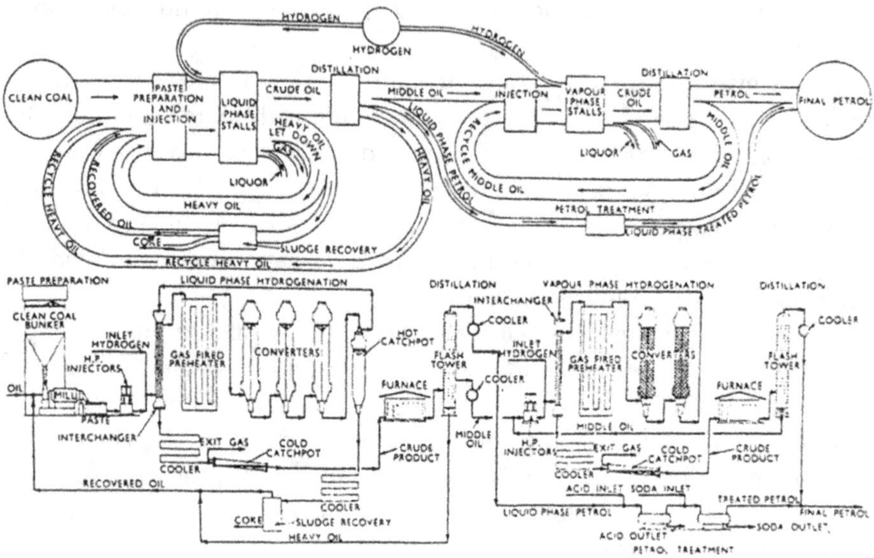

FIGURE 3.5 Schematic diagram for liquid and vapor phase hydrogenation plant
SOURCE: H. H. STORCH, "HYDROGENATION OF COAL AND TAR," IN *Chemistry of Coal Utilization*, ED. H. H. LOWRY, 3 VOLS. (NEW YORK: JOHN WILEY & SONS, 1945 AND 1963), VOL. 2, P. 1774

genation at 300 atm or 700 atm and 449° C. Splitting converted most of the middle oil to a high-quality gasoline.

The two-reaction vapor phase hydrogenation gave a 92–95% yield of motor gasoline with a boiling point of 180° C and a 62–64-octane number and an 80–83% yield of aviation gasoline with a boiling point of 150° C and a 68-octane number increasing to 86 after adding 1.15 mL lead tetraethyl/L of gasoline. A saturation stall hydrogenated 360–870 metric tons of light and middle oil per day, a splitting stall hydrogenated 350–430 metric tons of middle oil per day (500–630 metric tons of light and middle oil). Some unconverted middle oil for recycling, hydrogenation gases (ammonia, hydrogen sulfide), and liquefiable hydrocarbon gases, mainly propane and butane, also resulted. A Linde-Fränkl liquefaction unit, recovered all the gases, which amounted to 8% of the vapor phase products; compressors liquefied the hydrocarbon gases in cylinders as liquid gas.

Brown coal and bituminous coal hydrogenation were essentially the same process. Motor gasoline obtained from bituminous coal had a 69-octane number, aviation gasoline's octane number was 73–74 increasing to 90–91 after adding 1.15 mL lead tetraethyl/L gasoline. Brown coal with its much higher water content of about 50% gave a gasoline with a higher hydrogen content (more saturated) and therefore a lower quality. IG Farben produced high quality aviation gasoline high in aromatics (unsaturated cyclic compounds) by hydrogenating bituminous coal and blending the crude gasoline with about 20% iso-octane. Hydrogenation of coal tar and crude oil also followed the same liquid phase-vapor phase hydrogenation. The only significant differences from coal hydrogenation were that the vapor phase's saturation and splitting reactions used the same catalyst, usually molybdic acid or iron (II) sulfate, and the quantity of heavy oil residue remaining after hydrogenation was much less and of little importance.[45]

3.7 *Germany's Coal Hydrogenation Industry 1933–39*

The signing of the fuel agreement in December 1933 produced immediate results. Within three years Leuna nearly tripled its synthetic liquid fuel production, increasing it from 108,498 metric tons of motor gasoline in 1933 to 332,153 metric tons, including 65,955 metric tons aviation gasoline, in 1936. The increased production resulted largely from increasing the size of the converters from 800 mm to 1,200 mm diameter and the number of converter stalls. Leuna had one 4-converter stall in 1933, in 1935 it had five 4-converter stalls. Production also increased because Leuna's plant engineers more accurately controlled reaction conditions so that the average operational time of a stall increased yearly, from only 11 days in 1927, to 155 days by the mid-1930s, to one year in 1942. Pier's ongoing catalyst research at Ludwigshafen leading to the development of better catalysts such as the vapor phase catalyst 6434 in November 1936 and the introduction of 700 atm hydrogenation in industrial-size plants in 1937–38 also contributed to increased production.

During the same period Germany's coal dealers established several industrial associations such as Brabag, Hydrierwerk Scholven AG, and Ruhröhl GmbH which, prior to the outbreak of World War II on 1 September 1939, constructed seven coal and tar hydrogenation plants.[46] Production costs for motor and aviation gasoline at the plants ranged from RM 181–206/metric ton to RM 360–390/metric ton and depended considerably on the cost of the raw material, brown or bituminous coal at RM 10–21.5/metric ton, tar at RM 40/metric ton, or petroleum residues at RM 123/metric ton. Cost figures for the five plants constructed later, during the war years, varied little.

In the 1930s, prior to World War II's outbreak, industrialized nations, particularly Britain, Canada, and the United States, had studied carefully the articles

---

45 *High-Pressure* (note 3), Vol. 1; *Report on the Petroleum and Synthetic Oil Industry of Germany* (London: His Majesty's Stationery Office, 1947), pp. 51–59.

46 *High-Pressure* (note 3), Vol. 1.

and patents published on Germany's advancing synthetic fuel industry and had visited the synthetic fuel plants. Carleton Ellis (1876–1941), an American consulting chemist at Ellis laboratories in Montclair, New Jersey, visited IG Farben's facilities in Ludwigshafen-Oppau and Leuna in 1930. The war, however, ended the plant visits and limited the availability of Germany's synthetic fuel publications, although Pier and other German scientists continued publication into the 1940s.

During the years Germany was developing its synthetic fuel industry and became the world's leader in synthetic fuel research and development, the Roosevelt administration recognized the real possibility of American involvement in a second world war. As early as 1941, President Franklin Roosevelt (1882–1945) expressed concern over American petroleum reserves, and he established at that time the Office of Petroleum Coordinator for National Defense, to gather information on petroleum products and to make recommendations to insure an adequate petroleum supply. By June 1941, private industry joined the program, lending technical advice to the government. An executive order on 2 December 1942, created the Petroleum Administration for War (PAW) within the Office of Petroleum Coordinator, and from PAW originated the idea of sending technical experts to Germany to collect information on Germany's synthetic fuel industry. Harold Ickes (1874–1952), PAW administrator, in the summer of 1943 made a formal recommendation to the Joint Chiefs of Staff to establish a Technical Oil Mission (TOM) to Germany. The PAW expected private industry and the United States Bureau of Mines (BOM) to capitalize on Germany's considerable experience in synthetic fuel production. With the $30 million that Congress authorized in April 1944 for a five-year synthetic petroleum research and development program, the BOM planned to resume synthetic fuel research where Germany had stopped.

Seven months later, in late November 1944, with Germany's defeat a distinct possibility, the American and British governments and their joint chiefs of staff established the first of three intelligence agencies that would go to Germany to inspect the synthetic fuel plants, interrogate synthetic fuel scientists, and write detailed reports of their investigations. The Combined Intelligence Objectives Sub-Committee (CIOS), with headquarters in London and responsible to the Supreme Headquarters, Allied Expeditionary Forces, was the first of the intelligence agencies. Its proposed 59-member American component, TOM under the general direction of Cornelius S. Snodgrass (1900–74), director of Foreign Refining of the Petroleum Administration for War, by 11 December 1944 consisted of 26 scientists, 6 from the BOM, 2 from PAW, and 18 from industries such as Standard Oil, Pure Oil, Gulf Oil, and Humble Oil. TOM teamed with 40 British scientists from the Ministry of Fuel and Power to establish the CIOS.

Wilburn C. Schroeder of the BOM directed the TOM in its search for documents in eight research areas: (1) petroleum products and testing methods, (2) gasification and carbonization of solid fuels, (3) hydrogenation of coal, tar, and oil, (4) synthesis of oil from water gas, (5) recovery and separation methods for gases, liquids, and solids, (6) oil shale processes, (7) materials of construction, including metallurgy, and (8) by-products from the synthetic oil processes.

The first TOM (CIOS) members, 8 of the 26 TOM members who went to Europe, arrived in Germany in late February 1945 and with their British counterparts entered the German plants as soon as possible after the plants fell into Allied hands. Although inspection of the plants was useful, the most valuable information TOM obtained came from plant production documents, research documents, and interviews with German scientists and engineers such as Matthias Pier and Ernst Donath. The amount of captured documents utterly amazed the TOM. When TOM entered the IG Farben headquarters in Frankfurt, papers were knee-deep on the floors of the gigantic building and waist-deep in the stairwells. TOM collected the technical documents and sent them to London by the ton where CIOS members began translating and abstracting them.

As the flow inundated CIOS offices, its staff could only microfilm the documents at random, producing 141 reels. Later CIOS moved the entire program to the United States where it produced another 164 microfilm reels. Many documents that arrived in the United States found their way to loosely defined depositories, which often were simply the first convenient locations with available storage space. CIOS published 1,400 reports, designated confidential or restricted, and released the interviews of German scientists and engineers, written at the plant sites or shortly after in Paris and London before CIOS's mission ended on 1 July 1945. Many of these reports and interviews are on the TOM reels.

The US War Department established the second agency, Field Intelligence Agency Technical (FIAT), in Frankfurt in April 1945. FIAT, which included British Ministry of Fuel and Power scientists from its beginning, took over some of CIOS's activities and continued its operation until 30 June 1947. The third agency, the British Intelligence Objectives Sub-Committee (BIOS), established on 18 July 1945 in London, consisted solely of British scientists from the Ministry of Fuel and Power. Each agency operated as a 1–15 member army subunit, its members considered army personnel attached to a combat unit, and arrived at its designated target usually within a few hours after its army unit captured its target.

Beginning on 7 March through May 1945, when CIOS and FIAT, and BIOS later, visited the hydrogenation and F-T synthetic fuel plants, the three agencies had acquired in addition to the patents and published literature, information on the synthetic fuel plants' reaction conditions, catalysts, and operations; plant processing facilities, and probable plant output. British (Royal Air Force, RAF Bomber Command) aerial reconnaissance (photography) throughout 1942, American (United States Army Air Forces, USAAF) aerial reconnaissance beginning in 1944, and British intelligence reports, sometimes outdated because of German secrecy and five and one-half years of war, provided most of the information. Aerial reconnaissance, for example, indicated the approximate size of the plants, many of which occupied about 0.50 square miles; the number and size of the stalls at the hydrogenation plants and the number and size of the converters at the F-T plants; and thereby enabled the intelligence agencies to estimate each plant's daily and annual production. The interrogation of plant managers and scientists, such as Pier, all of whom cooperated, and examination of plant records confirmed and added important operational details previously unavailable, such as accurate production figures and costs, hydrogen and synthesis gas ($2\ H_2$:1 CO) costs, pressure vessel manufacturing, and catalyst efficiency. BIOS team member Thomas E. Warren of Canada's Mines Branch also visited four of the Ruhr's non-operating hydrogenation plants (Wesseling, Gelsenberg, Scholven, Welheim) and six of the non-operating F-T plants in September 1945. Warren confirmed information obtained from the earlier plant visits, again noting that before the war ended the ten plants he visited had removed all the control room instruments and sent them to central Germany in preparation for later application, very likely in underground hydrogenation and F-T plants that the German government proposed but never fully constructed.

The United States Strategic Bombing Survey (USSBS), which the secretary of war established on 3 November 1944 with a staff of 1,150 civilians, enlistees, and officers and headquarters in London, enacted a similar program of plant visits and interrogations and wrote in its 200 reports that it found German synthetic fuel documents in safe-deposit vaults, in private homes, in barns, in caves, in a hen house, and in two coffins. USSBS and TOM also microfilmed all relevant captured synthetic fuel documents. The statement from CIOS scientists, Captain C. C. Hall of the British Ministry of Fuel and Power and Dr. Vladimir Haensel (1914–2002), a German-born American chemical engineer with the US PAW, taken from their written report on the Sterkrade-Holten F-T plant, explains the necessity of the synthetic fuel plant inspections and interrogations: "to obtain technical details of the operation… of the research and development work… in the field of hydrocarbon synthesis."

The plant inspections also allowed the intelligence teams to see firsthand the damage the Allied bombings had inflicted on the synthetic fuel plants, enough damage so that by November-December 1944 most of the plants ceased production. From the start of World War II on 1 September 1939 until May 1944, the German synthetic fuel plants, suffered little wartime damage from land or air attacks. The destructive attacks began on 12 May 1944 when 649 bombers of the US Eighth Air Force bombed and extensively damaged five synthetic fuel plants in the Ruhr-west Germany and central Germany (Leuna, Lützkendorf, Zeitz, Böhlen, Brüx) and again on 28 May when 410 bombers attacked seven synthetic fuels plants. Bombs fell on the hydrogenation plants at Leuna, Lützkendorf, and Zeitz both times.

Hitler had promised to protect the synthetic fuel plants with fighter planes, but by July 1944 bombs struck every major synthetic fuel plant, and production fell from 316,000 tons /month when the attacks began, to 107,000 tons/month in June, to 17,000 tons/month in September. Production ended at Germany's western-central hydrogenation plants in Gelsenberg and Wesseling in September 1944, Welheim in October 1944, and at Scholven and Ludwigshafen in November 1944. Leuna, the most heavily defended German synthetic fuel plant, experienced additional attacks, one in July 1944, heavy attacks on 2 November 1944 (638 B-17s) and in January 1945, a total of 22 attacks (20 by US, 2 RAF), yet produced at 15% of capacity until nearly the war's end. A report of 25 February 1945 noted that bombing damaged 70 of 863 converters, 16 beyond repair. The USAAF Eighth Air Force and RAF flew 6,552 bomber sorties during the war and in one year dropped 18,328 tons of bombs on Leuna. US troops occupied Leuna on 15 April 1945, and on 21–26 April 1945 CIOS's 11-member team interrogated Dr. Schunck director and manager of the hydrogenation plant, Dr. Becker his assistant, Dr. Cron plant engineer, and Dr. Helmut Pichler plant costs expert, who provided details of the damage, in addition to other operational information. A second 10-member interrogation followed on 8–24 May, and a third interrogation by two CIOS members on an unspecified later period. Leuna plant officials also informed CIOS that they had evacuated engineering drawings, prints, and tracings of Leuna's synthetic ammonia apparatus to buildings in Bottendorf, 30 miles southwest of Merseburg, which CIOS visited on 10 May 1945.

Bombing ended production in November 1944 at eight of the Ruhr's nine F-T plants, only the Schwarzheide F-T plant near Dresden in eastern Germany continued production. The RAF in two attacks on 13–14 January 1945 and 8–9 February 1945 devastated the Pölitz hydrogenation plant nine months after the Eighth Air Force's bombing on 11 April 1944 that caused little damage. German industrialists established a Strosstrupp (Shocktroop) service in various cities to assist in repairing the damaged synthetic fuel and other essential plants, and ARSYN (Association for Hydrogenation, Synthesis, and Low Temperature Carbonization) with headquarters in Berlin assisted with bomb damage prevention and repair of damaged plants. Beginning on 21 June 1944 the synthetic fuel plants installed flak guns to reduce the damage from air attacks. The minimum number installed at several plants follows: Leipzig area (Lützkendorf, Magdeberg, Zeitz 1,000 total); Pölitz, Auschwitz, Hamburg 200 each; Brüx 170; Wesseling 150; Gelsenkirchen, Scholven 140 each; Heydebreck 130; Leuna 120; Blechhammer, Moosbierbaum 100 each; Böhlen 70. Blast walls and concrete housing surrounding vital plant machinery provided additional protection.

Allied intelligence agencies estimated that the bombings required 150,000–350,000 workers to repair the damaged synthetic fuel plants, many of which resumed partial production within days only to experience another round of bombings. The intelligence agencies also noted during their plant visits and interrogations that prior to and increasingly during the war years the synthetic fuel plants, like other German industrial organizations, used forced and slave workers from nearby concentration camps to repair the damaged plants and in some locations to construct the plants and provide the labor to keep the plants operational.[47]

Brabag constructed the first of the nearly identical, Nazi-era hydrogenation plants at Böhlen (Brabag I) south of Leipzig in eastern Germany in 1935-spring 1936 and Magdeburg (Brabag II) on the Elbe River in central Germany in fall 1936. Each plant cost RM 72 million, the high-pressure section having the highest cost at RM 17.1 million. Captured plant records and the 15-member CIOS interrogations of 12 Böhlen plant officials on 4–7 May 1945 before the European's war ending on 8 May, and extending through 13 May, identified Dr. von Falbert (Felbert) as Böhlen's plant manager, Herr Richard Lachner (Lackner) its chief engineer, Dr. Schneider its deputy manager, and Dr. Carl Schwitzer its high-pressure section manager. Böhlen employed 670 staff and 2,300 workers

47 In addition to the 59 TOM members, TOM administrators twice called for additional nominees. The first request called for 7 nominees, the second for 8 nominees. See Albert E. Miller, "The Story of the Technical Oil Mission, Preprint to the Twenty-fifth Annual Meeting of the American Petroleum Institute," Chicago, Illinois, 14 November 1945. Miller's paper has a list of the 26 TOM members who went to Germany and their affiliations. The information on the plant bombings comes from the United States Strategic Bombing Survey (note 2) and from the descriptions given in the CIOS, BIOS, and FIAT reports on plant visits cited in the references. Several other investigative groups (there were thirty groups altogether) including the United States Strategic Bombing Survey, Naval Technical Mission to Europe and an Air Force Team, went to Germany to collect technical documents. Little, if any, coordination existed among the groups.

but because of wartime bomb damage and reconstruction had increased its workforce to 4,500 men. Herr K. Fabricius managed the Magdeburg plant at the time the US Ninth Army captured the city on 18 April 1945. Two months later, on 1 July 1945, the US army left Magdeburg, and the Russian army began its occupation.

Both plants had two liquid phase stalls, each containing three converters ($T_{reaction}$ = 460–475° C), and two vapor phase stalls, each with four converters ($T_{reaction}$ = 370–410° C), all operating at 300 atm. A third liquid phase stall, third and fourth vapor phase stalls added in 1939 at Böhlen, and a third vapor phase stall added later at Magdeburg also operated at 300 atm. Two Winkler generators increased to three in 1939 each produced water gas (for hydrogen gas) from coke at the rate of 20,000 m^3/hr; five Linde-Fränkl units with a capacity of 2,300–3,000 m^3/unit, supplied oxygen. At both plants the hydrogenation of coal tar obtained from LTC of brown coal produced mainly motor and aviation gasoline, diesel oil, and liquid propane (LP) gas. Böhlen had a maximum production capacity of 250,000 metric tons/yr, Magdeburg 220,000 metric tons/yr, and each operated relatively trouble free from the start but produced a motor gasoline of only 59-octane number with vapor phase catalyst 5058. Pier's introduction of catalyst 6434 in November 1936 and its adoption in July 1937 improved the gasoline's quality, increasing its octane number to 66–68 and diesel's cetane number to 40. Both plants experienced extensive bombings from RAF and USAAF B-17s, Böhlen on 29–30 June 1944 and 7 July, Magdeburg beginning on 28 May 1944 and continuing through August-October 1944 and January-February-2 March 1945. The bombing raids extensively damaged the Böhlen and Magdeburg plants, particularly the RAF's Marburg night raid on 16 January 1945 and 648-bomber attack on 21 January, forcing both plants' shutdown in early 1945.[48]

Hydrierwerk Scholven, which the Prussian state-owned mining company Bergwerksgesellschaft Hibernia AG established in July 1935 with a capitalization of RM 10 million, increased to RM 110 million in 1939, constructed Germany's first bituminous coal hydrogenation plants, and the first of the three

48 J. F. Ellis, *Brabag I Plant Böhlen*, CIOS XXXII-92, Item No. 30, 42 pp.; "Coal Hydrogenation-Germany, Hydrocarbons Synthesis," *Petroleum Press Service*, 6 (1939): 529–32; *High-Pressure* (note 3), Vol. 1, pp. 80–82; *Report* (note 46), p. 53; *Germany (including Czechoslovakia). Liquid Fuels, Synthetic Oil Plants, Hydrogenation* (typed) Report 75687, reference numbers 4.01–4.06. Average exchange rate from 1934 to 1941 was RM 1 = 40¢, *Banking* (note 9). Lurgi Gesellschaft für Wärmetechnik installed a high-pressure (18 atm) F-T pilot plant at Böhlen by 1940, but CIOS found no plant personnel and operational data at the F-T plant which officials at the nearby Sächsische Werke in Böhlen said never operated satisfactorily.

Ruhr plants, in the fall of 1936. The decision to construct the plant in Scholven at the site of the Hibernia nitrogen plant, near Recklinghausen in the Ruhr, western Germany, came after two years of laboratory and pilot plant tests on Ruhr bituminous coal by Mathias Stinnes-Mülheimer Bergwerks-Verein AG, a large mining company. The tests showed the particular amenability of Ruhr bituminous coals to hydrogenation, giving motor gasoline, diesel oil, fuel oil, and aromatic products with which Ruhr coal industrialists hoped to establish a bituminous coal-based chemical industry similar to the existing bituminous coal tar chemical industry. Low freight costs and the availability of inexpensive coke oven gas for hydrogen production and plant power fuel added to Scholven's attractiveness. IG Farben designed the plant and provided engineers to assist with its construction and operation, under an IG Farben license.

Six CIOS personnel, three American and three British, on 6, 9 April, and Warren of BIOS on 12 September 1945 visited Scholven and interrogated its managing director Dr. Jost who "proved very difficult," chief chemist Dr. R. Urban, chief engineer Herr Starker, Dr. Schmidt, and business director Herr Brunning at the badly damaged plant. Scholven had a workforce of 2,000, but after the daytime air attack on 10 March 1945, 1,000 workers remained to repair the plant. Plant records showed that at the time of its opening, Scholven had three 300 atm liquid phase stalls ($T_{reaction}$ = 460° C) and three 300 atm vapor phase stalls ($T_{reaction}$ = 375° C), each with four converters, and an annual capacity of 125,000 metric tons of motor gasoline. Expansion in 1940 increased the number of stalls to six liquid phase stalls and four vapor phase (two saturation and two splitting) stalls each with three converters, giving it an annual capacity of 220,000 metric tons of high-quality motor or aviation gasoline and liquid petroleum (fuel) gas, of which 10–15% came from hydrogenating heavy oils and bituminous coal high-temperature tar. Coke generators (coke + steam → water gas reaction) and a Linde-Fränkl liquefaction unit that removed hydrogen from coke oven gas provided hydrogen gas. Scholven used tin (II) oxalate with ammonium chloride added in the liquid phase catalysis and tungsten sulfide catalysts 5058 and 6434 in the vapor phase to produce a gasoline of 72- and 76–78-octane number but experienced considerable initial operating difficulties because of the chloride catalyst's corrosive effect, especially on the return bends (elbows) in the preheaters' pipes. Improvements in structural materials, design changes that simplified replacing the return bends, and reducing the chloride concentration by one-half, from 1% by weight of paste to 0.5%, greatly improved plant operation. Nevertheless, bituminous coal hydrogenation at 300 atm remained problematic because the liquid phase products had a high asphalt content that clogged the converters and made residue processing more difficult. Scholven also had difficulty hydrogenating coals with a carbon content greater than 83%, which some bituminous coals approached. Scholven consequently was the only German bituminous coal hydrogenation plant to operate at 300 atm.

Allied bombers attacked Scholven 71 times, the first attack, a heavy raid on 15 March 1941 caused a 3–4-week shutdown, 96 B-17 bombers dropped 238 tons of bombs in November 1943, attacks in July and October 1944 that killed two workers and badly damaged the plant ended production. The bombing raids also destroyed Scholven's freight lines. CIOS reported that Scholven's dismantling had begun before its arrival and that plant officials had removed all plant instruments and stored them in the Hibernia mine at Gelsenkirchen. CIOS also found huge masses of documents in Scholven's main offices, plant buildings, and air raid shelters. The documents contained information on raw materials, plant costs, and hydrogen production, and eight files authorizing construction of a 700 atm plant, producing 250,000 metric tons/yr of motor fuel (Scholven III) in Gladbeck north of Scholven. The authorization dated from 1938, but construction ended in April 1942 because of a construction worker shortage.[49]

Three more coal hydrogenation plants began operation in the three-year interval between Hitler's announcement of his Four Year Plan in September 1936 and the outbreak of war in September 1939. They were the Ruhröl GmbH plant adjoining the Vereingte Welheim coal mine near Bottrop-Welheim-Gelsenkirchen in the Ruhr, western Germany; the Gelsenberg Benzin AG plant in Gelsenberg northwest of Gelsenkirchen; and the Brabag IV plant in Zeitz south of Leipzig in central Germany. CIOS visited the three plants in spring 1945.

Mathias Stinnes established Ruhröl AG and Ruhröl GmbH, the operating company, in 1936, and its Welheim plant was the first of four 700 atm hydrogenation plants. Seven CIOS personnel on 10–13 March 1945 and on 10, 11, and 13 April 1945; two FIAT members in spring 1945; and Warren of BIOS on 13 September 1945 visited the Welheim plant at which times they interrogated Dr. Erich Freese, plant manager; Dr. Schirrmacher, Dr. Karl Winkler, Dr. Günther Ibing, Dr. Walter Muschenborn, and Dr. Hermann Thomas. Interrogations and plant records showed that Welheim employed 2,118 men and women including 176 war prisoners, and had an additional 497 contractors and laborers working on

49 C. Cockram, *Scholven Hydrogenation Plant*, CIOS XXX-102, Item No. 30, 31 pp.; *High-Pressure* (note 3), Vol. 1, pp. 82–84; *Report* (note 46), p. 65; T. E. Warren, *Inspection of Hydrogenation and Fischer-Tropsch Plants in Western Germany during September 1945*, BIOS Final Report No. 82, Item No. 30 (London: British Intelligence Objectives Subcommittee, 1945), pp. 1–28 on pp. 10–11; *Germany* (note 48); "Coal Hydrogenation-Germany" (note 48).

new buildings. Brabag and Mineralölbau GmbH in Berlin designed the plant, its construction began in July 1936. The first 700 atm liquid phase stall with $T_{reaction}$ = 480° C reached operation by November 1937, and a 700 atm vapor phase stall with $T_{reaction}$ = 480–502° C followed in May 1939. With the addition of three more liquid phase stalls during the war and another vapor stall on 9 May 1944, Welheim had four liquid phase stalls, each with four converters, and two vapor phase stalls, each with three converters, giving it a maximum annual production of 130,000 metric tons of fuel oil, 30,000–35,000 metric tons of gasoline, and a total plant cost of RM 81 million in December 1944.

Alfred Pott (1882–1951) and Hans Broche (1896–1963) at Mathias Stinnes-Ruhröl in 1927 invented a solvent-coal extraction process as a lower costing alternative to the expensive liquid phase hydrogenation. The extraction required pumping a 1 part: 2 part powdered coal and tetralin-cresol paste into an extractor maintained at 400–420° C and 100 atm. The coal dissolved in 1.5 hours after which filtration removed the 20–47% undissolved (unextracted) solid material and distillation separated the liquid extraction products into middle and heavy oils for hydrogenation in the vapor phase. The extraction unit had a capacity of 80,050 metric tons/yr and cost RM 8,734,600. IG Farben reached an agreement with Pott and Broche in March 1936 to construct and test at Leuna and Ludwigshafen an extraction unit in which coal dissolved in middle oil would produce a coal extract (solution) for hydrogenation at IG's 700 atm plants, but because of construction delays and operating difficulties, the extraction unit never produced as expected until 1943, having manufactured only high-quality electrode carbon (coke) for the aluminum and steel industries.[50] Four Pintsch-Hillebrandt coke consuming generators (water gas reaction) and Linde-Fränkl liquefaction units that had an output of 20,000 m^3/hr/unit removed hydrogen from water gas or from hydrogenation waste gases provided hydrogen gas.

For the liquid phase hydrogenation at 700 atm and 480° C, a 20–30% tar pitch-oil mixture containing a small amount of iron (II) sulfate catalyst (0.25% of the mixture oil) gave 20% motor gasoline and middle oil; 30% high-quality fuel oil, gases, and residue. Ruhröl developed its own vapor phase catalyst (0.4% Mo-2% Cr-5% Zn-5% sulfur on terrana earth) for hydrogenation at 700 atm and 500° C and synthesized 30,000–35,000 metric tons/yr

50 C. Cockram, *Bottrop-Welheim Hydrogenation Plant*, CIOS XXX-104, Item No. 30, 31 pp.; W. F. Faragher and W. A. Horne, *Supplemental Report on the Ruhröl Hydrogenation* Plant, *Welheim Ruhr*, FIAT 90; *High-Pressure* (note 3), Vol. 1, pp. 84–85; Warren, *Inspection* (note 49), pp. 12–13, *Germany* (note 49); J. H. Howell and R. M. Crawford, *The Production of Synthetic Fuels* (note 19), pp. 65–70.

of aviation gasoline with a high aromatic content and 80-octane number (91–92 after adding 1.15 mL lead tetraethyl/L of gasoline). Ruhröl's vapor phase catalyst reduced gasoline production cost to RM 2,000/metric ton from RM 10,000–12,000/metric ton with catalyst 5058 (WS_2) and allowed the Welheim plant to use a single recycling system for both phases, thereby simplifying plant operation.

Welheim officials planned to destroy its operational records and documents in March 1944 but never carried out the destruction before the Allied intelligence teams arrived a year later and seized numerous records and documents that contained detailed plant drawings, line diagrams, analyses, plant outputs, and operating data. A heavy air attack on 20–21 July 1944 seriously damaged the Welheim plant. Freese estimated its restoration to full capacity would take 12 months and cost RM 20–25 million, but another air raid on 31 October 1944 destroyed the plant. A third heavy air raid on 6 November flooded the region and led to Welheim's abandonment and removing of plant equipment east to nearby Witten. Despite the heavy bombing Welheim's three-story vapor phase catalyst building remained undamaged. CIOS, the first intelligence team to visit Welheim, upon completing its interrogations and plant inspection concluded, that "From an economic point of view the Welheim plant must have been a poor proposition."[51]

Gelsenberg Benzin AG, established in December 1936 with a capitalization of RM 100 million, RM 55 million obtained on loan, constructed the first 700 atm bituminous coal hydrogenation plant. Pier and the German synthetic fuel scientists believed that the higher pressure made the hydrogenation process more reliable and simpler to operate. Gelsenberg's construction at the Nordstern coal mine near Gelsenkirchen in the Ruhr took two years, 1937-June 1939, required 7,000 workers, and cost RM 293,113,000. Mineralölbau GmbH did the overall planning and design, Pier's group in Ludwigshafen designed the high-pressure section and made available experienced technicians to assist with the construction and operation. CIOS's six-member team interrogated chief chemist Dr. Paul Jacob and plant chemist Dr. Erich Friehmelt on 12–16 April and Warren of BIOS followed four months later, interrogating Jacob and Friehmelt on 7–8 September 1945. Plant manager Dr. Pross was absent. Their interrogations and the huge mass of captured plant records provided information on reaction conditions, catalysts, and plant operations summarized in a 39-page report.

Gelsenberg had 20 Demag coke consuming generators for the production of water gas, and a RM 7.52 million Linde-Fränkl liquefaction unit that removed

51 Note 50; *High-Pressure* (note 3), Vol. 1, pp. 85–86.

hydrogen from the hydrogenation waste gases, supplied hydrogen gas for consumption at the rate of 135,000 m³/hr. Gelsenberg used a liquid phase catalyst of iron (II) sulfate (1.2%) with Bayermasse (1.5%) on dry coal, and catalysts 5058 and 6434 (terrana earth)for the vapor phase hydrogenation. It produced a good quality gasoline having a 72-octane number (boiling point 155–160° C) that increased to 90 after adding 1.15 mL lead tetraethyl/L of gasoline, making it suitable for motor and aviation use. Plant records showed that when Gelsenberg opened in 1939, it had an annual capacity of 150,000 metric tons of gasoline. Additions in October 1942, giving it a total of six 700 atm liquid phase stalls operating at 480° C and seven 300 atm vapor phase stalls (four saturation and three splitting stalls) operating at 400–442° C), each liquid and vapor phase stall having four converters, its annual capacity more than doubled to 340,000 metric tons of gasoline and 100,000 metric tons of propane and butane for automobile fuel.

Because of Gelsenberg's higher operating pressure and the design engineers' decision not to pass the coal-oil paste through heat exchangers to heat the paste before entering the preheaters and liquid phase stalls, Gelsenberg initially experienced major equipment and structural difficulties. The N8 and N9 steel alloys that worked well at 300 atm proved too weak for operation at 700 atm, the lack of heat exchangers required heating the preheater tubes to 520° C and caused them to burst. The introduction of heat exchangers, which kept the preheater tube temperatures below 520° C, the development of a more durable steel alloy N10 (Cr-Mo-V-W), and the discovery that decreasing the coal in the paste from 46 to 40% reduced the stress on the N10 alloy and made it less sensitive to high-pressure operation, greatly improved Gelsenberg's production.

Gelsenberg employed 5,000 workers, making it the largest of the Ruhr hydrogenation plants, and because of its importance experienced several RAF bombing attacks. The earliest attack on 1 October 1940 caused a three-month shutdown, a 13 June 1944 attack destroyed the plant. Repair work began, but attacks in July and on 11 and 27 September ended the reconstruction and forced Gelsenberg's permanent shutdown in September 1944. Allied bombing of rail lines also hampered production.[52]

---

52 C. Cockram, *Gelsenberg Hydrogenation Plant*, CIOS XXX-105, Item No. 30, 39 pp.; *High-Pressure* (note 3), Vol. 1, pp. 86–88; *Report* (note 46), pp. 66–68; Warren, *Inspection* (note 49), pp. 8–9; *Germany* (note 49). Gelsenberg Benzin AG consisted of five partners: Vereingte Stahlwerke AG, Dusseldorf; Gelsenkirchener Bergwerks AG; August Thyssen-Hütte AG, Duisburg-Hamborn; Bochumer Verein; and Dortmund-Hörder Hüttenverein, Dortmund; "Motorkraftstoff von Kohl," *Teer und Bitumen*, 35 (1937): 231–32.

Brabag's plant (Brabag IV) at Zeitz, south of Leipzig in central Germany, was the last of the prewar plants and the only one to use a low-temperature hydrogenation process (TTH) and a medium-temperature process (MTH) to treat brown coal tar obtained from LTC plants in nearby Espenhain, Neukieritzsch, and Böhlen. Construction started in 1937, operation began in 1939 before the war's outbreak, and continued until USAAF attacks on 12, 28 May (187 B-24s) and 10 August 1944 forced its closing. Two CIOS members visited Zeitz on 9 May 1945 and questioned Dr. Johannes Saeger, who managed the low-pressure plant section, and Dr. Horst Roth, chief engineer of the stall section, about plant construction and operation.

The plant's six liquid phase stalls, each containing three converters operating at 300 atm and 350–356° C for TTH and 375–425° C for MTH, had an annual capacity of 250,000 metric tons of gasoline, diesel oil, and lubricating oil. Three Winkler generators gasified coke with oxygen and produced 20,000 m^3/hr /generator of water gas ($CO + H_2$, the hydrogen source). Brabag developed TTH and MTH because they made possible the hydrogenation of brown coal tar and heavy oil in the liquid phase over a fixed bed catalyst and gave a product nearly identical to vapor phase hydrogenation. Both required filtering the heavy oil or tar to reduce the amount of solids to less than 0.1% the concentration of solids that rapidly poisoned the catalyst, which despite such treatment lasted only four to six months. After filtration, pumping into an injector at the rate of 1.0 metric ton/hr/m^3 of catalyst, and adding hydrogen at the rate of 2,500 m^3/metric ton of oil, the reactants passed into a heat exchanger(s) and pre-heater and then into a series of three or four converters, each containing 20 metric tons (pellets) of catalyst 5058 or 8376 [7846W (Al_2O_3 70%-WS_2 27%-NiS 3%), developed on 8 August 1941] in a fixed bed. TTH had a 95% conversion to a liquid that upon distillation gave 25% motor gasoline, 50% diesel oil of 40–43-cetane number, 10–15% lubricating oil, and 10–15% refined wax; MTH yielded 35% motor gasoline, 60% diesel oil of 40–43-cetane number, and 5–15% lubricating oil. CIOS reported that TTH and MTH had relatively low operating costs of RM 15/metric ton, but without carefully controlling the liquid phase hydrogenation it led to polymerization of intermediate unsaturated products that poisoned the catalyst and reduced its effective lifetime.[53]

53 *High-Pressure* (note 3), Vol. 1, pp. 89–90; *Report* (note 46); J. F. Ellis and R. J. Morley, *Report on Investigations by Fuels and Lubricants Team at the Brabag Works at Troglitz-Zeitz*, File No. CIOS XXXIII-24, Item No. 30 (7 October 1945).

FIGURE 3.6 Brabag IV Zeitz synthetic fuel plant (hydrogenation) after bombing May 1944
SOURCE: AIR UNIVERSITY HISTORY OFFICE, VIRIN: 190626-F-XX000-003 OFFICIAL UNITED STATES AIR FORCE WEBSITE

3.8 *Germany's Coal Hydrogenation Industry 1939–45*

By the outbreak of World War II on 1 September 1939, Germany had seven coal and tar high-pressure hydrogenation plants in operation (eight with Ludwigshafen). Most of the important breakthroughs in synthetic fuel research already had occurred. The war years represented therefore a development period for existing plants and a construction period for five new plants, bringing the final number of hydrogenation plants to twelve.

Wintershall AG built the first of the wartime synthetic fuel plants in Lützkendorf, west of Merseburg and Leipzig in the Geiseltal brown coal mining district of central Germany (Saxony). Wintershall originally produced potassium salts, but in 1931 it shifted to oil production and recovery. It completed construction of a F-T plant with a 30,000 metric ton /yr capacity of synthetic oil in late 1938 and that same year planned the construction of a small 50,000 metric ton/yr coal hydrogenation plant.

The hydrogenation plant, which began operation in early 1940 with Heinrich Schmalfeldt, the F-T plant director also serving as its director, had one 700 atm liquid phase stall with two converters operating at 350° C, one 700 atm vapor phase stall with three converters, and used IG Farben's catalyst 6434 in the vapor phase. It hydrogenated refined petroleum residues, such as propane asphalt and crude paraffin wax obtained from the plant's petroleum refinery, and bituminous coal tar to give gasoline of 73.5-octane number, diesel oil of 35.6-cetane number, and fuel oil. CIOS's ten personnel on 9, 11, and 17 May 1945 inspected both plants and interrogated six plant officials, particularly Dr. Schneeberger, who succeeded Schmalfeldt as Lützkendorf plant director in 1940; Herr Schültz (Dip.Ing), F-T plant manager; and Dr. Wielland, hydrogenation plant manager. CIOS's 21-page report included information on catalysts, reaction conditions, and plant equipment such as the generator Schmalfeldt had invented for the production of F-T synthesis gas from brown coal; Wielland's statements that the hydrogenation plant achieved a maximum production of 44,000 metric tons in 1944; and Schültz's revelation that Allied bombings had destroyed the Schmalfeldt gas generator thereby reducing production from the F-T plant.

Lützkendorf cost RM 117,405,389 to construct all plant facilities including the F-T and hydrogenation plants, which cost RM 18,182,250 and RM 18,606,851 respectively. It employed 4,000 workers and 380 staff members, but both plants had a poor reputation during their first 18 months of operation. The water gas ($CO + H_2$) that Wintershall generators produced and provided hydrogen gas for hydrogenation, gave a hydrogen gas of very poor quality, having a hydrogen purity of only 86%, improved to 91–92% purity. German synthetic fuel experts described both the hydrogenation and F-T plants as badly designed and engineered, poorly staffed having a shortage of skilled engineers and foremen, inexperienced workers, little discipline, low morale, and weakly managed. IG Farben in October 1939 sent 187 scientists, engineers, and workers from Leuna, seven miles away, to Lützenkendorf, to improve operation, and by spring 1940 some improvement had occurred. Nevertheless, in spring 1945 when CIOS visited the plant, it concluded that Lützkendorf's operational standards ranked lower than the other synthetic fuel plants it had inspected. CIOS described Lützkendorf as a classic example of how not to design, build, and operate a plant. By that time Lützkendorf had ceased all production because of extensive bomb damage from a 12 May 1944 attack by 800 USAAF bombers and later USAAF attacks on 28 May, 30 November 1944, 9 February 1945, and a 7 April 1945 attack by 233 B-17 bombers.[54]

A few months after Lützkendorf began production, Hydrierwerk Pölitz AG, established in June 1937, completed construction of its RM 200 million hydrogenation plant on the Oder River near the Baltic port of Stettin and the

54 L. L. Newman, J. G. Allen, and H. Schindler, *Wintershall AG, Lützkendorf*, CIOS XXXII-90, Item No. 30 (28 September 1945), 21 pp.; *The Wintershall-Schmalfeldt Process for the Manufacture of Synthesis Gas at Lützkendorf*, BIOS Final Report 1142, Item No. 30 (London: His Majesty's Stationery Office, 9, 11 May 1945); *High-Pressure* (note 3), Vol. 1, p. 88; *Report* (note 46), p. 54; *Germany* (note 49).

German-Polish border. Because of the plant's location in far eastern Germany the Russian army seized the region, and the Allied investigative teams never visited the Pölitz plant. CIOS obtained information on the plant's operation from Pier and Müller Conradi in Oppau on 25–31 March, from plant manager Dr. Hans Hartmann and Dr. Wissel hydrogenation plant manager who fled Pölitz shortly before the Russians arrived in April 1945, and from Dr. Zorn at Leuna on 15 May 1945. Pölitz was the largest of the twenty-one synthetic fuel plants, employing 5,000 workers mostly forced laborers (30,000 from 1940–45 of whom 13,000 died from hunger or disease), 1,000 scientists and engineers, and 4,000 other personnel, and the only plant not located near its raw material supply, but in a predominately agricultural region. The entire Pölitz complex occupied 6 square miles (1,500 hectares).

Hydrierwerk selected Pölitz because it planned to hydrogenate mainly imported petroleum residues from Columbia and Aruba and some domestic coal, all of it shipped there by water. IG Farben had reached an agreement with Standard Oil and Royal Dutch-Shell to hydrogenate petroleum residues to produce motor gasoline, and in 1937 it established Hydro GmbH in Berlin and Hydrierwerk Pölitz AG to construct and operate the plant. Deutsche-Amerikanische Petroleum Gesellschaft, Standard Oil's German subsidiary, also held a considerable interest in the plant. But with the outbreak of war, and petroleum residues from Columbia and Aruba no longer available, Pölitz hydrogenated Upper Silesian bituminous coal, bituminous and brown coal tar, pitch, and some petroleum residues particularly from Romania. The original design had four 700 atm liquid phase and four 300 atm vapor phase stalls, but the numbers increased to six liquid phase and eight vapor phase stalls by the time Pölitz opened in July 1940. Eight Pintsch-Hillebrandt coke consuming generators that gasified coke obtained by carbonizing Lower Silesian bituminous coal (water gas reaction), and a gas cracking plant (methane-steam) provided hydrogen gas. The hydrogenation plant had its own gas power plant, but because of its large size and power demand Hydrierwerk Pölitz had to construct another large power plant in the city of Pölitz.

The Pölitz hydrogenation plant produced 700,000 metric tons/yr mainly of aviation gasoline as well as motor gasoline, diesel oil, fuel oil, gas oil, and petroleum (fuel) gas. Bergius and Pier claimed that hydrogenation had a major advantage over F-T and all other coal-to-oil conversion processes because of the wide variety of raw materials suitable for hydrogenation. Pölitz, which hydrogenated several kinds of pitches, tars, and petroleum residues, demonstrated hydrogenation's wide adaptability, but its costly raw materials, RM 85/metric ton of bituminous coal tar and RM 123/metric ton of petroleum residues, led to a high production cost. Pölitz's aviation gasoline cost RM

360–390/metric ton, at Leuna where brown coal tar cost RM 80/metric ton, motor gasoline cost RM 190/metric ton.[55]

The first bombing of Pölitz occurred in August 1940, shutting the plant for 2 1/2 days; a second attack in October caused a 2 1/2 week shutdown. A third attack by the USAAF Eighth Air Force on 11 April 1944 did little damage, but a 29 May 1944 attack decreased production from 1,500 metric tons/month to 800 metric tons/month for the period 1 June 1944 to 8 February 1945. RAF attacks on 13–14 January and 8–9 February 1945 devastated Pölitz. The Russians forced 25,000 German workers to dismantle the synthetic fuel plant, salvaged whatever undamaged equipment remained and shipped it to Siberia for construction of a 1 million metric ton/yr coal hydrogenation plant for the production of aviation gasoline, and used other plant remnants for scrap iron.

The last three coal hydrogenation plants began operation less than three years after the start of World War II. The new plants ranked among the largest, but they were no more advanced than the existing plants. Union Rheinische Braunkohlen-Kraftstoff AG, established on 27 January 1937 by Rhineland's brown coal producers (Knapsack) with a capitalization of RM 45 million and an additional RM 45 million loan, constructed the first of the newer hydrogenation plants one mile south of Wesseling on the Rhine River's west bank, eight miles south of Cologne, near Bonn. The German government strongly supported the plant, which hydrogenated lignite, as an important part of Hitler's Four Year Plan.

Plant construction started on 4 April 1938, but the outbreak of World War II interrupted construction from 8 November 1939 to 1 February 1940; a late-1939 decision to construct a coal hydrogenation plant in the Soviet Union, which Union Rheinische never constructed, also delayed Wesseling's construction. Operation began in August 1941 after a three-year construction period. Experiments in Ludwigshafen and Leuna had shown that the Rhine brown coal (lignite) hydrogenated at Wesseling, because of its asphalt, high water (60%), and ash content (5%), had hydrogenating properties and a product quality that ranked between those of central German brown coal and Ruhr bituminous coal. For this reason Wesseling had to operate a 700 atm liquid phase hydrogenation to produce an asphalt-free heavy oil. It had only one major start-up problem: salt and large caviar deposits (inorganic products of coal

55 Ronald Holroyd, *IG Farbenindustrie AG Works*, US Bureau of Mines, Information Circular 7375 (Washington, DC: Bureau of Mines, August 1946), pp. 54–55; *High-Pressure* (note 3), Vol. 1, pp. 92–93; *Report* (note 46), pp. 60, 64–66, 68; *Germany* (note 49); "Liquid fuel-Germany," *Petroleum Times*, 44 (1940): 61.

FIGURE 3.7 Wesseling plant 1953
SOURCE: DEUTSCHES BUNDESARCHIV, B 145 BILD-F000918-0009 / UNTERBERG, ROLF /

decomposition or crystal spheres) on the inside walls of the liquid phase converters forced frequent stoppages. Plant engineers eliminated the deposits by neutralizing them, injecting oil, and desanding the converters' walls. Wesseling achieved full production in December 1942, and in 1944 it had a production cost of RM 403.71/metric ton of liquid product.

The completed Wesseling plant had four 700 atm liquid phase stalls each containing four converters operating at 478° C and four 325 atm vapor phase stalls each containing four converters operating at 367–425° C. It had an annual production capacity of 200,000 metric tons of motor gasoline and aviation gasoline of 70-octane number, 87 with 0.09 volume percent tetraethyl lead, and diesel oil of 35 and above cetane number. Wesseling used catalysts 5058, 8376 (7846W), and 6434 in the two stages of conversion. Eleven Pintsch-Hillebrandt generators that gasified brown coal (lignite) briquettes at 700° C producing water gas, and a gas cracking plant that cracked residual methane and ethane with steam, provided hydrogen gas. A 70–80 atm steam turbine power plant driving 6,600 volt generators supplied most of the plant's 40,000 kilowatts energy requirements. Although Wesseling hydrogenated mainly Rhine brown coal, it treated small amounts of tar, pitch, and shale oil from Wurttemberg. Its Rhine River location also enabled Wesseling to hydrogenate imported oil and for a time large quantities of Estonian shale oil which it refined in its distillation unit.

Allied bombing of Wesseling began on 29 May 1942. A very heavy attack, which plant workers described as "1,000 bombs," came two years later, on late evening-early morning 18–19 July 1944. The 500-pound bombs, of which 200 were duds, fell on the well-lit plant because plant workers failed to activate immediately the artificial fog camouflage. The attack came in three waves, lasted 20 minutes, and destroyed about 20% of production capacity, but 10 weeks of accelerated repair work restored Wesseling to 40% of production. Two more attacks followed on 3 and 11 October 1944, each attack dropping 60 1,000-pound bombs, ended Wesseling's synthetic fuel production on 6 October 1944. Plant officials had trained workers to shut down the plant in two minutes and take cover in the plant's many air raid shelters, which they did at the time of the final bombing, but three of the plant's 1,000 workers died in the first attack.

The plant sat idle on 17–19 March 1945 when a four-member CIOS and a two-member Naval Technical Mission to Europe team interrogated the three remaining plant officials: Wesseling general manager Dr. Mueller von Blumencron, whom the German military ordered to stay at the plant after the army left, chief chemist Dr. von Heinz Sustmann who spoke English, and chief operating engineer Dr. Ernst Peuckert. Six other plant officials and nearly all the plant workers had left at the time of the plant's permanent shutdown, only 35 Russian workers in the barracks and a few German guards remained at the plant. The three officials provided "fairly complete information on the plant" but little information on the use of Wesseling's gasoline and diesel oil products because German army authorities never disclosed this information to them. Workers had removed some plant equipment from the boiler room and laboratory by the time the investigative teams arrived, and supposedly were transporting hydrogen compressors to a secret underground installation called JACOB II, at Pinch near Dresden. The compressor removal never occurred, although plant officials secretly sent essential compressor parts to a tube-working plant in Alster, west of Bonn and documents to Fabrikant Projahn in Waldbrohl, 30 miles from Wesseling. Warren of BIOS conducted a follow up interrogation of plant officials on 4–5 September 1945 that confirmed the previously obtained information.[56]

---

56 L. H. Mulit and R. C. Aldrich, *The Wesseling Synthetic Fuel Plant*, CIOS XXVII-60, Item No. 30 (London: His Majesty's Stationery Office), 105 pp.; J. A. Oriel and I. H. Jones, *The High Pressure Hydrogenation Plant, especially for Brown Coals, Wesseling*, CIOS XXVIII-40, Item No. 30, 155 pp.; *High-Pressure* (note 3), Vol. 1, pp. 93–94; *Report* (note 46), p. 65; Warren, *Inspection* (note 49), pp. 6–7; *Germany* (note 49). Union Rheinische consisted of Rheinische AG für Braunkohlen Bergbau und Brikettfabrikation in Cologne, and Braunkohlen und Brikettwerke Roddergrube AG in Brühl; "Motorkraftstoff" (note 53). The Germans had plans to construct five underground hydrogenation plants, code named

Sudetenland Treibstoffwerke AG in Maltheuern, established in 1940 as a subsidiary of Reichswerke Hermann Göring, started construction on a brown coal tar hydrogenation plant in Brüx, Sudetenland (Bohemia, northern Czechoslovakia), in May 1939. It began operation in mid-1942 with managing director Dr. Damm, chief engineer Dr. Amon, and chief chemist Dr. Ottens directing a workforce of 3,500. Brabag designed the plant, Mineralölbau GmbH in Berlin constructed it. Brüx's early design called for an annual production capacity of 270,000 metric tons of aviation and motor gasoline and diesel oil. With wartime expansion in 1941, Brüx had four 300 atm liquid phase stalls, six 300 atm vapor phase stalls (three saturation and three splitting), each stall having four converters, and an annual production capacity of 400,000 metric tons. A LTC plant, which began operation in August 1941 with 55 (80 planned) Lurgi Spuelgas ovens, each having a capacity of 250–300 metric tons/d, carbonized 7 million metric tons of brown coal annually to produce the 700,000–750,000 metric tons of tar required for hydrogenation. The brown coal, costing RM 12/metric ton delivered to Brüx, had a composition similar to Rhine brown coal and produced liquids similar to those obtained from Rhine tar. Coke, the other carbonization product, when treated with steam in six large Winkler generators, each of 20,000 m^3/hr capacity, produced hydrogen gas; Linde-Fränkl units provided oxygen gas.

The Brüx plant experienced only routine start-up problems and produced an excess of tar, resulting in a projected, though never achieved, annual yield of 1,000,000 metric tons of gasoline and diesel oil. It produced only 100,000–120,000 tons of liquid products in 1944. Production costs were RM 260/metric ton of gasoline, 65-octane number for motor gasoline, 78-octane number increased to 87 with 0.09% by volume lead added for aviation gasoline; and RM 220/metric ton for 40–45-cetane number diesel oil.

Allied bombing of Brüx, which began in May 1944 and continued at regular intervals, greatly reduced production to 20,000 tons of liquid products through December 1944, although Damm, Amon, and Ottens believed that of the plant's two liquid phase hydrogenation stalls, they could have the damaged stall operational within a month. Plant shutdown and evacuation of staff members followed five months later when several busloads of staff members, including Damm and Amon, left Brüx on the evening of 7–8 May 1945 just before the Russian occupation and arrived in Leuna on the evening of 12–13

Schwalbe, two in the Elbe valley on both sides of the Elbe River, one near Buckeburg and Minden in a natural cave, one at Berga Kelbia in Thuringia, and one in the gypsum quarries at Niedersachswerfen in the Harz mountains.

May. Ottens joined them later. In Leuna, Damm informed CIOS that they had planned in March to evacuate to Schwarzenfeld, near Amberg, Bavaria, in view of a likely Russian occupation, but because of the Allied advance they were proceeding much farther north to Reichswerke Hermann Göring at Walenstadt near Brunswick to arrange for the distribution of Brüx's staff to other hydrogenation facilities. Damm's interrogation provided information on two other significant developments: (1) Brüx plant officials had shipped a large batch of documents to the Porceline Fabrik Buchtal AG at Schwarzenfeld and to the Kohlen Chemische Institut in the Hartz mountains and had hidden other documents in the central shaft of the brown coal mine in Brüx and (2) Reichswerke Hermann Göring had started construction of an underground hydrogenation plant at Bad-Schandau, 50–55 miles northwest of Brüx. The underground plant would have two liquid phase stalls and two vapor phase stalls with a production of 5,000 metric tons of gasoline of coal tar.[57]

The Upper Silesian bituminous coal industry, Reichswerke Hermann Göring, and the German government, which in 1939 merged to become the Oberschlesische Hydrierwerke AG, jointly operated the Blechhammer plant northwest of Auschwitz and Katowice in Upper Silesia (southern Poland), the last coal hydrogenation plant constructed. The Blechhammer complex included a large LTC plant for coal tar production and occupied about 0.54 square miles. Dr. Max Josenhaus directed the plant, Dr. Walter Krönig, chief chemist, assisted him. Josenhaus was one of only seven German synthetic fuel scientists, of 1,600 total scientists, engineers, and technicians, whom Operation Paperclip, the secret United States intelligence program of 1945–59, brought to the United States.

Blechhammer operated for only four months, beginning in April 1944. No American-British investigative team visited Blechhammer, but CIOS learned from interrogating Josenhaus in Naumburg, 20 miles southwest of Leuna, and Krönig in Vorhum near Peine (Brunswick) that according to a pre-arranged plan practically the entire staff left Blechhammer at the end of January 1945 (January 21) for Naumburg and Vorhum. An office in Naumburg would then distribute the staff to other hydrogenation plants. The evacuation occurred in

57 *High-Pressure* (note 3), Vol. 1, pp. 93–94; *Report* (note 46), p. 57; Ronald Holroyd, ed., *IG Farbenindustrie AG Works, Leuna*, CIOS XXXII-107, Item No. 30, (London: His Majesty's Stationery Office, April, May 1945), 140 pp., pp. 134–37; Holroyd, *IG Farbenindustrie AG Leuna Works, Merseburg, Germany*, US Bureau of Mines, Information Circular 7370 (Washington, DC: Bureau of Mines, 1946). Plant officials at damaged or abandoned synthetic fuel plants usually sent undamaged light equipment to other synthetic fuel plants for possible use. Brüx officials said nothing about the equipment or the workers they left at the plant.

TABLE 3.4 German coal hydrogenation plants 1927–45

Plant location	Operating company	Construction started	Construction completed in/Started operation	Raw material used
Ludwigshafen/ Oppau: Rhineland Palatinate, west Germany	BASF (IG Farben)			
Leuna: central Germany near Merseburg	Ammoniakwerk Merseburg GmbH subsidiary of IG Farben	November 1926	1 April 1927	brown coal 1927, then brown coal tar until 1931 when difficulties with brown coal overcome, crude oil
Böhlen: central Germany south of Leipzig, Brabag I	Braunkohlen Benzin AG (Brabag) founded 26 October 1934	1935	Spring 1936	brown coal tar
Magdeburg: central Germany, Brabag II	Brabag	1935	Fall 1936	brown coal tar
Scholven: Ruhr, west Germany near Gelsenkirchen	Hydrierwerk Scholven AG founded July 1935	Fall 1935	Fall 1936	bituminous coal
Welheim: Ruhr, west Germany near Gelsenkirchen	Ruhröl GmbH established 1935 by Mathias Stinnes	July 1936	1937	coke oven tar pitch
Gelsenberg: Ruhr, west Germany near Gelsenkirchen	Gelsenberg Benzin AG founded December 1936	1937	June 1938	bituminous coal
Zeitz: central Germany south of Leipzig, Brabag IV	Brabag	1937	1939 before WW II	brown coal tar
Pölitz: northeast Germany north of Stettin	Hydrierwerk Pölitz AG founded June 1937	1937	July 1940	petroleum residues, bituminous coal, coke oven tar, shale oil
Lützkendorf: central Germany west of Merseburg, Leipzig	Wintershall AG	1938	1940	petroleum residues, bituminous coal tar, crude oil
Wesseling: west Germany near Bonn	Union Rheinische Braunkohlen Kraftstoff AG founded Jan. 1937	1938	August 1941	Rhenish brown coal, shale oil
Brüx (Most): Bohemia, Sudetenland, Czechoslovakia	Sudetenlandische Treibstoffwerke AG founded 1940	May 1939	October 1942	Brüx tar

TABLE 3.4 German coal hydrogenation plants 1927–45 (*cont.*)

Plant location	Operating company	Construction started	Construction completed in/Started operation	Raw material used
Blechhammer: Upper Silesia (southwest Poland)	Oberschlesische Hydrierwerke AG founded 1939	1942	April 1944	bituminous coal and tar

SOURCE: TABLE COMPILED FROM *Report on the Petroleum and Synthetic Oil Industry of Germany* (LONDON: MINISTRY OF FUEL AND POWER, 1947), PP. 48–50; *High-Pressure Hydrogenation at Ludwigshafen-Heidelberg*, FIAT FINAL REPORT NO. 1317, 9 VOLS. (DAYTON, OHIO: JOINT INTELLIGENCE OBJECTIVES AGENCY, 1951)

secrecy one or two days before Russian troops occupied the area to prevent the Russians from forcing the staff to restart the Blechhammer plant.

Blechhammer initially planned to produce fuel oil for the German navy by hydrogenating Silesian bituminous coal to give gasoline and then mixing the gasoline with LTC tar and coke tar-pitch. But with the navy's greatly increased demand for fuel oil in 1940, the revised strategy called for hydrogenating bituminous coal tar to give fuel oil and gasoline. Pier's high-pressure experiment group in Ludwigshafen developed the revised plant design, and when Blechhammer opened in April 1944 it had four 700 atm liquid phase stalls, each having four converters, and four 300 atm vapor phase stalls (two saturation each with three converters and two splitting each with two converters) operating at 450° C. Blechhammer's operation differed little from the Gelsenberg plant although the liquid phase hydrogenation that Krönig introduced required the successive addition of three catalysts: iron (II) sulfide, Bayermasse (iron, aluminum, other metallic oxides), and sodium sulfide; and three vapor phase catalysts: 5058 and 8376 (7846W) for the saturation reaction and 6434 for the splitting reaction.

Blechhammer had annual production capacity of 240,000 metric tons of aviation gasoline, 260,000 metric tons of fuel oil, and 60,000 metric tons of petroleum (fuel) gas. It achieved a maximum annual production of 420,000 metric tons. A large LTC plant equipped with 10 Demag generators that began operation in October 1943 produced bituminous coal tar for hydrogenation and coke for hydrogen gas production. Blechhammer obtained additional hydrogen from its Linde-Fränkl liquefaction unit that separated hydrogen from the hydrogenation waste gases and from the nearby Heydebreck (Blechham-

TABLE 3.5 German coal hydrogenation plants: Production

Plant location	Process	Pressure	Temperature	Catalyst	Final products	Plant prod. and capacity 1944 tonnes/yr liquid products including LP gas
Ludwigshafen/ Oppau	liquid and vapor phase	200–300 atm	450° C	MoO_3, 6434	gasoline	not applicable
Leuna	liquid and vapor phase	250/250 atm	380–400° C	5058	gasoline, diesel oil, LP gas	620,000 640,000
Böhlen	liquid and	300/300 atm	460–75° C liquid 370–410° C vapor	5058, 6434	gasoline, diesel oil, LP gas	220,000 275,000
Magdeburg	liquid and vapor phase	300/300 atm	460–75° C liquid 370–410° C vapor	5058, 6434 vapor	gasoline, diesel oil, LP gas	220,000 275,000
Scholven	liquid and vapor phase	300/300 atm	460° C liquid 375° C vapor	$Sn(C_2O_4)_2$ liquid; 6434, 5058 vapor	gasoline, LP gas	220,000 240.000
Welheim	liquid and vapor phase	700/700 atm	480° C liquid 500° C vapor	$FeSO_4$ liq. Mo–Cr–Zn vapor	gasoline, fuel oil	130,000 145.000
Gelsenberg	liquid and vapor phase	700/300 atm	520° C	$FeSO_4$ liq. 5058, 6434, vapor	gasoline, LP gas	400,000 430.000
Zeitz	TTH process	300 atm	360° C liquid	5058, 8376 vapor	diesel oil, gasoline, lubricating oil, LP gas, wax	250,000 250,000
Pölitz	liquid and vapor phase	700/300 atm	no data	no data	gasoline, fuel oil (diesel oil), LP gas	700,000 750,000
Lützkendorf	liquid and vapor phase	700/700 atm	350° C liquid, vapor	6434 vapor	gasoline, diesel oil, fuel oil	50,000 12,000
Wesseling	liquid and vapor phase	700/300 atm	478° C liquid 367–425° C vapor	5058, 6434, 8376 liq., vap.	gasoline, diesel oil, LP gas	200,000 230,000
Brüx (Most)	liquid and vapor phase	300/300 atm	no data	no data	gasoline, diesel oil, LP gas	400,000 360,000

TABLE 3.5 German coal hydrogenation plants: Production (*cont.*)

Plant location	Process	Pressure	Temperature	Catalyst	Final products	Plant prod. and capacity 1944 tonnes/yr liquid products including LP gas
Blechhammer	liquid and vapor phase	700/300 atm	450° C liquid and vapor	8376, 6434, 5058 vapor	gasoline intended, fuel oil	60,000 65,000

SOURCE: TABLE COMPILED FROM *Report on the Petroleum and Synthetic Oil Industry of Germany* (LONDON: MINISTRY OF FUEL AND POWER, 1947), PP. 48–50; *High-Pressure Hydrogenation at Ludwigshafen-Heidelberg*, FIAT FINAL REPORT NO. 1317, 9 VOLS. (DAYTON, OHIO: JOINT INTELLIGENCE OBJECTIVES AGENCY, 1951), ESPECIALLY P. 112

mer South) higher alcohol and saturation plant that began operation in early 1944 and produced hydrogen from methane and ethane in a water gas reaction. Oberschlesische planned to double Blechhammer's capacity to 450,000 metric tons/yr of gasoline and 500,000 metric tons/yr of fuel oil and convert its vapor phase stalls to 700 atm in two scheduled expansions, but labor shortages prevented the expansions and no plant increase ever occurred. Bituminous coal delivered to Blechhammer cost RM 10–11/metric ton, the small amount of middle German brown coal tar the plant hydrogenated after July 1944 cost RM 80/metric ton (1944 prices). This gave a production cost of RM 250/metric ton for fuel oil and RM 450/metric ton for aviation gasoline with 65% of the cost attributed to capital charges. The aviation gasoline had a 78–79-octane number and with alkylate or iso-octane added increased to 102–104-octane number.

Blechhammer had a workforce of 4,116 forced laborers, all of whom Nazi officials evacuated in a 10-day foot march (21 January to 2 February 1945) from the Blechhammer concentration camp located three miles from the hydrogenation plant, to the Gross-Rosen concentration camp before the arrival of Russian troops who later dismantled the Blechhammer hydrogenation plant and used its equipment for scrap iron. Blechhammer experienced heavy bombing during the war years, by mid-1944 the USAAF Fifteenth Air Force had dropped 7,082 tons of bombs on Blechhammer. Heydebreck also experienced heavy bombing beginning in June 1944, followed by another attack in July and subsequent monthly attacks that reduced production and forced Heydebreck's closing in late 1944.[58]

4 Industrial Development of the F-T Synthesis

The F-T plants operated at atmospheric pressure (1 atm) or medium pressure (5–15 atm) and 180–200° C. They produced synthesis gas by reacting coke with steam in a water gas reaction and adjusting the proportions of carbon monoxide and hydrogen, and used a cobalt catalyst (100 Co-5 ThO2-8 MgO-200 kieselguhr) that Ruhrchemie's director of research Otto Roelen developed from 1933 to 1938. Roelen's catalyst became the standard F-T catalyst because of its greater activity and lower reaction temperature, but had a high preparation cost, RM 3.92/kg cobalt. For this reason, Ruhrchemie recovered the cobalt and thorium from the spent (used) catalyst by treatment with nitric acid and hydrogen gas at a cost of RM 2.97/kg cobalt, and re-used them in preparing fresh catalysts.[59] This gave a total catalyst cost of RM 6.89/kg cobalt or nearly 30% of the total F-T production cost.[60]

By 1937–38 the combined annual capacity of the four F-T plants increased to 300,000 metric tons (2.17 million barrels) and with the completion of five additional plants, total capacity rose to 740,000 metric tons (5.4 million barrels) at the outbreak of World War II in September 1939. Production at the nine F-T plants peaked at 526,000 metric tons (3.79 million barrels) in 1944.[61]

---

58 *High-Pressure* (note 3), Vol. 1, pp. 94–96; *Report* (note 46), pp. 68–69, gives Blechhammer's a capacity of 425,000 metric tons per year and a maximum production of 65,000 metric tons/yr (p. 48). *Germany* (note 49); Holroyd, *IG Farbenindustrie* (note 57), pp. 128–33 and US Bureau of Mines (note 57).

59 *Report* (note 46), pp. 82–90; "Germany's Home Production of Motor Fuels," *Gas World*, 104 (9 May 1936): 421; Franz Fischer, "The Conversion of Coal into Liquid Fuels," *Chemical Age*, 35 (24 October 1936): 353–55; "The Fischer Process," *Chemical Age*, 35 (31 October 1936): 367.

60 Henry H. Storch, Norma Golumbic, and Robert B. Anderson, *The Fischer-Tropsch and Related Syntheses* (New York: John Wiley & Sons, Inc., 1951), pp. 336–38; *Report* (note 46); G. Wilke, "Die Erzeugung and Reinigung von Synthese gas für die Benzinsynthese," *Chemische Fabrik*, 11 (1938): 563–68; "Substitute Motor Fuels in Europe," *Petroleum Press Service*, 5 (1938): 301–04; Arno C. Fieldner, "Frontiers of Fuel Technology," *Chemical and Engineering News*, 26, no. 23 (7 June 1948): 1700–01. Average exchange rate from 1934 to 1941 was RM 1 = 40¢, *Banking and Monetary Statistics, 1914–1941*, Federal Reserve System (Washington, DC: Government Printing Office, 1947), p. 671.

61 P. L. Pavcek, *Production of Fatty Acids from by-products of the Fischer-Tropsch Process*, CIOS XXVIII-35, Item No. 22 (no date), 2 pp.; *Report* (note 46), pp. 82–90. As much as 4.15 million barrels.

FIGURE 3.8
Otto Roelen
SOURCE: *Otto Roelen, Life and Work*, TECHNISCHE UNIVERSITÄT BRAUNSCHWEIG (N.D.). http://www.tu-braunschweig.de/Medien-DB/anchem/homogenekat-4bw.pdf

The older F-T plants operated at 1 atm whereas three of the five newer plants were medium pressure 5–15 atm syntheses. Converter design differed depending on the reaction pressure, but all the plants had inefficient externally cooled converters that dissipated the high heat of reaction (600 kilocalories per m^3 of synthesis gas consumed) and controlled the reaction temperature by arranging the cobalt catalyst pellets in a fixed bed within the converter and circulating pressurized water through the converter. Synthesis gas entered at the converter's top at the rate of 650–700 m^3 per hour per converter and flowed down through the catalyst bed, hydrocarbon products passed out the bottom. The medium pressure synthesis gave a slightly higher yield and extended the catalyst's life from 4–7 months to 6–9 months.

For the 1 atm synthesis the converter (tube and plate design) was a rectangular sheet-steel box 5 m long, 2.5 m high, 1.5 m wide, containing about 600 horizontal water-cooling tubes interlaced at right angles with 555 vertical steel plates or sheets. The complicated grid-like arrangement over which the synthesis gas flowed from top to bottom eliminated any localized heat buildup in the converter. Each steel plate was 1.6 mm thick, a space of 7.4 mm separated adjacent plates. The cooling tubes, each of 40 mm diameter, placed 40 mm apart, led to a boiler (steam drum) for recovery of the heat released in the synthesis. One boiler recovered the heat released from two converters. An empty converter weighed 50 metric tons. The catalyst pellets, which contained 900 kg of cobalt and filled the space between the tubes and plates, occupied a volume of 12 m^3 and weighed 3 metric tons.

The medium pressure converter (concentric double tube) had a simpler design. It consisted of a 50-metric ton vertical cylindrical steel shell 6.9 m high, 2.7 m internal diameter, 31 mm thick walls, and contained 2,100 vertical cooling tubes. Each cooling tube was 4.5 m long and double in construction,

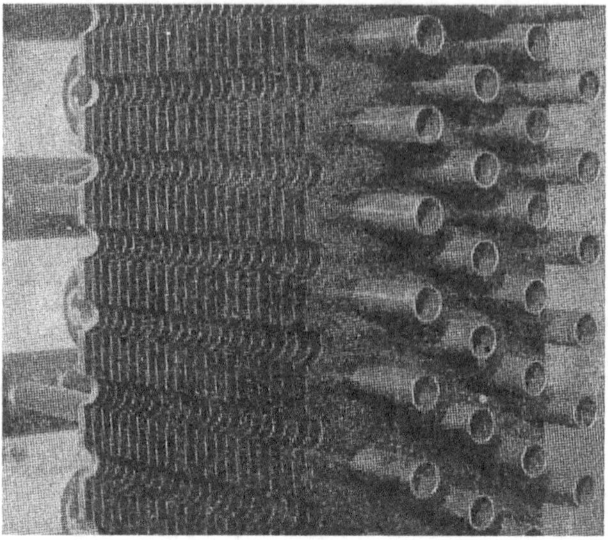

FIGURE 3.9 F-T tube and plate 1 atmosphere converter
SOURCE: HENRY H. STORCH, NORMA GOLUMBIC, AND ROBERT B. ANDERSON, *The Fischer-Tropsch and Related Syntheses* (NEW YORK: JOHN WILEY & SONS, INC., 1951), P. 348

consisting of an outer tube of 44–48 mm diameter fitted with a concentric inner tube of 22–24 mm diameter. A top and bottom weld (T-connections) between the converter's horizontal face and an outer tube connected an inner tube with a boiler that allowed cooling water to circulate from the boiler to the main space in the shell around the outer tubes and through the inner tube. One boiler recovered the heat released from four converters. The catalyst pellets filled the annular space between the concentric tubes and occupied a volume of 10 m^3.

In the 1 atm synthesis, water sprays in packed towers directly cooled the hot hydrocarbon vapors and gases (primary products or primary oils) leaving the bottom of the converter. The vapors condensed to give light oil C_5–C_{12} boiling point range 230–320° C, and hard and soft wax C_{20}–C_{30} boiling point range 320–460° C and above. The cooled gases (propane, butane) passed to an absorber for their removal and recovery with activated charcoal and subsequent liquefaction.

In the medium pressure synthesis about 3% of the primary products left the converter as hydrocarbon liquids. Passage through a tubular-type steel alloy condenser liquefied the hydrocarbon vapors. The remaining hydrocarbon gases, after expansion to atmospheric pressure, underwent recovery and removal with activated charcoal in an absorber.

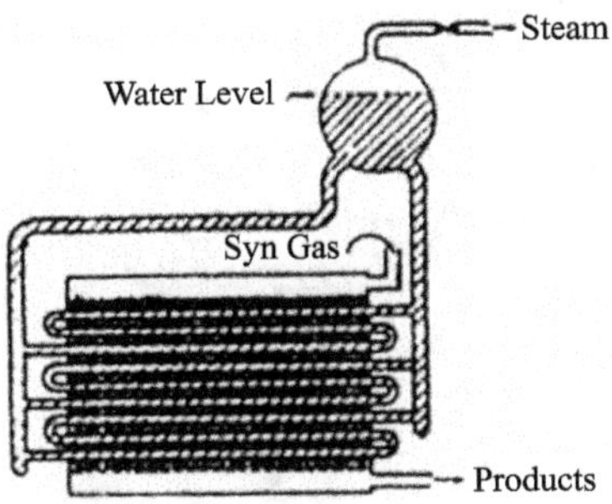

FIGURE 3.10 F-T tube and plate 1 atm converter
SOURCE: *Report on the Petroleum and Synthetic Oil Industry of Germany* (LONDON: MINISTRY OF FUEL AND POWER, 1947), P. 85. SYNTHETIC FUELS COLLECTION, TEXAS A&M ARCHIVES

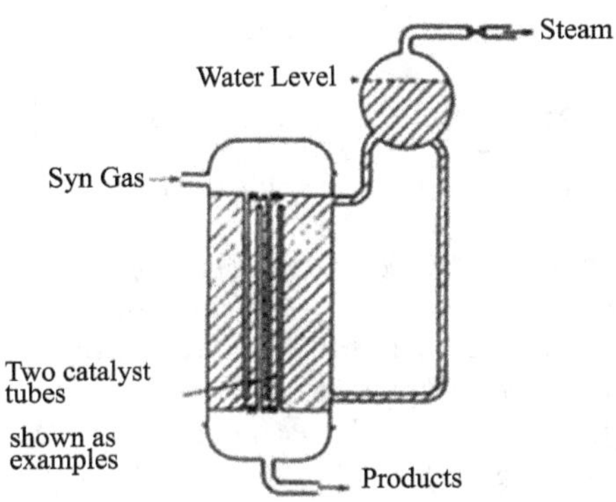

FIGURE 3.11 F-T concentric medium pressure (9–11 atm) double tube converter
SOURCE: HENRY H. STORCH, NORMA GOLUMBIC, AND ROBERT B. ANDERSON, *The Fischer-Tropsch and Related Syntheses* (NEW YORK: JOHN WILEY & SONS, 1951), P. 85

The biggest converter used in German F-T plants had a production capacity of only 2.5 metric tons/d (18 bbl/d) so that a small, 70 metric ton/d (504 bbl/d) plant had 25 or more converters, requiring considerable amounts of material and manpower for its construction and operation. All the plants operated their converters in stages. The 1 atm plants had two stages, operating two-thirds of the converters in the first stage and one-third in the second. Some of the plants placed the condensers and absorbers between the stages, others placed only condensers. All the plants had absorbers after the second stage converters and condensers. During the war's last two years the medium pressure plants switched from two stages to three stages, successively operating one-half, one-

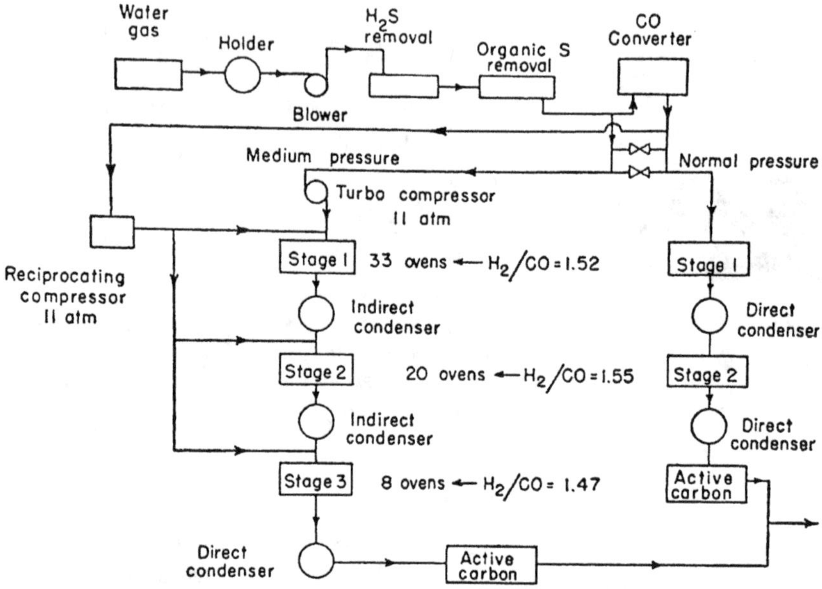

FIGURE 3.12 Schematic diagram for F-T 1 atm and middle pressure synthetic fuel plant, Ruhrchemie, Sterkrade

SOURCE: V. HAENSEL AND C. C. HALL, *The Fischer-Tropsch Plant of Ruhrchemie AG Sterkrade Holten*, CIOS-XXVII-69, ITEM NO. 30 (5 JULY 1945), 97 PP., ON P. 55; H. V. ATWELL, A. R. POWELL, AND H. H. STORCH, *United States Government Technical Oil Mission Fischer-Tropsch Report; TOM Report*, 5, JULY 1945, 66 PP.; *Fischer-Tropsch TAC Report SnMc-1, PB 2,051* (1945), 45 PP. REPRINTED IN HENRY H. STORCH, NORMA GOLUMBIC, AND ROBERT B. ANDERSON, *The Fischer-Tropsch and Related Syntheses* (NEW YORK: JOHN WILEY & SONS, 1951), P. 341

third, and one-sixth of their converters. They had condensers between each stage and absorbers after the final stage converters and condensers.

The 1 atm synthesis had an average plant yield of 130–165 g of liquid hydrocarbons per m^3 of synthesis gas, or about 80% of the theoretical maximum yield. Each converter produced 500–720 metric tons annually. The middle pressure synthesis had corresponding yields of 145–160 g/m^3 synthesis gas and 600–750 metric tons of products. The medium pressure synthesis also extended the catalyst's life from four-seven months to six-nine months.

Both syntheses used the same product refining process, fractional distillation. Low-grade gasoline which made up the light oil fraction, had a 45–53-octane number, which after blending with 20% benzol and adding 0.02–0.04% lead tetraethyl, increased to 70–78 and provided the German army with motor gasoline. High-grade diesel oil with a 78-cetane number (middle oil fraction) and some of the heavy oil fraction, after blending with 50% petroleum oil, served as aviation fuel for the German air force. Further treatment of most of the heavy oil at IG Farben's Leuna plant after its opening in 1927 gave the detergent *Mersol*; cracking and polymerizing the remaining heavy oil and some of the soft wax gave good quality lubricating oil. Oxidizing the rest of the soft wax produced fatty acids for conversion to soap and small quantities of edible fat. The German wax industry used most of the hard wax for electrical insulation, the manufacture of polishes, and as a paper filler.

The most efficient F-T plants recovered only 30% of the total heat energy input as primary products and another 25% as steam and residual gas. The net heat energy required to produce one metric ton of primary products equaled the heat energy produced by the combustion of 4.5 metric tons of coal (1 lb coal = 12,600 BTU).[62]

The first of the industrial-size F-T plants to produce synthetic fuel was the Steinkohlen-Bergwerk Rheinpreussen plant located in Mörs-Meerbeck (Homberg), near the Rheinpreussen coal mine in western Germany (Rhineland). Gütenhoffnungshütte, controlled by the Haniel Group, completed the plant that occupied an area 3,200 × 3,000 ft^2 in late 1936. Liquid fuel synthesis, using a Ruhrchemie (Oberhausen-Holten) cobalt catalyst that plant officials told CIOS they did not know its exact composition, occurred at 1 atm, a temperature of 190–195° C, in two stages with 60 of the plant's 96 tube and plate Ruhrchemie converters, each 1.5 m wide, 2.9 m high, 5 m long, operating in stage one, 30 operating in stage two, and 6 empty (idle). The catalyst had a four-month lifetime before requiring renewal. Synthesis gas passed into each converter at a rate of 1,000–1,200 m^3/hr. The converters had a 70–79% conversion in stage one and a 60–69% conversion in stage two. Rheinpreussen designed its own coal coking ovens (210 total) for the production of coke and coke (coal) oven gas, and prepared a hydrogen-carbon monoxide-methane mixture for cracking (reacting) with steam at 1,200° C in 11 Koppers gasifiers to increase the gas mixture's hydrogen content. Combining this mixture with twice as much water gas, produced by reacting coke with steam (981, 220 metric tons/d coke consumed produced 1,146,070 m^3/d water gas) gave the synthesis gas of proper proportions, 2 H_2: 1 CO. Rheinpreussen had an annual capacity of 25,000–30,000 metric tons (later increased to 70,000 metric tons)

62 *Report* (note 46), p. 91 (Table LV). The German Health Office officially approved the synthetic fat as fit for human consumption, but the Nazi government suppressed the findings of certain (unnamed) university scientists which threw considerable doubt on the fat's safety. Synthetic fat always contains esters of branched-chain fatty acids some of which are toxic (p. 94). C. C. Hall, "Oils and Waxes from Coal," *Chemical Age*, 55 (9 November 1946): 569–70.

of gasoline, 72-octane number for motor gasoline and 80–95-octane number for aviation fuel; diesel oil (primary oils) with a cetane number as high as 75–78; and paraffin wax. An alcohol plant produced another 3,000 metric tons of propyl and butyl alcohol. Rheinpreussen, which had a workforce of 1,000 and 25 researchers, achieved maximum production of 73,271 metric tons of liquid products in 1941. The CIOS report did not give the plant's construction cost.

American troops occupied the plant on the evening of 4 March 1945. Two CIOS members arrived at the plant on 6 March. They inspected the plant, collected samples of the cobalt catalyst and products, such as benzin and wax, and on 6, 9 March interrogated Wilhelm Breitenstein F-T assistant plant manager, Wilhelm Lohman F-T construction foreman, Herr Ziegler power engineer, and Fritz Häcklander coking plant manager. CIOS reports on Rheinpreussen published in March and May 1945 indicated that the plant suffered about 30% damage from RAF bombings on 19–20 July 1944 (destroyed 46,122 metric tons, its total stock, of treibgas); 21 August; 25 October; 2 November (destroyed 14 metric tons diesel fuel), and 8, 20–21 November (destroyed 103 metric tons benzin). The November attacks ended synthetic fuel production, but a final attack followed on 28 February 1945.

CIOS also learned that in January-February 1945, prior to the time of the army-CIOS arrival, Rheinpreussen plant officials had moved all technical and top administrative personnel, equipment, and many valuable plant records to locations east of the Rhine River, such as Alme in Sauerland, Hannover, Nachterstädt, Ludenscheid, and other locations. About 105 documents remained at the plant, and CIOS found them in basement storerooms, often scattered by the troops occupying the few habitable plant rooms, and in partly packed boxes indicating the hurriedness and incompleteness of the document removal. The documents contained correspondence, relocation sites of plant equipment, plant production and product quality, location and extent of plants bombed, and other relevant information. CIOS interrogators said they obtained more information from the documents than they obtained from plant officials. In visits to other F-T and hydrogenation plants CIOS personnel discovered that officials at those plants, such as Mörs-Meerbeck, had removed control room instruments and light equipment and sent them to different locations in Germany.[63]

63 H. V. Atwell and W. C. Schroeder, *Steinkohlen-Bergwerke Rheinpreussen Moers-Meerbeck*, CIOS XXVI-80, Item No. 30 (8 March 1945), 152 pp.; H. V. Atwell and W. C. Schroeder, *Report on Steinkohlen-Bergwerke Rheinpreussen Moers-Meerbeck (Homberg) Germany*, CIOS XXV-

FIGURE 3.13 Wilhelm Breitestein, Wilhelm Lohman, Fritz Häcklander at Mörs-Meerbeck plant (no date given for picture)
SOURCE: H. V. ATWELL AND W. C. SCHROEDER, *Steinkohlen-Bergwerke Rheinpreussen Moers-Meerbeck*, CIOS XXVI-80, ITEM NO. 30 (8 MARCH 1945), P. 11

The mining company Gewerkschaft Viktor AG (Klocknêrwerke AG), a subsidiary of Wintershall AG, constructed the second industrial-size F-T plant at a cost of RM 30 million in Castrop-Rauxel (Ruhr) also in late 1936. The plant site, occupying 3,300 × 2,350 ft^2, adjoined Gewerkschaft Viktor's coal mine and was the location of a colliery and a synthetic ammonia plant, that occupied 91 acres. Gewerkschaft Viktor designed its own coal coking ovens and a gasifier similar to a Koppers gasifier. It produced 300,000 m^3/d of synthesis gas by cracking coke oven gas with steam and mixing the synthesis gas with water gas obtained from coke in four Demag reactors that produced 600,000 m^3/d of water gas. The plant's 63 tube and plate converters operated in two stages at 1 atm, 53–55 in stage 1 and 20–22 in stage 2, a temperature of 176–191° C, and used Ruhrchemie's standard cobalt catalyst (100 Co-5 ThO2-8

6, Item No. 30 (15 May 1945), 150 pp.; H. V. Atwell and W. C. Schroeder, *Report on Investigation of Synthetic Lubricating Oil Plant, Rheinpruessen, Homberg*, CIOS XXIV-9, Item No. 30 (9 March 1945), 13 pp.; H. A. Schade, E. Foran, and R. C. Aldrich, *The Fischer-Tropsch Process*, CIOS VI-22, X-18, X-22, XV-5, Item No. 20 (26, 27 September 1944), 28 pp.; *Additional Information Concerning the Fischer-Tropsch Process and its Products*, BIOS Final Report No. 1722, Item No. 22 (London: His Majesty's Stationery Office (July, August 1946), 178 pages; *Report* (note 46), pp. 89; Warren, *Inspection* (note 49), pp. 16–18; *Germany* (note 49); "Coal Hydrogenation-Germany" (note 49).

MgO-200 kieselguhr), 3 tons in each converter. Gewerkschaft Viktor had an annual capacity of 30,000–40,000 metric tons of low-grade gasoline that had a 47–53-octane number and diesel oil used to synthesize Mersol detergents. It produced a maximum of 43,380 metric tons of liquid products in 1944, but bombing attacks by the RAF in September 1944, November 1944, January 1945 and the USAAF in March 1945 badly damaged the F-T plant. The five-member CIOS team arrived on 12 April 1945 and interrogated assistant plant manager Dr. Braune and F-T plant manager Dr. Volmer on 12 and 14 April. It intended to interrogate Dr. Leonard Alberts, Castrop-Rauxel plant manager who until late 1943 managed the Sterkrade-Holten F-T plant, but he had returned to his home in nearby Waltrop and underwent interrogation on 13 April 1945.[64]

Ruhrchemie's Ruhrbenzin AG constructed the third industrial-size F-T plant in Oberhausen-Holten (Sterkrade-Holten), (Ruhr). The complete plant complex occupied a triangular area 3,700 × 4,100 × 2,250 ft^2; the F-T plant measured 1,300 × 900 ft^2. Ruhrbenzin, established in September 1935 with a capitalization of RM 4.5–6 million and increased to RM 15 million in 1940, planned in 1936 to complete construction of a plant annually producing 30,000 metric tons (increased to 62,000 metric tons in 1942) of gasoline, diesel oil, and lubricating oil. Production did not begin until 1937, however. The plant differed from the Rheinpreussen and Viktor plants in having two independent synthesis systems: a two-stage 1 atm synthesis with 48 tube and plate converters and a three-stage 10–15 atm synthesis with 72 concentric double tube converters. Water gas, prepared from coke in 11 Demag reactors (2 more under construction November 1941–January 1942), one-third of which after an iron-catalyzed reaction with steam at 500° C, gave a mixture containing 61% hydrogen and 5% carbon monoxide; adding the mixture to the remaining two-thirds water gas provided synthesis gas for conversion to gasoline of 58–60-octane number, diesel oil of 80–85-cetane number, and lubricating oil. The plant had a single Linde-Fränkl liquefaction unit.

Compared to the eight F-T plants that eventually came into operation the Ruhrbenzin plant ranked low in efficiency. It lost RM 2.6 million in 1939 which plant officials attributed to constant experimentation with the plant's reaction conditions and operational procedures. An Allied aerial attack by 321 USAAF, RAF, and Royal Canadian Air Force (RCAF) bombers on 16–17 June 1944 did little damage because of thick cloud coverage, later attacks in August-October

64 C. C. Hall and A. R. Powell, *Report on Plant of Klocknerwerke AG, Castrop-Rauxel, Germany*, CIOS XXV-7, Item No. 30 (22 June 1945), 12 pp.; *Report* (note 46), p. 89; Warren, *Inspection* (note 49), pp. 23–24; *Germany* (note 48). Diesel oil cetane number not given. The US War Department brought Alberts, who had a Nazi affiliation, to the War College in Washington, DC through Project Paperclip.

1944 severely damaged the plant and permanently prevented it from resuming production.

When five CIOS personnel arrived at the Sterkrade plant on 6 April 1945, a month before the European war ended on 8 May, they learned from their interrogation of plant scientists and engineers that because of the bombing, Ruhrchemie in autumn 1944–early 1945 had evacuated 726 plant personnel to various locations, mainly to their homes, and that those evacuated had burned some plant records and taken other plant records with them. CIOS located a large number of records at a small moated, walled castle (country house), stacked from floor to ceiling in small rooms, and in barns and farm buildings in Reelkirchen, four miles south of Bloomberg. Additional records came from Haberlah Wiese, located 20 miles southwest of Braunschweig. A document found at Sterkrade contained a list of 77 department managers and workers.

The 12–13 Ruhrchemie executives, managers, scientists, and engineers that CIOS located and interrogated at the Sterkrade plant, their homes, or agency headquarters, included Dr. Friedrich Martin (1885–1951), president of Ruhrchemie's four-member directorate, whom CIOS found at a farm in Braunschweig and interrogated in Brussels and London; chemist Dr. Hagemann, a member of Ruhrchemie's directorate, interrogated in Brussels and London; Otto Roelen, Ruhrchemie's research director and cobalt catalyst specialist, located at a farm in Russian-occupied Salzwedle, north of Magdeburg, recovering from ill health before he returned to Sterkrade in October 1945, interrogated in London; Dr. Spanier, acting head of the Sterkrade plant; Dr. Schreiber, laboratory head; Dr. Velde (b. 1905), in charge of the research laboratories; and Dr. Landgraff, manager of the Sterkrade oxo plant.

Ruhrchemie in 1937 had licensed its F-T process to the Japanese company Mitsui Bussan KK in Rumoi, and in interrogations that focused on German-Japanese information exchange, CIOS learned that Mitsui had not allowed Dr. Schenk, a plant designer Ruhrchemie sent to Japan to assist in the planning and construction of Mitsui's Miike F-T plant, to return to Germany. The Miike F-T plant came on stream in 1940. CIOS interrogations also revealed that Lurgi in Frankfurt had experienced a similar problem at its Tokyo office where Lurgi-Tokyo kept Dr. C. Crayer, in semi-confinement from May to June 1939. Lurgi later sent Dr. Otto Hubmann (1891–1983), its director of fuel research, to assist Mitsui's oil processing and coal gasification programs which Hubmann found unimpressive and reported that Japan's fuel scientists were very secretive and greatly exaggerated their advances in fuel research and development. You could believe 50% of what they claimed.[65]

65 V. Haensel and C. C. Hall, *The Fischer-Tropsch Plant of Ruhrchemie AG Sterkrade Holten*, CIOS-XXVII-69, Item No. 30 (5 July 1945), 97 pp.; R. L. Hasche and R. H. Boundy, *Report*

FIGURE 3.14 Cobalt catalyst reduction vessel at Ruhrchemie, Sterkrade
SOURCE: V. HAENSEL AND C. C. HALL, *The Fischer-Tropsch Plant of Ruhrchemie AG Sterkrade Holten*, CIOS-XXVII-69, ITEM NO. 30 (5 JULY 1945), 97 PP., ON P. 30

Sterkrade-Holten became the research and development center for the catalytic studies of Roelen, Leonard Alberts, Walter Feisst, and others. Its catalyst plant supplied the six F-T plants in the Ruhr area with the standard cobalt catalyst, producing annually about 3,000 metric tons. Brabag's plant in Ruhland-Schwarzheide and beginning in 1938 the Wintershall plant in Lützkendorf also produced the standard cobalt catalyst.[66]

Brabag, which operated three coal hydrogenation plants, completed construction of Brabag III, the fourth F-T plant in Ruhland-Schwarzheide, Dresden area eastern Germany (Saxony), in 1937. It occupied an area 5,100 ×

---

on Inspection of the OXO Plant at Ruhrchemie Oberhausen Holten, CIOS XXVIII-18, Item No. 22 (18 June 1945), 4 pp.; O. G. Thompson and H. V. Atwell, *Metallgesellschaft-Lurgi, Frankfurt-am Main*, CIOS XXXI-23, Item No. 30 (14 September 1945), *Ruhrchemie AG, Sterkrade-Holten, Near Oberhausen*, Ruhr, BIOS Final Report 1038 (London: His Majesty's Stationery Office, 7–8 March 1946); *Report* (note 46), p. 89; Warren, *Inspection* (note 49), pp. 14–15; *Germany* (note 49).

66 Note 65; *Report* (note 46), p. 85.

3,000 ft^2. Brabag III was a two-stage 1 atm plant and had an annual capacity of 25,000–30,000 metric tons of gasoline and diesel oil. It used the standard Ruhrchemie cobalt catalyst and produced gasoline and diesel oil with high enough octane and cetane numbers. Later expansion, which increased the number of tube and plate converters to 262 and maximum annual production to 162,000 metric tons (200,000 metric tons capacity), made it Germany's largest F-T plant. Brown coal briquettes, gasified in Didier-Bubiag retorts, each with a capacity of 638,000 m^3/d (22 million cubic feet), and four Koppers gasifiers, each with a capacity of 26,100 m^3/hr (900,000 cubic feet), provided 20% and 80% of the synthesis gas. Purification of the synthesis gas by passing it though towers containing pellets of iron oxide and sodium carbonate to remove sulfur and other impurities was relatively simple because of the brown coal's low sulfur content.

On 11 September 1944, 36 USAAF Eighth Air Force B-17 bombers sent to destroy the Schwarzheide F-T plant lost 14 aircraft in aerial combat with German fighter planes. The remaining B-17s dropped 53 tons of bombs on the F-T plant causing significant damage and production to fall from 179,665 metric tons in 1943 to 158,500 metric tons in 1944. Other B-17 raids followed from 7 October 1944 through 22–23 March 1945. Because of Brabag III's location in eastern Germany, the Russian army occupied the plant area, and no American or British intelligence teams visited Brabag III. Ruhrchemie documents provided some operational details, CIOS also obtained information came from Brabag III officials, Erwin Sauter, A. Wagner, W. Sapper, and catalyst specialist Karl Meyer.[67]

In addition to the 1930s ongoing F-T catalytic research, both Ruhrchemie at its research center in Oberhausen-Holten and Fischer at the KWI investigated the F-T medium pressure synthesis hoping to improve F-T efficiency and economics. The studies showed that medium pressure gave a slightly higher yield of gasoline and diesel oil per m^3 of synthesis gas, extended the catalyst's life from 4–7 months to 6–9 months without any reactivation, and yielded a higher proportion, about 45% versus 18%, of heavier hydrocarbons such as soft and hard wax for the production of lubricating oil and chemicals. The middle pressure synthesis also had a higher operating cost. Consequently, unlike the 1 atm Brabag III plant, only two of the five F-T plants constructed in 1938 and 1939

67 Octane and cetane numbers not given. *Report* (note 46), p. 90; C. C. Hall, S. R. Craxford, and D. Gall, *Interrogation of Dr. Otto Roelen of Ruhrchemie AG*, BIOS Final Report No. 447, Item No. 30 (London: His Majesty's Stationery Office, 1946); *Interrogation of Dr. Roelen on OXO Synthesis*, BIOS Interrogation Report No. 736 (London: Ministry of Fuel and Power, 20–21 December 1945); "Coal Hydrogenation-Germany" (note 49); *Germany* (note 49).

before World War II began were medium pressure syntheses. A third plant was a combination atmospheric-medium pressure synthesis.[68]

The first of the newer F-T plants was the Wintershall subsidiary, Mitteldeutsche Treibstoff plant, constructed in Lützkendorf, central Germany (Saxony), in late 1938. Mitteldeutsche had 132 tube and plate converters that operated in two stages at 1 atm, a maximum of 77 converters operated at one time. Three Linde-Fränkl units provided oxygen gas, four Winkler generators supplied hydrogen gas. Mitteldeutsche had a workforce of 430 and occupied an area 5,200 × 1,740 ft^2.

The plant performed poorly except for its last two years of operation in 1943–44 when annual production reached 22,000 and 29,320 metric tons of gasoline with a 56.5-octane number and diesel oil with a 68.2-cetane number, or about 40% of its maximum. A synthesis gas problem, impurity of hydrogen gas extracted from synthesis gas; removal of suspended coal dust from synthesis and producer gas; and the presence of a high amount of organic sulfur ring compounds and gum forming compounds in the products, caused the plant's poor performance.

Mitteldeutsche installed the first industrial-size Schmalfeldt generator that plant director Heinrich Schmalfeldt had designed for the production of synthesis gas from the direct gasification of powdered brown coal. The coal had a very high sulfur content, and until plant engineers added activated charcoal absorbers to the purification system to remove the sulfur and eliminate the catalyst's poisoning (a standard procedure in F-T plants), the cobalt catalyst (100 Co-10 MgO-5 ThO_2) lasted only two months instead of the usual 4–7 months. Plant director since 1940 Dr. Schneeberger, plant director and power plant manager Dr. Schneider, gas plant manager Herr Dassau, and F-T plant manager Herr Schültz provided plant operating details when six CIOS personnel visited the F-T and hydrogenation plants on 9, 11, and 17 May 1945.[69] From 12 May 1944 to 8 April 1945, 9 RAF and USAAF air raids, with as many as 233

68 *Medium Pressure Synthesis with Iron Fixed-Bed Catalysts, and Operation of the Fischer-Tropsch Synthesis in the Liquid Phase*, BIOS Report 1712 (London: Technical Information & Documents Unit, 5 December 1947); *Report* (note 46), p. 87 (Table L); Hall, "Oils" (note 63).

69 H. Hollings, *Wintershall AG Lützkendorf Near Mülchen, Germany*, CIOS XXXII-90, Item No. 30 (Washington, DC: Bureau of Mines, 28 September 1945), 21 pp.; *The Schmalfeldt Process for making Synthesis Gas from Methane*, FIAT Final Report 1305 (30 September 1947); *The Wintershall-Schmalfeldt Process for the Manufacture of Synthesis Gas at Lützkendorf*, BIOS Final Report 1142, Item No. 30 (London: His Majesty's Stationery Office, 9, 11 May 1945); *Report* (note 46), p. 90; *Germany* (note 49).

aircraft in a raid, bombed Lützkendorf, but the raids did not reduce its production.

Friedrich Krupp AG in Essen joined the expanding group of synthetic fuel producers in 1937 when it established Krupp Treibstoffwerk GmbH in Wanne-Eickel, Essen (Ruhr) with a capitalization of RM 20 million and a RM 10 million loan. A six-member CIOS team on 15–16 April 1945 visited the plant, badly damaged from RAF attacks on 7 August 1941, by 277 bombers on 9 November 1944, by 323 bombers on 2–3 February 1945, and finally on 4 March. CIOS interrogated Dr. Erich Combles general manager and assistant general manager Dr. H. Fischer who directed the 900 workers operating the only F-T combination atmospheric-medium pressure plant.

Krupp-Lurgi gasifiers of 40 metric tons/d capacity converted coke, obtained mainly from high-temperature coal carbonization, to water gas, one-third of which underwent catalytic conversion to synthesis gas. Synthesis gas then passed through one set of 72 tube and plate converters at 1 atm for conversion to gasoline and diesel oil; residual synthesis gas, after flowing through standard tubular condensers, activated charcoal absorbers, and compressed to 10–15 atm, moved through a second set of 24 medium pressure converters to complete the conversion. The Krupp plant used the standard cobalt catalyst.

Of the 24 medium pressure converters, 16 were of a new design called tauschenrohren, in which single tubes of 72 mm internal diameter and fitted with fins of sheet steel, replaced the standard concentric double tube converter. The new converter design increased catalyst capacity by 5% but left carbon deposits in the converter. Maximum production of gasoline 60–61-octane number and diesel oil 85–95-cetane number at the plant reached 54,020 metric tons in 1943, falling to 39,802 metric tons in 1944. Krupp had a maximum annual capacity of 130,000 metric tons.[70]

Chemische Werke Essener Steinkohle AG in Essen (Ruhr), established in early 1937 as a partnership of Essener Steinkohlen Bergwerke AG and Harpener Bergbau AG in Dortmund with a capitalization of RM 12 million and a RM 10 million loan, constructed the second largest F-T plant with 600 workers and the most efficient 1 atm plant in Kamen-Dortmund (Bergamen), Ruhr. CIOS's five-member team interrogated Dr. Hermann Schwenke managing director, Dr. Thies chief chemist, plant manager Herr Gabriel, and assistant manager E. Tengelmann at the plant site on 19–20 April 1945.

70 C. C. Hall, *Krupp Treibstoffe Werke GmbH. Wanne Eickel. Fuels and Lubricants*, CIOS XXV-25, Item No. 30 (Washington, DC: Department of Commerce, Office of the Production Board, 15 April 1945), 8 pp.; *Report* (note 46), p. 89; Warren, *Inspection* (note 49), pp. 21–22; *Germany* (note 49); "Motorkraftstoff" (note 53), pp. 231–32.

Gasifying daily 800–850 metric tons of coke in 12 Demag water gas generators and cracking the resulting 550,000 m³ coke oven gas, produced 1,800,000 m³ of synthesis gas for conversion to gasoline and diesel oil in 124 tube and plate converters operating at 185–190° C in two stages, 2/3 in stage one, 1/3 in stage two. Each converter contained one metric ton of the standard cobalt catalyst. Essener Steinkohle plant's high efficiency, according to CIOS, appeared to depend on the purity of its synthesis gas, the equal distribution of the cobalt catalyst between the two stages, and the frequency of reactivating the catalyst by treating it with nitric acid and hydrogen gas. Gabriel and Tengelmann believed, however, that the constant composition of the synthesis gas and the plant's freedom from interruptions and breakdowns, which most likely resulted because of all the above factors, sufficiently explained the plant's successful operation from the time of its start up in 1939. Essener Steinkohle achieved a maximum annual production of 86,580 metric tons of gasoline 61–62-octane number and diesel oil 70-cetane number in 1944. RAF air raids on 9 September 1944 and on 24 February 1945 with 340 bombers heavily damaged the Essener plant and ended its operation.[71]

The last of F-T plants constructed were the medium pressure operations of Hoesch-Benzin GmbH in Dortmund (Ruhr) which began operation in 1939 and Schaffgotsch Benzin GmbH in Deschowitz-Beuthen, Odertal (Upper Silesia) which began operation in 1941. Hoesch-Benzin, a subsidiary of Bergwerksgesellschaft Trier GmbH (owned by Hoesch-Köhn-Neussen AG) and capitalized at RM 3 million, had a workforce of 800 under the direction of plant manager Dr. H. Weitenhiller and plant superintendent Dr. Werres. It occupied an area 2,100 × 870 ft². The medium pressure (10 atm) Hoesch plant converted coke to water gas in six water gas generators and then cracked the water gas with additional steam to produce synthesis gas. Its 65 concentric double tube converters, operating at 180–205° C, with standard cobalt catalyst, converted synthesis gas to 55–56-octane number gasoline and 70+, 100-cetane number diesel oil in two stages, using 40 converters in the first stage, 16–20 in the second stage, and added a third stage during the war. Hoesch, of all the F-T plants, had the highest operating efficiency, measured by monthly production per converter, its production reaching a maximum of 51,000 metric tons/yr in 1944. FIAT personnel, one from the US Bureau of Mines and two from British Ministry of Fuel and Power inspected the plant on 22 April 1945, 50% of which

71 C. C. Hall, *Chemische Werke Essener Steinkohle AG Bergkamen, Germany*, CIOS XXVII-54, Item No. 30 (19, 20 April 1945), 5 pp.; *Report* (note 46), pp. 89–90 (Table LIV); Warren, *Inspection* (note 49), p. 16; *Germany* (note 49); "Coal Hydrogenation-Germany" (note 49); "Motorkraftstoff" (note 53).

bombing attacks on 11 September 1944; 11, 12, 15 November 1944; and 22 April 1945 had destroyed. The RAF 11 November attack dropped a single 4,000-pound bomb, six 1,000-pound bombs, and seven 500-pound bombs from 16,000 ft contributing to the significant plant damage.[72]

Plant manager Alfred Pott, formerly director-general of Ruhrgas AG, supervised the Schaffgotsch Benzin plant operation. Pintsch generators produced synthesis gas from hard coke and coke oven gas, and until mid-1943 synthesis gas conversion to gasoline and diesel oil occurred in two stages. The addition of a third stage at that time resulted in a plant similar to the Hoesch-Benzin plant. Schaffgotsch had 68 converters, 50 of them wide-tube 22–23 mm diameter converters that contained single catalyst tubes rather than the concentric double tubes used in the other medium pressure F-T plants. Its engineers claimed that their modified design increased catalyst capacity by 10% and that their converters functioned particularly well in the second and third stages despite having to drill interior carbon deposits in order to remove the standard cobalt catalyst for reactivation. Schaffgotsch, which occupied an area 5,150 × 2,150 ft^2, achieved a maximum annual production of 39,200 metric tons of gasoline and diesel oil in 1944; its annual capacity was 80,000 metric tons.

USAAF B-24s bombed Schaffgotsch on 7 July 1944, 22 August, 13 September, 2 November, and 17–18 December 1944, seriously damaging the plant. As occurred with the Brabag III plant, the Russian army occupied the Upper Silesian region, and no CIOS, BIOS, or FIAT teams visited the Schaffgotsch plant. Ruhrchemie documents provided most of the operational information.[73]

F-T plant construction ended with the outbreak of war, resulting in standardization of plant apparatus and operation, although during 1943–45 research continued on finding cheaper iron catalysts to replace scarce wartime supplies of cobalt compounds, and on designing better converters. Ruhrchemie, which conducted 2,000 experiments, Rheinpreussen, KWI, IG Farben, Lurgi, and Brabag developed six iron catalysts, all gave satisfactory results in comparative tests done in September 1943 at Brabag's Schwarzheide plant. The Reichsamt für Wirtschaftsausbau, which arranged for the tests, never decided on the best iron catalyst, concluding only that all six were inferior to cobalt catalysts. None of the industrial-size plants used iron catalysts.

---

72 W. Schroeder and R. Neumann, *Fischer-Tropsch Plant of Hoesch Benzin AG at Dortmund, Germany*, FIAT Final Report No. 239 (Washington, DC: Joint Intelligence Objectives Agency, 2 October 1945); *Report* (note 46), p. 90; Warren, *Inspection* (note 49); "Motorkraftstoff" (note 53).

73 Octane and cetane numbers not specified. *Report* (note 46), pp. 90–91; *Germany* (note 49).

TABLE 3.6 German Fischer-Tropsch plants

Plant	Location	Coal (raw material), catalyst	Product. 1944 (metric tons)	Products	Pressure (atm) Atmos 1 Med 5–15	Started operation
Ruhrbenzin AG (Ruhrchemie AG)	Oberhausen-Holten (Sterkrade-Holten) Ruhr, w..Germany	bituminous cobalt	62,000	primary oils gasoline, diesel, lubricating	atmospheric medium	construction started Nov. 1935, in operation 1937
Steinkohlen-Bergwerk Rheinpreussen	Mörs-Meerbeck (Homberg), w. Germany	bituminous cobalt	19,700	primary oils, hard, soft paraffin wax oils for fatty acids	atmospheric	construction started by Nov. 1935, operating late 1936
Gewerkschaft Viktor, Klocknêrke-Wintershall AG	Castrop-Rauxel, Ruhr, w. Germany	bituminous cobalt	40,380	primary oils (gasoline and diesel oil)	atmospheric	construction started by Nov. 1935, operating second half 1936
Braunkohle Benzin AG (Brabag)	Ruhland-Schwarzheide, north of Dresden, e. Germany	lignite cobalt	158,500	primary oils	atmospheric	construction started by Nov. 1935, in operation 1937
Mitteldeutsche Treibstoff und Öl Werke, Wintershall AG	Lützkendorf, near Leipzig, Merseberg	lignite cobalt	29,320	primary oils	atmospheric	1938
Krupp Treibstoffwerke GmbH	Wanne-Eickel, Ruhr, w. Germany	bituminous cobalt	54,000 (1943) 39,802 (1944)	primary oils	atmospheric and medium	late 1938
Chemische Werke Essener Steinkohle AG	Dortmund (Bergkamen), Ruhr, w. Germany	bituminous cobalt	86,580	primary oils	atmospheric	1939
Hosech-Benzin GmbH	Dortmund, Ruhr, w. Germany	bituminous cobalt	51,000	primary oils	medium	March 1939
Schaffgotsch Benzin GmbH	Deschowitz-Beuthen Odertal, Upper Silesia (sw Poland)	lignite cobalt	39,200	primary oils	medium	1939

SOURCE: COMPILED FROM *Report on the Petroleum and Synthetic Oil Industry of Germany* (LONDON: MINISTRY OF FUEL AND POWER, 1947), PP. 88–93, AND THE *Joint Intelligence Objectives Agency, High-Pressure Hydrogenation at Ludwigshafen-Heidelberg*, FIAT, FINAL REPORT 1317, 9 VOLS. (DAYTON, OHIO: JOINT INTELLIGENCE OBJECTIVES AGENCY, 1951)

TABLE 3.7 Converter design summaries

Gas as the heat transfer medium	Oil as the heat transfer medium
IG Farben's fixed bed hot gas recycle process	Fluidized bed oil slurry process: IG Farben, Ruhrchemie, Rheinprussen Fixed bed oil circulation process: IG Farben

SOURCE: COURTESY OF AUTHOR

The three new converter designs developed during the war operated at 20 atm, used iron catalysts, and were internally cooled compared to the inefficient externally cooled fixed bed converters in the existing F-T plants. Their design summaries appear below.

The first of these new converters removed the heat the synthesis released in the gas stream by recirculating residual gas through a wide shallow bed containing a powdered iron-1% borax catalyst. IG Farben developed this converter design and successfully tested it on a small scale with 5 L of catalyst for 10 months. Large-scale tests were unsuccessful because of the catalyst's overheating. The converter had high energy requirements and cost more to operate than the older design converters, and for these reasons IG Farben abandoned its development before the end of the war. The gasoline produced had a 68–70-octane number that additional refining increased to 75–78- and 84-octane number.

Both of the remaining converter designs used oil as the heat transfer medium. The fluidized bed or oil slurry process forced water gas through a ceramic plate at the bottom of a cylindrical converter that contained a catalyst of iron with carbonate or borate suspended in a high boiling heavy synthetic oil. The tests were small scale and aimed at the production of C20-C70 olefins in the gas oil boiling point range (232–426° C) for use in chemical syntheses.

The other oil-cooled converter had the iron catalyst (iron oxide and other metallic oxides) arranged in a fixed bed and removed the heat of reaction by circulating oil through the catalyst bed. IG Farben tested this converter for an extensive time in a pilot plant of 8–10 metric tons capacity, synthesizing a gasoline with a 62–65-octane number.[74]

Emphasis also shifted at this time from the production of fuels and lubricants to the production of olefins, (unsaturated hydrocarbons), waxes,

74 *Report* (note 46), pp. 96–100.

alcohols, and other organic compounds. The oxo synthesis from the German "oxiering," meaning ketonization, was the most important result of this research. In the oxo synthesis, straight-chain olefins such as C_2H_4 and C_3H_6, reacted with carbon monoxide and hydrogen at 110–150° C, 150 atm, and with a cobalt catalyst to form an aldehyde that had one more carbon atom in the chain. Hydrogenating the aldehyde under the same conditions gave the corresponding alcohol. Roelen of Ruhrchemie patented the process in 1938, and in 1940 Ruhrchemie and IG Farben cooperated in its development. Their objective was the production of long-chain alcohols (C12 to C18) for conversion to detergents (sulfate esters), but the process had general applicability to all olefin-like compounds. Ruhrchemie, IG Farben, and Henkel et Cie, organized a new company called Oxo-Gesellschaft and in 1944 completed construction of an oxo plant in Sterkrade-Holten that had an annual capacity of 12,000 metric tons of fatty alcohols and a production cost of 78.23 pfennig/kg. Allied bombing in August-October 1944 permanently prevented the plant from beginning production.[75]

Information on F-T plant construction and operating costs has come from two main sources: captured German synthetic fuel documents and their summaries and interrogation of German synthetic fuel scientists, such as Martin, Ruhrchemie's managing director; Heinrich Bütefisch (1894–1969), an IG Farben director and government economic advisor on wartime petroleum production; and F-T plant managers and operating personnel.[76] The collective information indicated high plant capital and production costs. A F-T plant cost (capital cost) approximately RM 30 million. Production cost, including catalyst, water gas manufacture, synthesis of primary products and all other costs was 23.5–26 pfennig/kg (RM 240–330/metric ton, \$13.2–18.4/barrel, 31–44¢/gallon) for both the 1 atm and medium pressure operation.

Hoesch-Benzin's medium pressure plant with an annual production of 40,000 metric tons had a capital cost of RM 26 million (RM 650/metric ton per year) and a production cost in 1942 of 25.81 pfennig/kg of products. The Essener Steinkohle 1 atm plant with 80,000 metric tons annual production

75 *Interrogation of Dr. Roelen on OXO Synthesis*, BIOS Interrogation Report No. 736 (note 68).

76 *Synthetic Oil Production in Germany, Interrogation of Dr. Bütefisch*, BIOS Final Report 1697, Item No. 30 (London: His Majesty's Stationery Office, 1946). Bütefisch was production head at IG Farben's Bunawerke, Monowitz, part of IG Auschwitz. He received a six-year prison sentence in 1948 at the Nuremberg Military Tribunals, December 1946–April 1949, for war crimes; also W. F. Faragher and W. A. Horne, *Interrogation of Dr. Helmuth Pichler and Prof. Karl Ziegler at Kaiser Wilhelm Institut für Kohlenforschung, Mulheim*, FIAT Final Report No. 276.

had a capital cost of RM 32 million (RM 400/metric ton/year) and a production cost of 23.71 pfennig/kg of synthetic products. Ruhrchemie's combined atmospheric-medium pressure plant with an annual production of 42,000 metric tons had a capital cost of RM 15 million in 1940 (RM 380/metric ton/year) and a production cost in 1939–40 of 23.57 pfennig/kg of products. The nine F-T plants provided 12–15% of Germany's total synthetic fuel production during their nine years of operation.[77]

5 Summary of Synthetic Fuel Industrial Development in Germany 1927–45

Despite substantial government support and Hitler's 1936 call for petroleum-independence, the synthesis of petroleum from coal and tar never completely solved Germany's liquid fuel problem. Bureaucratic confusion, material shortages, and later Allied bombing limited its effectiveness. But production increased dramatically under the Four Year Plan and its renewal in October 1940. In 1933, Germany had only three small synthetic fuel plants in operation: Ludwigshafen, Leuna, and Ruhrchemie Oberhausen-Holten, the last a F-T plant, that produced mainly diesel oil and petrochemicals. At that time, it consumed about one-half of Great Britain's petroleum consumption, one-fourth of Russia's, and one-twentieth that of the United States. Yet, even at such low consumption, domestic resources remained inadequate. Its consumption of liquid fuels (including 274,000 metric tons of lubricating oil) in 1932 totaled 2,755,000 metric tons, 73% of which (2,020,000 metric tons) Germany imported mainly from the United States. For gasoline consumption the situation was the same: Germany consumed 1,460,000 metric tons, two-thirds of which (930,000 metric tons) it imported. On 1 September 1939 when World War II broke out Germany had in operation seven coal hydrogenation (plus Ludwigshafen) and eight F-T plants, and they had begun to contribute increasingly to Germany's domestic liquid fuel supply. When plant construction ceased in 1942 twelve coal hydrogenation and nine F-T plants converted coal and coal tar into gasoline, diesel oil, and other petroleum products.

From the first coal hydrogenation plant that began operation at Leuna on 1 April 1927, the twelve coal hydrogenation plants in early 1944 reached a peak production of over 3 million metric tons (21.6 million barrels) of synthetic fuel. Two million metric tons (14.6 million barrels) after adding lead tetraethyl were

77 *Report* (note 46), pp. 92–95, 101–102.

TABLE 3.8 Summary of German petroleum availability from various sources at beginning of 1944

Type of petroleum production	Annual rate of production in metric tons						Total
	Hydrogenation	Fischer-Tropsch synthesis	Refining of German and Austrian petroleum	Brown coal and bituminous coal tar distillation	Benzole	Imports from Romania and Hungary	
Aviation fuel	1,900,000	–	–	–	50,000	100,000	2,050,000
Motor spirit	350,000	270,000	160,000	35,000	330,000	600,000	1,745,000
Diesel oil	680,000	135,000	670,000	110,000	–	480,000	2,075,000
Fuel oil	240,000	–	120,000	750,000	–	–	1,110,000
Lubricating oil	40,000	20,000	780,000	–	–	–	840,000
Miscellaneous	40,000	160,000	40,000	50,000	–	–	290,000

SOURCE: *Report on the Petroleum and Synthetic Oil Industry of Germany* (LONDON: MINISTRY OF FUEL AND POWER, 1947), P. 50

high quality motor and aviation gasoline approaching 100-octane number. In World War II these plants provided 95% of the German air force's aviation gasoline and 50% of Germany's total liquid fuel requirements.[78] Extensive Allied bombing of the Leuna, Böhlen, Zeitz, Lützkendorf, and Brüx plants in May 1944 significantly reduced production from 236,000 metric tons to 107,000 metric tons in June. With the repeated attacks on the Ruhr plants at Welheim, Scholven, and Gelsenberg, production fell dramatically to 17,000 metric tons in August, then stopped completely in March 1945.

The nine F-T plants contributed another 526,482 metric tons of primary products to the war effort, or 12–15% of Germany's total liquid fuel requirements. Their production also fell significantly because of Allied bombing, decreasing from 43,000 metric tons in the first four months of 1944 to 27,000 metric tons in June, to 7,000 metric tons in December, and to 4,000 metric tons in March 1945.[79]

Hydrogenating coal or tar had a high average cost, 19–26 pfennig/kg (RM 190–260/metric ton) or the equivalent of 24¢/US gal ($10/bbl) of gasoline. The average cost of primary products at the F-T plants compared favorably at 23.71–25.81 pfennig/kg (RM 240–330/metric ton). Synthetic fuel produc-

78 *Report* (note 46), pp. 1–2, 88; Strategic Bombing Survey (note 2).
79 Strategic Bombing Survey (note 2).

TABLE 3.9 Synthetic fuels in Germany 1911–45

Period	Activity
Fischer-Tropsch: Research and development falls into two divisions	
1922–35	Laboratory research and pilot plant development
1936–45	Industrial-size development, most development occurring prior to World War II
High-pressure coal hydrogenation: Research and development falls into four divisions	
1911–20	Bergius, laboratory research and pilot plant development BASF-IG Farben laboratory research and pilot plant development
1920–33	BASF and IG Farben research and development at Ludwigshafen and Leuna; most important breakthroughs occur such as introduction of two-phase conversion and sulfur resistant catalysts
1933–39	IG Farben use of 700 atm pressure (1936) First hydrogenation of bituminous coal (1936) New hydrogenation companies established Seven plants constructed New steel alloys introduced Research and development continue
1939–45	Five plants constructed Mainly a period of development, not much research other than catalysis

SOURCE: COURTESY OF AUTHOR

tion costs averaged more than twice the price of imported gasoline, but for Germany, with only a limited supply of natural petroleum, no alternative remained during the war other than the construction of synthetic fuel plants. In this way Germany utilized its naturally abundant supplies of bituminous and brown coal.[80]

80 *Report* (note 46), pp. 1–2, 50, 61, 67, 68, 93. The prewar exchange rate is from *Statistical Abstract of the United States* 1939 (Washington, DC: United States Government Printing Office, 1940): 208. RM 1 = 25¢.

FIGURE 3.15 Bombed hydrogenation synthetic fuel plant (unidentified)
SOURCE: SYNTHETIC FUELS COLLECTION, TEXAS A&M ARCHIVES

FIGURE 3.16 Bombed hydrogenation synthetic fuel plant (unidentified)
SOURCE: SYNTHETIC FUELS COLLECTION, TEXAS A&M ARCHIVES

6 Labor Force in the Synthetic Fuel Plants

Faced with a growing labor shortage as the war dragged on, German industrial firms, including synthetic fuel producers IG Farben, Brabag, Sudetenlandische Treibstoffewerke AG, Hydrierwerke Pölitz AG, and Oberschlesische Hydriewerke AG, increasingly supplemented their labor force with paid coerced (forced) laborers and (or) concentration camp inmates (slave laborers) of many nationalities. French, Belgian, Polish, British, Serbian, Czech, Hungarian, and Russian laborers, Jews and non-Jews, worked in the prewar plants in Ludwigshafen in west Germany and Leuna in east Germany and in several of the synthetic fuel plants constructed after the war had started. Lützkendorf in central Germany (early 1940), Pölitz in northeast Germany (July 1940), Wesseling in west Germany near Bonn (August 1941), Brüx in Bohemia, Sudetenland (October 1942), and the Blechhammer plant (April 1944) and Heydebreck saturation plant in Upper Silesia (April 1944) used forced laborers (prisoners) and (or) concentration camp inmates. Blechhammer, located near the town of Kolze (Poland), initially a labor camp for Jews, had 4,116 forced laborers on 17 January 1945. IG Farben's labor force contained about 9% forced laborers and concentration camp inmates by 1941, increasing to 16% in 1942, and to 30% of all workers in its synthetic fuel plants near the war's end. In addition to the forced laborers and concentration camp inmates, some free foreign workers came from Germany's allies, mainly Italy and Romania. The USSBS estimated that the synthetic fuel plants used about 34,000 concentration camp inmates, representing 30% of their total workforce, and that women accounted for 13% of the concentration camp inmates.

All skilled and unskilled foreign workers in a specific industry (automotive, coal, steel) earned the same wage as an unskilled German worker in that industry, about 64.1 pfennig/hr or RM 38 for a 60-hour week, but received RM 18–25 after deductions for taxes, room and board. A skilled German worker received 81 pfennig/hr or about RM 49 for a 60-hour week.[81] At Blechhammer, one of the few synthetic fuels plants for which Allied investigative teams provided labor costs, CIOS reported a wage of 55–70 pfennig/hr for unskilled labor and 80 pfennig-RM 1/hr for skilled labor. This was considerably less than the RM 1.30/hr average wage that synthetic fuel plant officials told postwar Allied investigating teams they paid all workers at the coal hydrogenation and F-T plants, and the wage they used to calculate synthetic fuel production costs.[82]

The never-completed, German-government financed IG Farben synthetic fuel plant at Auschwitz (Auschwitz III) or Oswiecim, in south Poland west of Cracow, relates a different story. Free German and Polish workers as well as forced eastern European workers contributed to its construction, but the largest group of workers was the approximately 300,000, concentration camp inmates that included Germans, Greeks, Dutch, Czechs, Hungarians, Poles, and Russians, most of whom were Jews. IG Farben paid all its unskilled workers 30 pfennig/hr (RM 0.30/hr) or RM 3 for a 10-hour workday and all skilled workers 40 pfennig/hr (RM 0.40/hr) or RM 4/d after deductions for taxes, room and board. This was the same wage as the 1944 average industrial wage for unskilled and skilled foreign workers for a six-day week (RM 18–25) with one difference. Free and forced foreign workers received these wages, but the total wage each concentration camp inmate earned went instead to the SS (Schutz-Staffel) for taxes and expenses (room and board and clothing). In effect, IG Farben paid the SS for the labor it provided.[83]

Auschwitz, however, never produced a drop of synthetic fuel. Construction started in 1941, and it remained largely unfinished at the time Soviet troops overran it on 27 January 1945. Its scheduled production of 24,000 metric

81 Edward Homze, *Foreign Labor in Nazi Germany* (Princeton, NJ: Princeton University Press, 1967), p. 171.

82 *Report* (note 46), p. 61 (Table XXXV); Ronald Holroyd, ed., *IG Farbenindustrie AG Works, Leuna*, CIOS XXXII-107 (note 58), p. 133.

83 Homze, *Foreign Labor* (note 82), p. 171; Joseph Borkin, *The Crime and Punishment of IG Farben* (New York: The Free Press, 1978), p. 117. RM 10 = DM 1 = 40¢ US, currency conversion law of 1948.

FIGURE 3.17 IG Farben plant near Auschwitz
SOURCE: BUNDESARCHIV, BILD 146-2007-0058 / CC-BY-SA 3.0

tons/yr, making it the smallest of the coal hydrogenation plants, represented only a fraction of the Leuna plant which produced at its rated capacity of 620,000 metric tons of synthetic liquids per year. Dr. Müller Conradi and Dr. Schierenbeck in Oppau and Dr. Durfeld, Auschwitz's managing director now at Leuna, provided information on Auschwitz. Auschwitz cost 25,000–30,000 lives and RM 900 million for all operations including the never-completed synthetic rubber plant, but it was a miserable failure.[84]

To determine an accurate production cost in the operational synthetic fuel plants, or in any of the wartime that plants used forced laborers and concentration camp inmates, remains very complicated, however, mainly because of difficult-to-measure factors and incomplete data. First of all, even though concentration camp inmates received no wages, the cost of their guarding, housing, even their near starvation feeding involved some expense. The production cost calculation also requires knowing the number of years each plant was in operation; what percentage of workers in each plant was forced or concentration camp; how long each worker worked at the plant; and worker efficiency. Fritz Sauckel, Reich Commissioner for Labor, not surprisingly, ranked Polish forced workers one-half as efficient, and concentration camp inmates one-third as efficient as German workers. Total German synthetic fuel production

84 *Report* (note 46), p. 48 (Table XXVI).

in fact fell to its lowest in the last months of World War II when the number of coerced and concentration camp inmates reached a maximum.

Other factors to consider are plant operation time versus shutdown time because of bombings and equipment malfunctions and what reduction in production cost resulted completely from technical improvements in each plant.[85] Some of this information, such as the composition of the workforce in a few of the plants, is available. The Leuna plant as of 1 October 1944 employed 34.9% foreign workers; in Ludwigshafen, which was a research facility and only a small producer of synthetic fuel, the foreign work force numbered 36.6%. Auschwitz, by the same date, had 55.1% foreign workers, 26.6% concentration camp inmates both foreign and German, and 18.3% free German workers.[86]

Postwar court settlements, such as the 1957 Braunschweig court case settlement between a Jewish concentration camp inmate and the Bussing Company of Braunschweig, have provided additional information on wartime labor. Bussing manufactured trucks for the German army, and during the war it used foreign inmates from the Neuengamme concentration camp. Because the inmates received no compensation for their wartime work the court set the wage at RM 1/hr (before deductions) for a 10-hour workday, arguing that RM 1/hr was the scale established according to wartime wage controls.[87]

In another postwar settlement IG Farben and the Conference on Jewish Material Claims Against Germany (a consolidation of twenty-three major Jewish organizations) reached an agreement to provide compensation for IG Farben's use of unpaid concentration camp inmates. By 1958, IG Farben had arranged to pay DM 27 million to the Jewish Material Claims Conference.[88] Its settlement followed an earlier 1952 agreement between the Federal Republic of Germany and the Material Claims Conference in which the German government paid DM 450 million ($105 million) to the Material Claims Conference and also sent DM 3 billion ($700 million) worth of goods such as petroleum and steel to Israel over a ten-year period. The German government estimated that its payments would continue beyond the year 2000 and its total payments would reach DM 100 billion ($40 billion).[89]

---

85 Jurgen Kuczynski, *Germany: Economic and Labor Conditions under Fascism* (New York: Greenwood Press, 1968), p. 213.

86 Homze, *Foreign Labor* (note 82), pp. 238–39.

87 Homze (note 82), pp. 238–39.

88 Benjamin B. Ferencz, *Less than Slaves: Jewish Forced Labor and the Quest for Compensation* (Cambridge, MA: Harvard University Press, 1979), p. 52.

89 Ferencz (note 88), pp. xvi–xvii.

7 Conclusion

Germany had the first technologically successful synthetic fuel industry producing 18 million metric tons (130 million barrels) from coal and tar hydrogenation and another 3 million metric tons from the F-T synthesis in the period 1939–1945. After the war ended German industry did not continue synthetic fuel production because the Potsdam (Babelsberg) Conference of 16 July 1945 prohibited it.[90] The Allies maintained that Germany's Nazi government had created the industry for strategic reasons under its policy of autarchy and that in postwar Germany economically better uses existed for its coal than synthetic fuel production. Four years later on 14 April 1949, the Frankfurt Agreement ordered dismantling of the four coal hydrogenation plants in the Western zones, all of which were in the British zone.[91] But shortly after the formal establishment of the West German government on 23 May 1949, a new agreement, the Petersberg (Bonn) Agreement of 22 November 1949, quickly halted the dismantling process in an effort to provide employment for several thousand workers.[92] The West German government completely removed the ban on coal hydrogenation in 1951, although by this time Ruhröl GmbH (Mathias Stinnes) had deactivated the Welheim plant, and the plants in Scholven, Gelsenberg, and Wesseling after design modifications, were hydrogenating and refining crude oil rather than hydrogenating coal.

The Russians dismantled the Magdeburg plant located in their zone and the three plants in Poland at Pölitz, Blechhammer, and Auschwitz. They used parts from the Magdeburg and Auschwitz plants to reconstruct a plant in Siberia that had an annual production capacity of one million metric tons of aviation fuel and a second plant in Kemerow-Westbirien that also produced aviation fuel from coal. The Pölitz and Blechhammer plants provided scrap iron. Three other plants in their zone, at Leuna, Böhlen, Zeitz, and the Sudetenland plant at Brüx, which the Russians gave to Czechoslovakia, continued with coal and tar hydrogenation, and after modification, refined petroleum

90 "Principles to Govern the Treatment of Germany in the Initial Control Period," No. 848, The Conference of Berlin (The Potsdam Conference), 1945, *Foreign Relations of the United States, Diplomatic Papers*, 17 Vols. (Washington, DC: Government Printing Office, 1949–68), Vol. 2 (1960), pp. 750–53, on p. 752.

91 "Multilateral: German Industries," *United States Treaties and other International Agreements*, Part I, 1951, 35 Vols. (Washington, DC: Government Printing Office, 1950–84), Vol. 2 (1952), pp. 962–72, on p. 963.

92 "Multilateral: Incorporation of Germany into European Community of Nations," *United States Treaties and other International Agreements*, Part II, 1952, 35 Vols. (Washington, DC: Government Printing Office, 1958–84), Vol. 3 (1954), pp. 2714–2722, on p. 2716.

into the early 1960s. Some dismantling and conversion to synthetic ammonia production for fertilizers occurred at the Leuna plant which by 1947 the Russians had renamed the Leuna Chemical Works of the Soviet Company for Mineral Fertilizers. The last of the coal hydrogenation plants in the Soviet Zone at Lützkendorf did not resume production after the war.

Three of the F-T plants continued operation after the war. Schwarzheide in the Soviet Zone, which had a labor force of 3,600, produced gasoline for Russian civilian and military construction. Gewerkschaft Viktor in Castrop-Rauxel and Krupp Treibstoffwerk in Wanne-Eickel in the British zone, as of February 1946 were producing oils and waxes from fatty acids and using hem to make soaps and margarine.[93] The six other plants remained inoperative.

The German synthetic fuel industry succeeded technologically because in the 1920s Pier at IG Farben developed suitable catalysts for the hydrogenation of coal and divided the process into separate liquid and vapor phase hydrogenations, improving both economics and yield. A short time later Fischer and his co-workers at the KWI prepared the catalysts and established the reaction conditions that made the F-T synthesis a success. But neither coal-to-oil conversion process could produce a synthetic liquid fuel at a cost competitive with natural petroleum. Coal hydrogenation and the F-T synthesis persevered and survived because they provided the only path Germany could follow in its search for petroleum independence.

---

93 Birkenfeld, *Der synthetische Treibstoff* (note 1): 213–15; E. E. Donath, "Hydrogenation of Coal and Tar," in *Chemistry of Coal Utilization*, ed. H. H. Lowry, 3 Vols. (New York: John Wiley & Sons, 1945 and 1963), Supplementary Volume, pp. 1041–1080, on pp. 1042–1044; "German Synthetic Petrol and the Moscow Conference," *Petroleum Times*, 51 (1947): 430, 446; "Present Position and Future Role of Most (Brüx) Synthetic Oil Plant," *Petroleum Times*, 50 (1946): 852; Strategic Bombing Survey (note 2).

CHAPTER 4

From Birmingham to Billingham: High-Pressure Coal Hydrogenation in Great Britain

> Britain cannot afford not to know how to get oil from coal quickly in time of need.
>
> *Journal of the Society of Chemistry and Industry*, 53 (1934): 1034

1 Introductory

Friedrich Bergius's successful conversion of coal to petroleum sent waves of enthusiasm through the industrial world of the 1920s. In his small factory at Rheinau on the Rhine near Mannheim, Germany, Bergius had converted soft coal into petroleum by reacting the coal with hydrogen gas at high pressure and temperature. The announcement of his success captured the attention of scientists and engineers in Britain, Italy, Switzerland, Hungary, South Africa, Australia, and Japan, countries rich in coal but lacking petroleum.[1] Even the

1 H. I. Waterman and J. N. J. Perquin, "Hydrogenation by the Bergius Method," *J. Inst. Petroleum Technologists*, 10 (1924): 670–77; Jose M. Pertierra, "The Liquid Products of the Berginization of Coal," *Annales de la Sociedad Espanola de Fisica y Quimica*, 28 (1930): 792–806 and "Hydrogenation of Coal," *Annales de la Sociedad Espanola de Fisica y Quimica*, 28 (1930): 389–415; Saburo Tashiro, "Hydrogenation of Neutral Oil of Low-Temperature Tar," *J. Fuel Soc. Japan*, 7 (1928): 67–70; Yoshikiyo Oshima and Saburo Tashiro, "Hydrogenation of Japanese Coals," *J. Fuel Soc. Japan*, 7 (1928) 70–73; Shigeru Komatso, "Liquefaction of Coal," *J. Fuel Soc. Japan*, 7 (1928): 128–29; K. Smolenski and W. Badzyoski, "Liquefaction of Naphtha Asphalt By The Bergius Method," *Przemysl Chemiczny*, 12 (1928): 117–36; Hans E. Fierz-David, "The Liquefaction of Wood and Cellulose and some General Remarks on the Liquefaction of Coal," *J. Soc. Chem. Ind.*, 44 (1925): 942–44; "Coal Hydrogenation in Hungary," *J. Soc. Chem. Ind.*, 46 (1927): 1131; "Hydrogenation of Australian Coal," *Fuel*, 8 (1929): 52; "Oil from Australian Lignite," *Chemical Age*, 21 (1929): 233–34; "Motor Fuel from Coal in S. Africa," *J. Soc. Chem. Ind.*, 48 (1929): 408; "Coal Liquefaction in South Africa," *Chemical Age*, 20 (1929): 233–34; "Coal Liquefaction in South Africa," *Chemical Age*, 20 (1929): 373; Andre Kling and M. Daniel Florentin, "Hydrogenation at High Pressure and High Temperature With or Without Catalyzing Agents," *Proceedings of the Second International Congress on Bituminous Coal*, 1928, 2 Vols. (Pittsburgh: Carnegie Institute of Technology, 1929), Vol. 2, pp. 523–41.

 | DOI:10.1163/9789004690912_005

United States, which had an abundance of both coal and petroleum, sent Arno C. Fieldner (1881–1966), chief chemist at the Bureau of Mines in Pittsburgh, to observe Bergius's operation. The Bureau had an interest in producing a synthetic motor fuel, and it wanted to test the suitability of hydrogenating American coals at some future time when the United States might have to supplement its petroleum reserves.

Global supplies of petroleum abounded in the 1920s and 1930s, especially with the discovery of large reserves in East Texas and Oklahoma. Yet a growing concern existed for petroleum's availability in the future.[2] Petroleum derivatives already had begun to replace coal in commercial heating, in stationary-power generation, and in fueling many ships.[3] The British navy that emerged

2 Global production of petroleum increased by 101,000,000 barrels during 1933 to give a total production of 1,453,000,000 barrels. Global consumption increased by 18,000,000 barrels to 1,340,000,000 barrels in 1933, an overproduction of 113,000,000 barrels. "World Production of Petroleum," *J. Soc. Chem. Ind.*, 53 (1934): 106.

3 The tonnage of ships equipped with Diesel engines rose from 960,000 tons in 1920 to over 8,000,000 in 1930. In 1920 the tonnage of oil burning steamships was 9,400,000 tons; in 1930 tonnage increased to 20,000,000 tons (47%). Sir John Cadman, address delivered at the Centenary Meeting of the British Association for the Advancement of Science (BAAS), 28 September 1931, p. i; published with the title, "British Fuel Problem: Oil," *J. Soc. Chem. Ind.*, 50 (1931): 908–12. Another useful comparison is the following: in 1914, oil-burning ships transported 3.4% of the world's tonnage of shipping; in 1931 they transported over 40%. Sir David Milne-Watson (1869–1945), "British Fuel Problem: Coal," *J. Soc. Chem. Ind.*, 50 (1931): 905–08. Milne-Watson read this paper at the same BAAS meeting.

The table below compares Britain's consumption of coal, fuel oil, and gasoline for the years 1913 and 1932. W. Idris Jones, "Hydrogenation of Coal," *J. Soc. Chem. Ind.*, 53 (1934): 321–26.

Britain's consumption of coal, fuel oil, and gasoline

Fuel	1913 Millions of long tons	1932 Millions of long tons
Consumption of coal	184	150
Consumption of fuel oil	0.655	ca. 3
Consumption of gasoline	0.4	4

In this same article, Jones wrote: "The world production of oil in 1932 was 200 × 10^6 T. This indicates that the oil supply will suffice for only 35 years, even assuming that 100% of the crude oil is utilized" (p. 321). Tons in this chapter refer to long (Imperial) tons, where 1 long ton = 2,240 pounds. Gallons are Imperial gallons. Also 1 tonne (metric ton) = 0.984 long tons = 1.102 US short tons, and 2,204 pounds.

from World War I was to a large extent an oil-fired navy. The considerable increase in the number of automobiles produced at this time also guaranteed a greater dependence on petroleum. In 1920 a world census showed that 9 million automobiles were in use, a similar census ten years later put the number at 35 million. During 1929 alone the automobile industry manufactured 6.2 million motor vehicles.[4]

Of all the countries that watched closely the development of Bergius's high-pressure coal hydrogenation process, Britain watched most closely. Britain imported 90% of its petroleum and by 1931 consumed annually about 7 million long tons of petroleum which included 3 million long tons of gasoline.[5] The remaining 10% consumed came from the distillation of Scottish shale and from the high-temperature (HTC) and low-temperature carbonization (LTC), or decomposition, of coal.[6] But these reactions gave only small yields: 25 Imperial gallons of petroleum per long ton of oil shale distilled, about 3 Imperial gallons per long ton (Imperial gal/long ton) from HTC or LTC of coal.[7] British scientists also conducted small-scale research on the Fischer-Tropsch synthesis in the 1920s–30s, but coal hydrogenation seemed much more promising for

4 In 1926 the United States Department of Commerce estimated that 24,589,249 motor vehicles operated globally. The United States led with 19,954,347; followed by Britain, 815,957; France, 735,000; Canada, 715,962; and Germany, 323,000. A. C. Fieldner and R. E. Brown, "Future Trends in Automotive Fuels," *Industrial and Engineering Chemistry*, 18 (1926): 1009–14.

5 Cadman, "British Fuel Problem: Oil" (note 3), p. 911; G. W. Himus, "Some Recent and Impending Developments in Fuel Technology," *The Industrial Chemist*, 10 (1934): 491–94.

6 Jones, "Hydrogenation of Coal" (note 3), p. 321; Milne-Watson, "British Fuel Problem: Coal," (note 3), p. 905. In 1934 Britain produced 154.1×10^6 gallons or 0.51×10^6 long tons of gasoline from these three processes (1 long ton = 300 Imperial gallons, 1 barrel = 35 Imperial gallons); in 1932 Britain consumed about 7×10^6 long tons of petroleum. *The British Fuel and Power Industries, P.E.P.* (Political and Economic Planning), (London: The Stanhope Press, October, 1947), pp. 109–10. The F-T process also converts coal to petroleum.

7 Milne-Watson, "British Fuel Problem: Coal" (note 3), p. 905; Himus, "Some Recent and Impending Developments in Fuel Technology" (note 5), p. 493. The reaction temperatures for low and high temperature carbonization are, respectively, 600° C and l 100° C. The essential difference between coal and petroleum is the hydrogen content, 5% in coal vs. 15% in petroleum. In the distillation and carbonization processes, no hydrogen gas addition occurs while heating the shale or coal. The hydrogen required to convert coal to petroleum must come from within the coal structure, at the expense of existing molecules. In the hydrogenation process, hydrogen gas is added and reacts with the coal structure. Consequently a much higher conversion of coal to petroleum results from the hydrogenation process. See also, R. Lessing, "The Carbonization of Coal," *Fuel*, 1 (1922): 137–49; A C. Fieldner, "The Low Temperature Carbonization of Coal," *Fuel*, 5 (1926): 203–04; Philip C. Pope, "The Economics of Low-Temperature Carbonization," *Fuel*, 5 (1926): 510–11; W. A. Bristow, "Liquid Fuel from Coal Obtained by Low-Temperature Carbonization," *J. Inst. Fuel*, 10 (1936): 43–47.

the production of petroleum and thereby relieving Britain of her dependence on imported supplies.

A high-pressure coal hydrogenation research program began at the University of Birmingham in the early 1920s, and shortly after, it underwent considerable expansion at the Government's East Greenwich Fuel Research Station. Alfred W. Nash (1886–1942) at Birmingham summed up the British outlook toward coal to oil conversion at that time: "The importance of the discovery [of coal hydrogenation] will be evident in this country where free oil on a large scale has still to be found, but where huge deposits of coal suitable for hydrogenation exist."[8]

Within a few years Imperial Chemical Industries (ICI) took an active interest in coal hydrogenation. It conducted a much larger research program than Birmingham or the Fuel Research Station, and it led in 1935 to the construction of an industrial-size high-pressure coal hydrogenation plant. The plant, located on a 1,000-acre site at Billingham on the River Tees north bank, only 10 miles from the Durham coalfields of northeast Britain, remained in operation until 1958 when the high cost of synthesizing petroleum forced its closing. Britain's modest yet successful synthetic petroleum program ended 35 years after its beginning in the University of Birmingham's research laboratories.

2 Fischer-Tropsch Synthesis and the Beginning of Coal Hydrogenation in Britain: Birmingham

F-T development in Britain lagged far behind the continent's, with only four small experimental plants in operation. One of them was at the University of Birmingham where in a series of experiments in 1920–27 Oliver C. Elvins and Nash obtained a 2–3% maximum conversion to liquid hydrocarbons. The low yield and unusually high reaction temperature of 280–350° C at 1 atm resulted from the inactivity of their catalyst (60 Co_2O_3 – 45 MnO_2 – 30 CuO).[9] The

8 Alfred W. Nash, "The Berginisation of Coal and Oil," *J. Inst. Petroleum Technologists*, 10 (1924): 329–34. Quotation is on p. 334. C. H. Lander, "Our Available Coal Supplies and Their Utilization," *Fuel*, 6 (1927): 521–25. The situation in Britain changed with the discovery of the North Sea deposits in 1970–71 that yield about 1.5 million bbl/d.

9 Oliver C. Elvins and Alfred W. Nash, "The Reduction of Carbon Monoxide," *Nature*, 118, no. 2961 (31 July 1926): 154; Elvins and Nash, "Synthetic Fuel from Carbon Monoxide and Hydrogen," *Fuel*, 5 (1926): 263–65; Alfred W. Nash, "Recent Developments in the Formation of Synthetic Fuel From Carbon Monoxide and Hydrogen," *J. Soc. Chem. Ind.*, 45 (1926): 876–98; Oliver C. Elvins, "Hydrocarbon Synthesis from Carbon Monoxide and Hydrogen," *J. Soc. Chem. Ind.*, 46 (1927): 473T–78T.

other British F-T plants were the 4 Imperial gal/d and the 150 Imperial gal/d plants at the East Greenwich Fuel Research Station[10] and Synthetic Oils Ltd. 136–180 Imperial gal/d plant at Bedlay Colliery, near Glasgow.[11]

The Fuel Research Station experiments ran from 1935 to 1939 and tested a cobalt-thorium catalyst of the same composition as Fischer's. Its researchers intended to synthesize lubricating oils, but the products included gasoline of 46-octane number and diesel oil of 100-cetane number.[12] Synthetic Oils operated the only privately owned F-T plant in Britain. Its plant had been producing synthetic oil since 1936, adopting the Robinson-Brindley process which used water gas ($CO + H_2$) as the synthesis gas, and like the Fuel Research Station produced a variety of liquid hydrocarbons including lubricating oils, gasoline of 68-octane number, and diesel oil of 81–86-cetane number.[13]

With the growing threat of war in the late 1930s and concern for additional petroleum supplies, the British government established several committees, such as the British Falmouth Subcommittee of Committee of Imperial Defense in 1937 and the Committee on Fischer-Tropsch and Similar Synthetic Processes in 1940, to examine Britain's domestic sources of liquid fuels. The Falmouth Subcommittee dismissed the Scottish oil shale industry and LTC as too small and insignificant either economically or strategically, but F-T's potential impressed both committees, and they recommended construction of a pilot plant.[14] Economic data that the Falmouth Subcommittee obtained from Ruhrchemie and British fuel experts gave a construction cost of £1.9 million ($9.5 million, RM 37.8 million) for a plant producing annually 35,000 long tons (11.1 million Imperial gal) of synthetic products and an average production

10 W. W. Myddleton, "The Production of Hydrocarbon Oils from Industrial Gases: The Process of Synthetic Oils Ltd.," *J. Inst. Fuel*, 11 (1938): 477–84; Report of the Fuel Research Board for the Year Ended March 1937 (London: His Majesty's Stationery Office, 1938), pp. 141–47.

11 Myddleton (note 10).

12 Storch, Golumbic, and Anderson (note 3), pp. 129–31. The Fuel Research Station used the CFR motor method to calculate octane numbers.

13 Myddleton (note 10).

14 D. J. Payton-Smith, *Oil* (London: Her Majesty's Stationery Office, 1971), pp. 22–24, 94–95. The 1937 Falmouth Committee, a subcommittee of the Committee of Imperial Defense under Lord Falmouth (Viscount Falmouth, Evelyn Hugh Boscawen, 1887–1962), a member of the Advisory Council of the Department of Scientific and Industrial Research, adopted a somewhat cautious attitude toward the Fischer-Tropsch process, acknowledging that it would provide jobs for plant workers and miners but viewing it as uneconomic. Therefore no government support arose for the construction of industrial-size F-T plants in Britain and no plants operated in Britain in 1939. *The British Fuel and Power Industries*, p. 211.

cost of its marketable products as 13 d/Imperial gal (d = pence, 26¢/Imperial gal, 21.7¢/gal, 33 pfennig/kg), a price comparable to that of coal hydrogenation.

Britain's Labour Party saw the F-T (and coal hydrogenation) process as a way of putting coal miners back to work, and in 1938 it issued a 79-page report detailing construction and production costs for a 35,000 long ton/yr plant and a smaller 10,000 long ton/yr plant that agreed with the Falmouth Subcommittee's figures. It calculated a construction cost of £225,000 ($1,125,000, RM 4.5 million) for the 10,000 long ton/yr plant (excluding the coke ovens) and a cost of 8 d/Imperial gal (16¢/Imperial gal, 13.3¢/gal, 20.21 pfennig/kg) for the finished gasoline.[15] Shortages of steel and labor prevented any F-T plant construction during the war years, and with little interest and economic incentive to develop .a synthetic fuel industry in the postwar period, the British synthetic fuel program simply died, except for some laboratory research the Fuel Research Station conducted in the early 1950s.[16]

British research on high-pressure coal hydrogenation began late in 1923 a few months after J. Ivon Graham (1888–1983) and Hugh George Shatwell (1889–1959) from the University of Birmingham's Mining Research Laboratory visited Bergius's Rheinau-Mannheim plant. Their trip had a single objective: to

15 Payton-Smith, *Oil* (note 14), pp. 22–24; H. H. Storch, "Synthesis of Hydrocarbons from Water Gas," in *Chemistry of Coal Utilization*, ed. H. H. Lowry, 3 Vols. (New York: John Wiley & Sons 1945 and 1963) Vol. 2, pp. 1839–45. The motor fuel's estimated average cost was 22 pfennigs/kg. According to Arno C. Fieldner and American engineers the wartime cost of either the F-T or the hydrogenation process was 20–30¢/gal. See Fieldner, "Frontiers of Fuel Technology," *Chemical and Engineering News*, 26, no. 23 (7 June 1948): 1700–01. With the coke ovens included, the costs of the two plants were comparable. For currency conversion, the symbol for pence is d, after 1971 p is the symbol for pence; 1 RM = 100 pfennig, 1 RM = 25¢ (1934–41 prewar rate of exchange). Exchange rate is from *Statistical Abstract of the United States 1939* (Washington, DC: Government Printing Office, 1939), p. 208. For units conversions use below.

1 bbl = 42 US gal, 34.97 Imperial gal, or 159 L
1 US gal = 3.78 L
1 Imperial gal = 1.2 US gal
For crude oil 1 bbl = 300 lb or 136 kg

16 Payton-Smith, *Oil* (note 14). C. C. Hall and J. Rennie at the Fuel Research Station tested a fused iron catalyst at 300 psi (20.4 atm) and 300–330° C in a fluidized bed reactor. They obtained a 96% CO conversion and a yield of 104 grams of liquid hydrocarbons and oxygenated compounds per cubic meter of synthesis (feed) gas. C. C. Hall and J. Rennie, "A Study of the Fischer-Tropsch Synthesis Using Fluidised Iron Catalysts," *Proceedings of the Third World Petroleum Congress*, The Hague, 11 Vols., Vol. (Section) 4 (Leiden: E. J. Brill, 1951), pp. 25–37.

observe the high-pressure hydrogenation of coal, coal tars, and heavy petroleum oils and to consider its applicability to British coals.[17] What Graham and Shatwell saw clearly impressed them, for Bergius had expanded his coal hydrogenation program from a small laboratory process in 1911 to a plant-size operation. In 1923 the Rheinau-Mannheim plant had the capability of hydrogenating 18,000 metric tons of heavy petroleum oil per year, and scheduled for completion in 1924 was an additional unit for the hydrogenation of 1 metric ton of coal per day. The Deutsch-Bergin Aktien-Gesellschaft, which Bergius formed in 1914 to develop a coal hydrogenation industry in Germany, directed the Rheinau-Mannheim operation. The plant employed about 235 men, including 30 chemists and engineers.[18]

Graham and Shatwell's visit to Germany occurred during a period of decline in the British coal mining industry. Coal consumption had fallen steadily for several years, the result of petroleum fuels gradually replacing coal. With the coal mining industry in need of revitalization, coal hydrogenation seemed ideal. Its establishment as a new industry, requiring 10 million long tons of coal per year, would put miners back to work, simultaneously enabling Britain to become independent of natural petroleum.[19]

In addition to research that could lead to the development of a British high-pressure coal hydrogenation industry, the Birmingham program included theoretical studies on coal's composition and structure. Graham and Shatwell pointed out that for too long chemists had simply classified coals according to their macroscopic constituents, clarain (ordinary bright coal), vitrain (glassy, black coal), durain (dull, rough coal), and fusain (soft, powdery coal), but hydrogenation now provided a new method of investigation.[20] Birmingham would carry out a program far less empirical than Bergius's study. It

---

17 While the major discussion in this paper is on coal hydrogenation, both Bergius and the British scientists investigated the hydrogenation of tars and heavy oils, processes which are practically identical with coal hydrogenation. To deal with the industrial development of coal hydrogenation in Britain, Bergius and several British investors in 1924 established the British Bergius Syndicate. They had the patent rights for the process throughout the British Empire.

18 J. Ivon Graham and H. G. Shatwell, "The Artificial Production of Coal and the Hydrogenation of Coal and Oil," *Fuel*, 2 (1923): 54–57. See also Friedrich Bergius, "An Historical Account of Hydrogenation," *Proceedings of the World Petroleum Congress*, 2 Vols. (London: Office of the Congress, 1934), Vol. 2, pp. 282–89 and "The Transformation of Coal into Oil," *Proceedings of the International Conference on Bituminous Coal*, 1926 (Pittsburgh: Carnegie Institute of Technology, 1927), pp. 103–31.

19 "Society of Chemical Industry, Official Notices," *J. Soc. Chem. Ind.*, 48 (1929): 37.

20 Graham and Shatwell, "The Artificial Production of Coal" (note 18), p. 57. Marie Stopes introduced the macroscopic classification of coal in her 1919 paper "On the Four Visi-

would establish the reaction conditions that most readily induced absorption of hydrogen and provide a clearer conception of the reaction's course.[21] Support for the Birmingham program came from the recently formed British Colliery Owners' Research Association which in 1924 gave Graham and Shatwell a three-year grant of £2,500 ($13,422). The Association increased this to £3,000 for another three years beginning in 1927.[22]

The Birmingham program consisted of three different groups of experiments: (1) hydrogenating hard and soft coals in the dry, powdered state and not suspended in a heavy oil liquid medium, as Bergius had done, (2) hydrogenating these coals in phenol as the liquid medium and (3) examining the influence of different metallic catalysts on the reaction.[23] Graham and Shatwell's results along with those of other university collaborators, D. G. Skinner from the Mining Research Laboratory, Arthur R. Bowen from the Department of Oil Engineering and Refining, and Alfred W. Nash, chairman of Oil Engineering and Refining, appeared in the journals *Fuel* and *Journal of the Institute of Fuel* in the period 1923–36.[24]

---

ble Ingredients in Banded Bituminous Coal," *Proc. Royal Society* B, 90 (1919): 470–87. For research on the structure and composition of coal in the 1920s and early 1930s see Franz Fischer and Hans Schrader, "New Contributions on the Origin and Chemical Structure of Coal," *Fuel*, 1 (1922): 113–19; D. T. Jones and R. V. Wheeler, "On the Formation and Chemical Constitution of Coal," *Fuel*, 1 (1922): 91–93; Marie Stopes, "The Constitution of Coal," *Fuel*, 1 (1922): 93–101; R. Lessing, "Coal Ash and Clean Coal," *Fuel*, 5 (1926): 17–22; S. Uchida, "The Primary Decomposition of Coal," *Fuel*, 5 (1926): 221–24; J. G. King and D. MacDougall, "The Determination of Carbon in Coal," *Fuel*, 5 (1926): 33–35; R. V. Wheeler, "The Chemistry of Coal," *Fuel*, 6 (1927): 506–11; William A. Bone, "The Chemical Constitution of Coal," *J. Inst. Fuel*, 5 (1931): 21–29; Clarence A. Seyler, "Some Problems of the Classification and Nomenclature of Coal," *J. Inst. Fuel*, 5 (1932): 248–59.

21 H. G. Shatwell and J. Ivon Graham, "The Hydrogenation and Liquefaction of Coal-I," *Fuel*, 4 (1925): 25–30, 75–81, 127–31, on pp. 27–28.

22 "Investigation on Coal Hydrogenation," *Chemical Age*, 16 (1927): 600.

23 In one of the early experiments at Birmingham they used a heavy mineral oil as the liquid medium. For all later experiments the Birmingham investigators selected phenol because the reaction conditions did not affect it, and they could readily remove it from the products by solution in sodium hydroxide. They recognized that the solution process also removed any phenol-like products formed during hydrogenation. Shatwell and Graham, "The Hydrogenation and Liquefaction of Coal-I" (note 21), pp. 27–28.

24 Graham and Shatwell, "The Artificial Production of Coal" (note 18), pp. 54–57; H. G. Shatwell, "The Hydrogenation of Hydrocarbon Oils," *Fuel*, 2 (1923): 229–32; Nash, "The Berginisation of Coal and Oil" (note 8), pp. 329–34; Alfred W. Nash and H. G. Shatwell, "Fuel Oil Resources of the Future," *J. Inst. Petroleum Technologists*, 10 (1924): 855–71, 884–91; H. G. Shatwell, "Hydrogenation By the Bergius Process," *J. Soc. Chem. Ind.*, 44 (1925): 471–77; Shatwell and Graham, "The Hydrogenation and Liquefaction of Coal-I"

All the Birmingham experiments were batch type, carried out initially in apparatus of two kinds: one for small-size tests at maximum temperature and pressure of 450° C and 200 atm, the other for larger-size experiments and capable of withstanding a temperature and pressure of 500° C and 500 atm. The small-size apparatus consisted of a steel tube 61 cm long (3.8 cm internal, 5.1 cm external diameter) with a capacity of 400 cm^3. A strong detachable steel cap sealed one end of the tube, a screwed-in steel cover sealed the other end. The cover had a small opening, a small flexible copper tube that held a pressure gauge and gas valve fit into the opening. The cap, tube, gauge, and valve connections were cone-shaped metal-to-metal facings. An electric furnace heated the tube, a nitrogen-filled gas thermometer wired to the tube's outer wall measured the furnace temperature, a rheostat included in the electric circuit permitted temperature adjustment. In each experiment, the investigator placed powdered coal in a wide glass tube plugged lightly at each end with glass wool and then inserted the glass tube into the horizontally held steel tube. This arrangement prevented the coal from making contact with the outer tube, eliminating any possible catalytic effect, but still permitted incoming hydrogen gas easy access to the coal.

The reaction container for the larger-size hydrogenation experiments was a horizontally arranged, nickel-steel cylindrical autoclave with walls 7.6 cm thick and a 4.25 L (4,250 cm^3) capacity. The autoclave rotated at the rate of sixty revolutions per minute during each experiment. Gas jets heated the

(note 21), pp. 25–30, 75–81, 127–31; H. G. Shatwell and A. R. Bowen, "The Hydrogenation and Liquefaction Coal-Part II," *Fuel*, 4 (1925): 252–55; D. G. Skinner and J. Ivon Graham "The Hydrogenation and Liquefaction of Coal, Part III," *Fuel*, 4 (1925): 474–85; A. R. Bowen and A. W. Nash, "The Thermal Decomposition of Cellulose and Lignin in Presence of Catalysts and Hydrogen Under Pressure," *Fuel*, 5 (1926): 138–42; Arthur R. Bowen and Alfred W. Nash, "The Thermal Decomposition of Coal in Presence of Catalysts and Under Hydrogen Pressure," *Fuel*, 5 (1926): 361–64; D. G. Skinner and J. Ivon Graham, "The Hydrogenation and Liquefaction of Coal-Part IV," *Fuel*, 6 (1927): 74–81; D. G. Skinner and J. Ivon Graham, "The Hydrogenation and Liquefaction of Coal-V," *Fuel*, 7 (1928): 543–55; J. Ivon Graham, "Some Aspects of the Hydrogenation of Coal," *Proceedings of the Second International Conference on Bituminous Coal*, 1928, 2 Vols. (Pittsburgh: Carnegie Institute of Technology, 1929), Vol. 2, pp. 456–84; J. Ivon Graham, "The Hydrogenation of Coal," *The Chemical Age*, 20 (1929): 25–26; J. Ivon Graham and D. G. Skinner, "The Action of Hydrogen on Coal," *J. Soc. Chem. Ind.*, 48 (1929): 129–36; J. Ivon Graham and D. G. Skinner, "Further Investigations Connected With the Action of Hydrogen Upon Coal," *Proceedings of the Third International Conference on Bituminous Coal*, 1 931, 2 Vols. (Pittsburgh: Carnegie Institute of Technology, 1932), Vol. 2, pp. 17–27; F. C. Hall, W. R. Wiggins, and A. W. Nash, "The Production of Lubricating Oils From Coal Products," *J. Inst. Fuel*, 9 (1936): 153–57.

autoclave, an asbestos-lined cover which completely enclosed the autoclave prevented excessive heat loss and allowed rapid and easy heat control. A gauge attached to the autoclave's side arm gave pressure readings, a mercury thermometer, inserted in a thermometer pocket, extended to the autoclave's center and measured reaction temperatures. The autoclave had a removable cover at one end that permitted the autoclave's ready cleaning. A small hole drilled through one wall and plugged when not using the autoclave, allowed addition of reactants and discharge of liquid products without removing the cover after every experiment. The side arm with the pressure gauge contained a valve for admitting hydrogen gas and withdrawing the gaseous products. All joints for the pressure gauge, thermometer pocket, and plug were cone-shaped metal-to-metal facings.[25]

The initial experiments carried out at Birmingham in 1924 used powdered coal samples, ground to pass through a 60-mesh sieve, and ranged in weight from 20 to 200 g. Reaction time was 3–9 hours at temperatures and pressures of about 415° C and 150 atm. Temperatures and pressure chosen fell within the limits Bergius found sufficient for hydrogenation. The Birmingham investigators kept the reaction conditions fairly constant in order to compare under similar conditions the suitability of different British coals for hydrogenation. In all experiments carried out after 1925 the Birmingham investigators adopted the following standardized conditions.

TABLE 4.1 Standardized hydrogenation reaction conditions at Birmingham

Reaction time	8 hours
Weight of coal	200 grams
Weight of phenol	100 grams
Pressure	150 atm
Temperature	430° C
Type of autoclave	3 L capacity, gas heated, equipped with mercury thermometer, steel construction (3% nickel, 0.5% carbon)

SOURCE: COURTESY OF AUTHOR

Graham and Shatwell's initial experiments quickly established that British hard and soft coals in the powdered dry state (not suspended in a liquid medium) gave an insignificant amount of oil. In three separate trials, using

25 Shatwell and Graham, "The Hydrogenation and Liquefaction of Coal-1" (note 21), pp. 27–28.

both steel cylinder and rotating autoclave, they converted only 3.98, 4.46, and 11% of a bituminous coal to oil.[26] They somewhat expected the result, for Bergius had shown ten years earlier that hydrogenation occurred readily only after suspending the powdered coal in a liquid medium, such as crude petroleum oil, to form a paste.

They expected greater success with the second group of experiments, the hydrogenation of coals suspended in phenol. Graham, Shatwell, and Skinner used the larger, rotating autoclave for these experiments (approximately twenty-nine of them) working with 200-g samples of coal in 400 g of phenol. While some variation occurred in the amount of coal converted to oil, they obtained a maximum yield of about 40% with bituminous coals.[27] In an exceptional experiment Shatwell and Bowen attained a 61.4% conversion, or 140–50 Imperial gallons of oil/long ton of ash-free dry coal.[28] Overall, they obtained yields much smaller than the yields Bergius had achieved. For coals containing less than 85% carbon, which included the bituminous coals tested at Birmingham, Bergius claimed complete conversion (100%) leaving only the ash of the original coal. Graham and Skinner believed their use of a pressure considerably lower than Bergius, 150 atm vs. 200 atm, probably caused their smaller yields.[29] Graham and Shatwell nevertheless succeeded in hydrogenating a hard coal with 94% carbon, obtaining a 26.1% conversion to oil. Their success seemed significant for it contradicted Bergius's results that showed no hydrogenation of coals with over 85% carbon.[30]

One question that required an answer at this time was why the hydrogenation of a coal occurred more readily when suspended in phenol or in some other liquid medium. Graham and Skinner believed that with no medium present, the tars first produced in the reaction quickly covered the surface of the coal particles and prevented the incoming hydrogen gas from making

26 Average reaction time was about eight hours but one experiment with a dry, powdered coal and no solvent ran for twenty-nine hours. Nash and Shatwell, "Fuel Oil Resources of the Future" (note 24), p. 865; Shatwell and Graham, "The Hydrogenation and Liquefaction of Coal-I" (note 21), pp. 26, 77.

27 Shatwell and Graham, "The Hydrogenation and Liquefaction of Coal-I" (note 21), pp. 75–81; Skinner and Graham, "The Hydrogenation and Liquefaction of Coal, Part III" (note 24), pp. 474–85.

28 Shatwell and Bowen, "The Hydrogenation and Liquefaction of Coal, Part II," (note 24), pp. 252–55.

29 Shatwell and Graham, "The Hydrogenation and Liquefaction of Coal-I" (note 21), pp. 25–31.

30 In another experiment Graham and Shatwell tried but could not hydrogenate a coal containing 95% carbon. "The Hydrogenation and Liquefaction of Coal-I" (note 21), pp. 127–31.

contact with the particles. The presence of a solvent, however, removed the tars, exposing fresh coal surfaces to the hydrogen gas.[31]

In agreement with Bergius, the Birmingham investigators also concluded from these experiments that the hydrogenation of coal consisted of three distinct reactions. In the first reaction, hydrogen gas removed the oxygen, sulfur, and nitrogen present in coal forming water, hydrogen sulfide, and ammonia respectively. Next, a cracking or decomposition of the different molecular groups making up the coal occurred, followed finally by addition of hydrogen atoms to the resulting unsaturated groups to form compounds of low molecular weight. The addition of hydrogen atoms in the third reaction prevented polymerization of the unsaturated groups that led to coke formation.[32]

By 1925, the Birmingham research group had established the feasibility of hydrogenating British soft coals suspended in a liquid medium such as phenol, but the effectiveness of different metals to catalyze the reaction remained for investigation. Some confusion existed about the effectiveness of metallic catalysts because Bergius in his original patents of 1913 and in later publications claimed that the hydrogenation of coal did not require a catalyst. Yet in all of his experiments, Bergius always added 5% iron (III) oxide by weight of ash-free dry coal to the coal-oil paste ostensibly to remove any sulfur present in the oily products.[33] Hans E. Fierz-David (1882–1953) at the University of Zurich also questioned the need of a catalyst. In separate reactions with bituminous coal in the presence of either 10% nickel or 10% iron, Fierz-David in 1925 obtained only a 20% oil yield. He believed that sulfur present in the coal might have poisoned the catalyst, destroying its potential benefits, and that some coking of the oil may have further decreased the yield.[34] Fierz-David did not stir or agitate the autoclave during reaction, a procedure Bergius had shown essential for hydrogenation to occur appreciably.[35]

31 Shatwell and Graham, "The Hydrogenation and Liquefaction of Coal-I" (note 21), p. 27; Shatwell, "Hydrogenation by the Bergius Process" (note 24), p. 471.

32 Shatwell and Graham, "The Hydrogenation and Liquefaction of Coal-I" (note 21), p. 76; Graham, "Some Aspects of the Hydrogenation of Coal" (note 24), pp. 483–84.

33 Friedrich Bergius, German Patent 301,231 of 9 August 1913; British Patent 18,232 of 1 August 1914; US Patent 1,251,954 of 1 January 1918 (filed 3 August 1914). For discussion of Bergius's later research, see Bergius, "Chemical Reactions Under High Pressures," Nobel Lecture, 21 May 1932, in *Nobel Lectures: Chemistry 1922–1941* (New York: Elsevier Inc., 1966), pp. 244–76 and "An Historical Account of Hydrogenation" (note 18), pp. 282–89.

34 Hans E. Fierz-David, "The Liquefaction of Wood and Cellulose and Some General Remarks on the Liquefaction of Coal" (note 1), p. 942; Hans E. Fierz-David and Max Hannig, "Über die Destillation von Cellulose, Holz und anlichen Stoffen unter Wasserstoffdruck mit Katalysatoren," *Helvetica Chimica Acta*, 8 (1925): 900–23.

35 Franz Fischer and Wilhelm Frey, "Über die Hydrierung von Kohlen, insbesondere von Braunkohlenhalbkoks nach Bergius," *Brennstoff-Chemie*, 6 (1925): 69–79.

Other investigations carried out in the late 1920s seemed to indicate the positive influence of various metallic catalysts on the hydrogenation reaction. Bretislav Hlavica's research in 1928 at the Fuel Technology Institute in Prague showed that of zinc, nickel, cobalt, and copper oxides, zinc oxide (ZnO) most effectively catalyzed the hydrogenation. A one-third reduction in reaction time occurred, a higher quality oil resulted, and oil yields increased from 30 to 100%. The catalyst had no influence on the optimum reaction temperature and pressure, and according to Hlavica the catalyst affected chiefly the coal's decomposition, not its subsequent hydrogenation.[36]

At IG Farben in Ludwigshafen, Carl Krauch and Matthias Pier had been investigating the catalytic hydrogenation of coal, coal tars, and heavy petroleum oils since 1925. The IG Farben researchers discovered a new class of metallic oxides and sulfides that did not lose their catalytic activity easily and were not susceptible to sulfur and other catalyst poisons. The compounds included molybdenum oxide, red mud (essentially iron (III) oxide, iron (II) sulfate, and tungsten sulfide.[37] The success of these catalysts led to IG Farben's constructing at Leuna in 1927 the first industrial-size plant for the hydrogenation of German lignite coals. By 1928, Leuna produced annually 70,000 tons of gasoline of which 40,000 tons came from coal.[38]

At Birmingham, the research program on high-pressure hydrogenation would clarify the catalyst's role. In a series of experiments published in August

36 Bretislav Hlavica, "Hydrierung von Kohle bei Gegenwart von Katalysatoren," *Brennstoff-Chemie*, 9 (1928): 229–31. Richard Lessing in London showed how important the presence of inorganic constituents and catalysts was on coal carbonization. "Clean Ash and Clean Coal, III," *Fuel*, 5 (1926): 117–24.

37 Matthias Pier, US Patent 1,955,829 of 24 April 1934 (filed 27 August 1930); Pier and Karl Winkler, US Patent 1,946,108 of 6 February 1934 (filed 3 January 1929); Pier, "The Hydrogenation of Bituminous Coal," *Fuel*, 14 (1935): 136–46 and "The Hydrogenation of Bituminous Coal," *Chemistry and Industry*, 54 (1935): 284–88; Carl Krauch, "Catalysis-Applied to the Conversion of Hydrocarbons," *National Petroleum News*, 21 (24 July 1929): 69–72; 21 (31 July 1929): 72–74; 21 (7 August 1929): 75–78; and "Catalysis Applied to the Conversion of Hydrocarbons," *Proceedings of the Second International Conference on Bituminous Coal*, 1928, 2 Vols. (Pittsburgh: Carnegie Institute of Technology, 1929), Vol. 1, pp. 32–47; Krauch and Pier, "Kohleveredlung und katalytische Druckhydrierung," *Z. angew. Chem.*, 44 (1931): 953–58; *High-Pressure Hydrogenation at Ludwigshafen-Heidelberg*, FIAT Final Report No. 1317, Vol. I, published for Joint Intelligence Objectives Agency by Central Air Documents Office, Dayton, Ohio, 1951, pp. 24–26; Alfred von Nagel, *Methanol Treibstoffe, Schriftenreihe des Firmenarchivs des BASF*, Band 5 (Ludwigshafen: 1970), pp. 41–47; E. E. Donath and Maria Höring, "Early Coal Hydrogenation Catalysis," *Fuel Processing Technology*, 1 (1977): 3–20 and "Kohle- und Ölhydrierung," *Ullmann's Enzyklopädie der technischen Chemie*, 3rd ed., (Munich: Urban & Schwarzenberg, 1958), Vol. 10, pp. 483–569, on pp. 486–535.

38 Graham and Skinner, "The Action of Hydrogen on Coal" (note 24), p. 135.

1926, Bowen and Nash investigated the catalytic hydrogenation of a bituminous coal from Arley in the dry powdered, ash-free state with no liquid medium (solvent). To one-half of the samples they added iron (III) oxide, to the other half nickel (II) oxide (NiO), each in concentrations varying from 5 to 20% by weight of coal. Both oxides had a definite catalytic effect, increasing significantly the yield of oil obtained from bituminous coal under similar temperature and pressure conditions.[39] Nickel (II) oxide, 10% by weight, proved the better catalyst giving a maximum conversion of 45.64% oil compared to 30.95% with 10% iron (III) oxide. Doubling the catalyst's weight gave no increased yield.[40] Both investigations yielded far less than what Shatwell and Bowen had obtained earlier (61.4%) for the uncatalyzed hydrogenation of the identical coal suspended in phenol.[41]

In subsequent investigations in 1928 Graham and Skinner, now using phenol as the solvent, again studied the catalytic effects of iron and nickel oxides, as well as two additional catalysts, nickel oxalate (NiC_2O_4) and molybdenum oxide introduced as ammonium molybdate. Each catalyst mixed thoroughly with the coal amounted to 5% of the ash-free dry coal. Graham and Skinner's results in Table 2 showed molybdenum oxide gave the highest yield, leaving only 5.5% of the coal as residue. In addition to oil the reaction produced lighter alcohols and gases such as carbon monoxide, carbon dioxide, and methane.[42]

With the experiments on the catalytic hydrogenation of British coals completed in 1928, the Birmingham program came to a close. Working with about 30 different kinds of coals under fairly standardized conditions, 200 g coal, 100 g phenol as solvent, temperature 430° C, pressure 150 atm, time 8 hours, the Birmingham investigators found coals of the para-bituminous type (Seyler's classification) easiest to hydrogenate.[43] They showed that high-pressure hydrogenation yielded substantially greater amounts of oil, in some

39 Shatwell and Graham, "The Hydrogenation and Liquefaction of Coal-I" (note 21), pp. 26, 77.

40 Bowen and Nash, "The Thermal Decomposition of Coal in Presence of Catalysts and under Hydrogen Pressure" (note 24), pp. 361–64.

41 Shatwell and Bowen, "The Hydrogenation and Liquefaction of Coal, Part II" (note 24), p. 252.

42 Skinner and Graham, "The Hydrogenation and Liquefaction of Coal-V" (note 24), pp. 543–55, on p. 550.

43 Note 42, pp. 543–50. Clarence Arthur Seyler first proposed his classification of coal in 1900–1901 in papers read before the South Wales Institute of Engineers, but it began to receive widespread use only in the 1920s. His system essentially classified coals according to their carbon and hydrogen content. The series anthracite, carbonaceous, bituminous (meta-, ortho-, para-), lignite, decreased in carbon content and increased in hydrogen content. See Wheeler, "The Chemistry of Coal;" Bone, "The Chemical Constitution of

TABLE 4.2 Graham and Skinner's experiments with catalytic effects

Type of coal	Catalyst (5% by weight)	Tar oil yield (%)	Total liquid, gas, gaseous products (%)	Residue (%)
North Staffordshire Seam A (85.81% carbon)		43.7	81.8	18.2
	Fe_2O_3	44.1	83.0	17.0
	NiO	49.3	90.4	9.6
	NiC_2O_4	53.1	92.8	7.2
	NH_4MoO_4	58.8	94.5	5.5

SOURCE: SKINNER AND GRAHAM, "THE HYDROGENATION AND LIQUEFACTION OF COAL-V," *Fuel* 7 (1928): 543–55, ON P. 550

cases four to six times greater than the destructive distillation (carbonization) of coal at nearly the same temperature, 450° C. The work at Birmingham seemed to disprove Bergius's claim that hydrogenation of coals with greater than 85% carbon content did not occur and to eliminate the need for pressures above 150 atm. Bergius had worked with pressures of 200 atm, but at Birmingham pressures as high as 245 atm failed to convert additional coal into oil.

Despite its success, the Birmingham program remained a batch process with most of the experiments carried out in a rotating autoclave. Therefore, no removal of products occurred until the end of the reaction, and the products sometimes underwent secondary reactions. The industrial development of high-pressure coal hydrogenation, however, required construction of semi-continuous and continuous operation units. In a semi-continuous unit, either the reactants or products flowed continuously. But because of the difficulty in pumping continuously the coal-oil paste into a small stationary autoclave, the first British semi-continuous units usually introduced the paste in batches, and removed continuously the gaseous and liquid products. In a continuous unit

Coal;" and Seyler, "Some Problems of the Classification and Nomenclature of Coal" (all note 20). For discussion of Seyler see "Clarence Arthur Seyler, Melchett Medalist for 1941," *J. Inst. Fuel*, 15 (1941): 13–14.

both reactants and products flowed continuously.[44] Batch-autoclave and semi-continuous units, although easily operated, provided little experimental data on the effect of varying hydrogen gas and coal-paste input and on the physical properties of the paste that might affect its pumping into a continuous unit. Bergius already had constructed a continuous unit in his Rheinau-Mannheim plant, the first British semi-continuous and continuous units appeared at the government's East Greenwich Fuel Research Station in 1926. The continuous unit at East Greenwich represented a small industrial plant.[45]

3 Coal Hydrogenation at the Fuel Research Station

By the beginning of World War I, the British government had recognized that despite the present abundance of natural fuels, supplies had limits, and Britain, therefore, needed a program of conservation and supplementation. To study Britain's fuel requirements, the government's Privy Council, one of four advisory councils to the monarchy, in 1915 appointed a committee of its members, and in 1916 this committee established the Department of Scientific and Industrial Research. The Department created a Fuel Research Board the following year, directing the Board to promote and coordinate two programs of fuel research: (1) a survey of Britain's coal resources and (2) an experimental investigation of more economical and efficient methods of utilizing coal and its solid, liquid, and gaseous products.[46] In 1919 the Board erected a Fuel Research Station at East Greenwich where it conducted studies on coal hydrogenation from 1923 until the beginning of World War II in 1939.[47]

44 L. Horton, J. G. King, and F. A. Williams, "The Progressive Action of Hydrogen on Coal," *J. Inst. Fuel*, 7 (1933): 85–97; W. R. K. Wu and H. H. Storch, "Hydrogenation of Coal and Tar," *Bureau of Mines, Bulletin 633*, 1968, p. 27.

45 L. Horton, F. A. Williams, and J. G. King, "The Action of Hydrogen Upon Coal," *Fuel Research Technical Paper No. 42* (London: His Majesty's Stationery Office, 1935), 57 pp.; A. T. Barber and A H. Taylor, "High Pressure Plant for Experimental Hydrogenation Processes," *Proc. Inst. Mech. Eng.* (London), 128 (1934): 5–75; Barber and Taylor, "High Pressure Plant for Experimental Hydrogenation Processes," *Engineering*, 138 (1934): 576–78, 635–36; N. Booth and F. A. Williams, "Experiments on the Continuous Hydrogenation of Coal-I," *J. Inst. Fuel*, 11 (1937): 40–49; Part II, 11 (1938): 493–502; Wu and Storch, "Hydrogenation of Coal and Tar" (note 44), p. 27.

46 D. J. Payton-Smith, *Oil* (note 14), pp. 19–25; C. H. Lander, "Notes on the Work of the British Fuel Research Board," *Fuel*, 2 (1923): 191–94.

47 The Fuel Research Station also investigated the high- and low-temperature carbonization of coal and the Fischer-Tropsch synthesis of hydrocarbons. See F. S. Sinnatt and J. G. King, "A Study of the Tars and Oils Obtained from Coal," *J. Soc. Chem. Ind.*, 44 (1925): 413–24T;

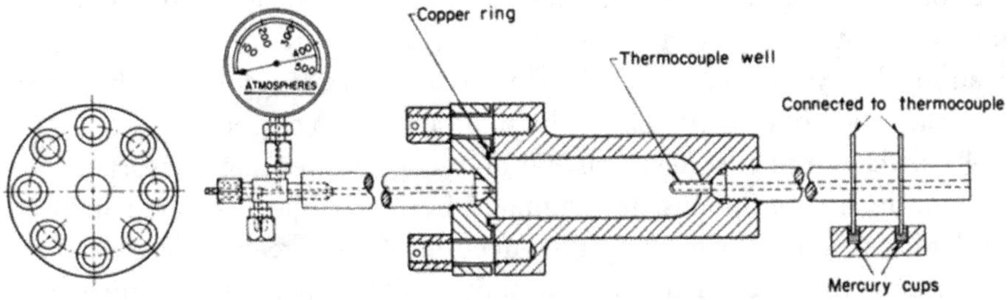

FIGURE 4.1 Section of early type of converter made at the Fuel Research Station
SOURCE: L. HORTON, F. A. WILLIAMS, AND J. G. KING, "THE ACTION OF HYDROGEN UPON COAL," *Fuel Research Technical Paper No. 42* (1935): 8

During its first years of operation, the Fuel Research Board worked closely with the British Bergius Syndicate. The syndicate consisted of fifteen British companies and individual investors that Bergius and William R. Ormandy, a research chemist and consultant to Distillers Company Ltd., organized in 1924. Members included Brunner, Mond and Company, South Metropolitan Gas Company, United Steel Companies, and Distillers Company, Ltd.[48] The syndicate held the option on the Bergius patent rights for the British Empire, and by the end of 1924, under Ormandy's direction, it had successfully hydrogenated British coals at Bergius's Rheinau-Mannheim plant. Ormandy informed the Department of Scientific and Industrial Research of the syndicate's success and urged it to cooperate in the investigations on British coals.[49] The department's Secretary and Director of Fuel Research, Frank S. Sinnatt (1880–1943), therefore visited Bergius in January 1925 and by early 1926 completed an agreement with the syndicate and its parent, the International Bergin Company which held all the patents on the Bergius process.

According to the agreement, the British Bergius Syndicate continued to hold the option on patents in Britain. A committee consisting of two department representatives, two syndicate representatives, and Bergius would supervise further research on British coals. The government would receive full infor-

F. S. Sinnatt, J. G. King, and W. H. Linnell, "A Study of the Tars and Oils Obtained from Coal-Part II," *J. Soc. Chem. Ind.*, 45 (1926): 385–93T; J. G. King, "The Production of Liquid Fuels from Coal," *J. Soc. Chem. Ind.*, 46 (1927): 181–86T; "Report of the Fuel Research Board," *Chemical Age*, 19 (1928): 318–19; Payton-Smith, *Oil* (note 14), p. 20.

48 "British Interests in Bergius Coal-Oil," *Chemical Age*, 15 (1926): 150; "The Hydrogenation of Coal," *Chemical Age*, 19 (1928): 310.

49 "The Late Dr. W. R. Ormandy," *J. Inst. Fuel*, 15 (1941): 68–69.

mation on the latest findings at Rheinau-Mannheim and incorporate the findings in a continuous coal hydrogenation unit constructed at the Fuel Research Station. The hydrogenation unit began operation in late 1926 at which time the hydrogenation of British coals at Rheinau-Mannheim ceased.[50]

Research at the Fuel Research Station followed three main lines: (1) experiments in the continuous unit supplied by the Bergius Research Institute, (2) batch experiments to study the effect of different reaction variables, and (3) experiments in a small-size continuous unit.[51] Experiments in the continuous unit commenced in February 1927 under the direction of James G. King, chief chemist at the Fuel Research Station; L. Horton, and F. A. Williams. The unit consisted of three horizontal converters of special chromium-molybdenum steel, each 190 cm long, 20 cm bore, 5-cm thick walls, and a 55-L capacity. Each converter had a metal mechanical stirrer, thermocouple, and gas-heated molten lead bath. The paste undergoing reaction contained in the proportions 20:8:1 powdered coal; a tar to serve as the carrier or vehicle; and a catalyst of luxmasse, an impure form of iron (III) oxide with a titanium oxide impurity obtained upon extracting alumina from bauxite. The unit had a rated hydrogenation capacity of 1,000 kg (1 long ton) of bituminous coal/d.[52] A Lane-Rogers generator with an output of 50 cm^3/hr (40,000 ft^3 in 24 hours) provided the hydrogen gas using the steam-iron process.[53]

50 Horton, Williams, and King, "The Action of Hydrogen Upon Coal" (note 45), pp. 3–4, 15.

51 The Fuel Research Station carried out some of the early experiments (1924) in autoclaves and semicontinuous units, but the results did not differ much from the work at Birmingham. See Sinnatt and King, "A Study of the Tars and Oils Obtained from Coal" (note 47), pp. 423–24; Sinnatt, King, and Linnell, "A Study of the Tars and Oils Produced from Coal" (note 47), pp. 392–93; Barber and Taylor, "High Pressure Plant for Experimental Hydrogenation Processes" (note 45), pp. 5–75; Horton, Williams, and King, "The Action of Hydrogen Upon Coal" (note 45), pp. 30–57; Horton, King, and Williams, "The Progressive Action of Hydrogen on Coal" (note 44), pp. 85–97; Wu and Storch, "Hydrogenation of Coal and Tar" (note 44), pp. 25–30.

52 King, "The Production of Liquid Fuels from Coal" (note 47), pp. 183–85; A. Crawford, F. A. Williams, J. G. King, and F. S. Sinnatt, "The Action of Hydrogen upon Coal-I," *Fuel Research Technical Paper No. 29* (London: His Majesty's Stationery Office, 1931), 37 pp.; Horton, Williams, and King, "The Action of Hydrogen upon Coal" (note 45), pp. 30–57. The special chromium-molybdenum steel contained carbon 0.25%, molybdenum 0.7%, chromium 1.25%. The first converters were of mild carbon steel construction.

53 W. P. Rogers, British Patents 249,925 of 1 January 1925 and 251,124 of 29 May 1925; Howard Lane, British Patents 17,591 of 29 July 1909 and 11,878 of 29 January 1910. See also T. P. Hilditch, *Catalytic Processes in Applied Chemistry* (New York: D. Van Nostrand, 1931), p. 215.

$$3\,Fe + 4\,H_2O \rightarrow Fe_3O_4 + 4\,H_2$$

Iron + Steam → Iron Oxide + Hydrogen

Hydrogen Production

$$Fe_3O_4 + 4\,H_2 \rightarrow 3\,Fe + 4\,H_2O$$

Iron Oxide + Hydrogen → Iron + Steam

$$Fe_3O_4 + 4\,CO \rightarrow 3\,Fe + 4\,CO_2$$

Iron Oxide + Carbon Monoxide → Iron + Carbon Dioxide

Iron Regeneration

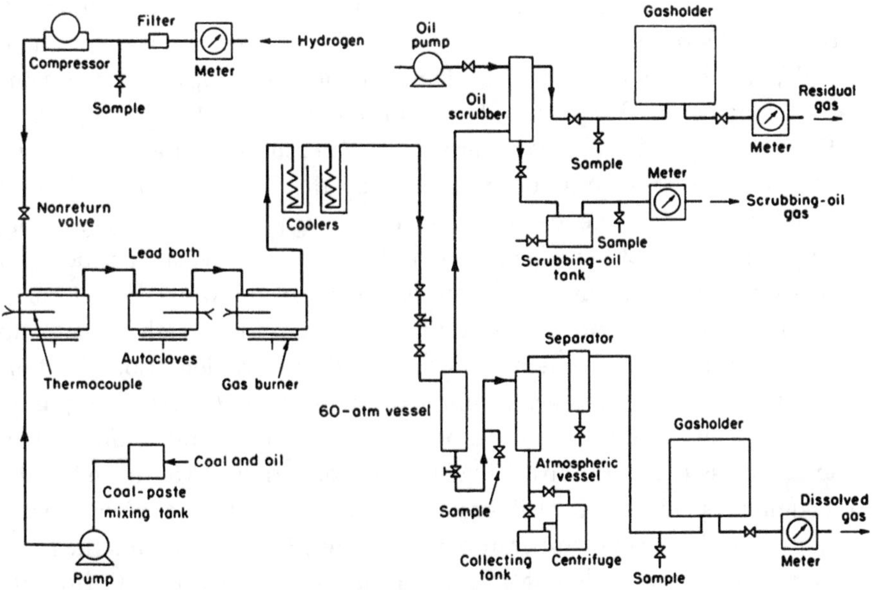

FIGURE 4.2 Bergius continuous unit
SOURCE: L. HORTON, F. A. WILLIAMS, AND J. G. KING, "THE ACTION OF HYDROGEN UPON COAL," *Fuel Research Technical Paper No. 42* (1935): 30 FACING

In general, the Fuel Research Station employed reaction conditions similar to those Bergius employed at Rheinau: working pressure 200 atm; temperatures in the three converters, respectively, 460° C, 480° C, 480° C; paste feed-rate of 20 kg/hr. The paste entered one end of the first converter at the bottom, hydrogen gas passed through a filter to a compressor under a pressure of 200 atm and entered the top of the converter at the same end as the paste. A dip pipe projected into the converter at its other end. When the level of the liquefied paste reached the top of the dip pipe, the incoming hydrogen gas blew a mixture of gas and liquid into the second converter, which in turn filled to the level of its dip pipe, before overflowing into the third converter. Finally, the liquid and gaseous products and any unused hydrogen gas passed from the third converter through two cooling coils to a control valve where the pressure fell from 200 to 60 atm. From the control valve the products flowed into a large cold vessel of about 0.25 m^3 capacity, called the 60-atm vessel, where the liquid, a light oil, and gas separated.

The gas, still at 60 atm pressure, moved to a scrubbing tower for washing with oil, and after reduction to atmospheric pressure, to a gas holder. The light oil, drawn off at intervals, passed into another container (the atmospheric vessel) held at atmospheric pressure. It remained there for several minutes, to permit any dissolved gas to escape to another gas holder and the frothing to subside, and then flowed into a receiving tank for weighing and centrifuging. Centrifuging removed any ash and undissolved solids remaining in the oil.

King, Horton, and Williams experienced numerous difficulties very early in the operation of the continuous unit. Initially, they encountered mechanical difficulties because of the reaction's high operating pressure and temperature; the corrosive effect of hydrogen, phenols, hydrogen sulfide, and ammonia on the construction materials soon followed. Hydrogen embrittled the mild steels (Fe_3C) first used by removing carbon from them to form methane gas; substituting chromium-molybdenum steels solved this problem. Complications arising from the hydrogen gas compressor's corroding and the paste pump's clogging because the paste sometimes set in the pump and formed a hard coke-like mass occurred most frequently. Introduction of galvanized piping and stainless steel rings in the compressor eliminated the corrosion; careful regulation of the paste's temperature immediately before it entered the pump prevented any paste hardening. In spite of these difficulties, King, Horton, and Williams by 1932 obtained a 50–75% conversion of coal to oil, corresponding to about 140 Imperial gal/long ton of moisture, ash-free coal. Hydrogen gas consumption equaled 4–6% the weight of coal consumed in the reaction.[54]

---

54 Horton, Williams, and King, "The Action of Hydrogen Upon Coal" (note 45), pp. 30–57; Horton, King, and Williams, "The Progressive Action of Hydrogen on Coal" (note 44), pp. 91–97; *Report of the Fuel Research Board, 1932* (London: His Majesty's Stationery Office, 1933), pp. 52–56.

4 Influence of Variables on the Conversion Process

The second series of experiments carried out at the Fuel Research Station by 1932 examined the influence of temperature, pasting carrier, and catalyst on the hydrogenation reaction. A typical British bituminous coal with a low ash content and ground to pass a 72-mesh sieve served as the standard raw material. In each experiment, a 2-L autoclave contained 150 g of coal and hydrogen gas at 100 atm pressure and underwent a three-step reaction sequence: (1) heating the autoclave to the reaction temperature, usually 450° C, (2) holding it at this temperature for two hours and then cooling, (3) discharging, measuring, and analyzing the products.

King, Horton, and Williams showed that in the absence of carrier and catalyst no reaction occurred below 400° C and that the reaction proceeded rapidly only above 440° C. They obtained the smallest amount of insoluble residue at 445–450° C; above 450° C the residue increased because of cracking reactions. Hydrogen consumption rose with temperature because more gaseous products (methane and ethane) formed at the expense of high-boiling liquids. Over the temperature range 440–470° C the yield of distillable oils varied little.

The conversion of coal to oil markedly increased on pasting the coal with a carrier, such as a low-temperature tar, even without a catalyst. Aromatic compounds, particularly m-cresol and tetralin, were effective. King, Horton, and Williams concluded that the advantage of using a carrier was principally in the greater rate of hydrogen gas absorption, especially with a catalyst added.[55]

The experiments on catalytic activity, with and without a pasting carrier, showed the effectiveness of about twenty chemical compounds. Luxmasse, ammonium molybdate (NH_4MoO_4), and tin (II) hydroxide [$Sn(OH)_2$] present as 2.5% by weight of coal and with no carrier added, lowered the initial reaction temperature and increased the reaction rate. They gave coal-to-oil yields of 66.1%, 59.2% and 71.9%, respectively, when the reaction ran for two hours at 450° C, and an initial hydrogen pressure of 100 atm. Luxmasse, which Bergius had originally added to remove any sulfur from coal, provided a particularly interesting case. The impure iron (III) oxide form catalyzed better than pure iron (III) oxide, because of its titanium oxide impurity.[56]

---

55 *Report of the Fuel Research Board, 1932* (note 54), pp. 52–56. The staff at the Fuel Research Station used both gas and electrically heated autoclaves but got better results with electric heat, presumably because it provided better temperature control, on p. 55.

56 Horton, King, and Williams, "The Progressive Action of Hydrogen on Coal" (note 44), pp. 85–97; *Report of the Fuel Research Board, 1933* (London: His Majesty's Stationery Office,

Other tin compounds, the oxide and sulfide, catalyzed effectively when present in 0.1% concentration. Zinc oxide (70.4% yield); nickel oleate (67.6% yield); germanium oxide (62.3% yield) catalyzed effectively, particularly in 2.5% concentration. With few exceptions, alkaline substances behaved as negative catalysts; halogens (chlorine and iodine) provided moderately good catalysis; and mixtures of 0.5% halogen with 0.1% tin hydroxide gave better results than either alone. Chlorine compounds left the lowest yield of insoluble material, while iodine compounds gave the best yield of distillable oils. The Fuel Research Station experiments showed that most of the catalysts increased both production rate and yield of oil when used with a pasting carrier. Increasing the amount of catalyst also increased somewhat the rate and yield.[57]

In addition to the batch experiments carried out in the two-liter autoclave, King, Horton, and Williams in 1933 worked with a semicontinuous unit. In these experiments, incoming hydrogen gas, kept at constant pressure, circulated over the coal-paste in the converter and the resulting volatile products along with excess hydrogen gas flowed continuously from the unit. Separation of the products from hydrogen gas followed upon passing both through a condenser. The semicontinuous unit provided a way to measure the hydrogenation reaction rate and the variation in the products' composition with time.[58]

5 Development of a Small-Size Continuous Plant

The batch experiments described above proved extremely valuable in making available quickly, information on catalysts and coal pasting carriers. Investigating the rate of hydrogen gas and coal-paste input on the coal-to-oil yield, as well as additional studies on temperature and pressure effects, required working with a continuous operating unit. For this reason, the British Government in 1936 constructed at the Fuel Research Station a small-size continuous unit capable of treating about 454 g (one pound) of coal per hour. As in previous experiments, the Fuel Research Station reacted a bituminous coal, ground

1934), pp. 94–103; *Report of the Fuel Research Board, 1934* (London: His Majesty's Stationery Office, 1934), pp. 98–106.

57 *Report of the Fuel Research Board, 1935* (London: His Majesty's Stationery Office, 1935), pp. 123–34; King, Horton, and Williams, "The Progressive Action of Hydrogen on Coal" (note 44), pp. 86–90.

58 Horton, King, and Williams, "The Progressive Action of Hydrogen on Coal" (note 44), pp. 85–91; *Report of the Fuel Research Board, 1933* (note 56), pp. 94–103.

to 72-mesh sieve size, mixed with 0.1% by weight tin hydroxide catalyst, and pasted with an equal weight of an oil produced from coal by hydrogenation. The paste and hydrogen entered at the bottom of a vertically arranged converter, gaseous products and unused hydrogen departed through a tube dipping into the space above the resulting oil. The oil and any remaining solids left the converter in the form of a sludge which flowed down a tube into a sludge receiver and then to a centrifuge for separation of ash and solids from oil.

In investigating the effect of varying hydrogen gas input at 400° C, 200 atm, and a paste input of about 1.0 kg/hr, Norman Booth and Williams showed in 1937 that the hydrogen rate had little significance within their experimental input limits of 1,275–1,875 L/hr. Other experiments at 440° C, with paste input of 1.0 kg/hr and hydrogen input of 1,600 L/hr illustrated that a total pressure of 180 atm (equivalent to a hydrogen partial pressure of 145 atm) sufficed for hydrogenation to proceed. Indeed, their experiments revealed that as the pressure increased above 200 atm the coal to oil yield actually decreased slightly.

In studying the effect of paste input on yield at 440° C, 200 atm, and a hydrogen input of 1,600 L/hr, Booth and Williams found that at the lowest input, 0.75 kg/hr, they obtained barely enough heavy oil to paste fresh coal. A rate of 1.0 kg/hr seemed normal and had a mean reaction time of about four hours. At higher rates the yield fell again. To examine the influence of temperature on the coal-to-oil yield, Booth and Williams selected their best experimentally established values of pressure, paste, and hydrogen input: 200 atm, 1 kg/hr, and 1,600 L/hr respectively. For the temperature range explored 420–460° C, they achieved a maximum oil yield at the lowest temperature, 420° C.[59]

By 1939 the Government had largely completed its three-part coal hydrogenation program at the Fuel Research Station. Over a sixteen-year period, the Research Station had successfully tested a number of different coals, experimented with many different catalysts, and established a set of optimum conditions for temperature, pressure, and other reaction variables.[60]

---

59 N. Booth and F. A. Williams, "Experiments on the Continuous Hydrogenation of Coal," *J. Inst. Fuel*, 11 (1937): 42–49.

60 *Ibid.*; J. G. King, "The Technique of Hydrogenation of Coal and its Products," *J. Inst. Fuel*, 10 (1936): 323–33; N. Booth and F. A. Williams, "Experiments on the Continuous Hydrogenation of Coal, Part II," *J. Inst. Fuel*, 1 (1938): 493–502; N. Booth, F. A. Williams, and J. G. King, "The Action of Hydrogen Upon Coal, Part III," *Fuel Research Technical Papers*, No. 29 (1931) and No. 42 (1935) both of which have the title "The Action of Hydrogen Upon Coal;" N. Booth, "The Catalysis of the Liquid Phase Stage of Coal Hydrogenation," *Chemistry and*

They obtained high-quality liquids from coal, the gasoline had an 88-octane number. If Britain intended to establish a high-pressure coal hydrogenation industry, the next step required a move from the Fuel Research Station's relatively small pilot plant operation to the construction of an industrial-size plant. The government's plan never included a next step, for long before 1939 it had turned to Imperial Chemical Industries (ICI) for the industrial development of high-pressure coal hydrogenation.

6 ICI and the Billingham Plant

ICI's experimental work, which in 1935 resulted in the construction of an industrial-size plant at its Billingham works in northeast England, represented the third and final stage in the history of Britain's high-pressure coal hydrogenation industry. Under the direction of Kenneth Gordon, ICI's research manager; and his assistants Herbert A. Humphrey (1868–1951), Melchett Medal winner, engineer, and inventor of the Humphrey pump (1909), an internal combustion, piston pump for large water projects; Philip J. Askey; Ronald Holroyd, later Sir and ICI's chairman; Howard W. Strong, and Charles Cockram, ICI had begun the Billingham program early in 1927 after acquiring control of the British Bergius Syndicate that same year. ICI thereby held the Bergius patent rights for the British Empire which the syndicate had obtained in 1924 from its parent organization, the International Bergin Company.[61]

The initial research at Billingham duplicated to some extent Bergius's earlier work and that at Birmingham and the Fuel Research Station: the experiments were batch type carried out in gas-heated, steel autoclaves. By 1928,

Industry, 63 (1944): 1–3. Booth completed this paper in 1939 but did not publish it until 1944.

61 British Interests in Bergius Coal-Oil," *Chemical Age*, 15 (1926): 150; *Bergius-Vertrage*, Summary Survey, 1–8, Bergius Papers, BASF Archives, Ludwigshafen, Germany; William J. Reader, *Imperial Chemical Industries, A History*, 2 Vols. (London: Oxford University Press, 1975), Vol. 2, p. 164. The British government acquired the Billingham site in 1917 when a German submarine blockade threatened to cut off nitrate imports from Chile. It planned to construct a factory to manufacture synthetic nitrogen products but made little progress. Brunner, Mond and Company took over Billingham after the war in 1920 and in December 1923 it opened a synthetic ammonia plant. In October 1926, Brunner Mond, United Alkali Company, Ltd., Nobel Industries Ltd., and British Dyestuffs Corporation Ltd., four of Britain's principal chemical companies formed Imperial Chemical Industries. See E. A. Blench, "The Billingham Enterprise," *Chemistry and Industry*, 30 (1958): 926–33; 31 (1958): 960–68.

however, the Billingham researchers had introduced and patented one significant modification: they arranged the converters (autoclaves) vertically rather than horizontally, and in this way the hydrogen gas bubbling up through the coal paste eliminated the need for mechanical stirring.[62] Their initial research had two aims: (1) to examine the thermodynamics of coal hydrogenation and (2) to study the influence of different catalysts on coal hydrogenation.

Bergius's early work on the thermodynamics of coal hydrogenation had shown the reaction to be highly exothermic,[63] and this fact convinced Gordon that in a large continuous unit, the reaction required no external heating once having heated the coal paste and hydrogen gas. ICI, therefore, built and tested three large steel converters, one of them 20 ft (6.1 m) high, 4 in (10.2 cm) in diameter, and by 1928 had in operation a successful self-heated coal conversion pilot plant. ICI already perfected the converter design used in the pilot plant as a result of its earlier industrial-size hydrogenation of nitrogen gas to produce synthetic ammonia at its Billingham works.[64]

The study on different catalysts produced another triumph. In 1929, Gordon and his co-workers found that elemental tin proved the most promising catalyst for the hydrogenation of British bituminous coal. Initially they used tin in the form of a tin-plated iron grid set inside the converter.[65] They quickly replaced this arrangement upon obtaining better results by injecting tin (II) oxalate or acetate (1% by weight of coal) into the converter.[66] Indeed, the discovery of tin's catalytic activity led ICI early in 1929 to begin the erection of a much larger pilot plant designed to hydrogenate daily 10 long tons of bituminous coal. By December the plant had begun operation and continued to operate until shut down at the end of 1931.

The hydrogenation process in Billingham's pilot plant took place in two stages, a procedure Pier, IG Farben's research director, had developed in 1926

62 H. A. Humphrey and ICI Ltd., British Patent 308,995 of 30 November 1927; Synthetic Ammonia and Nitrates, Ltd., and A. Rule, British Patent 309,258, of 12 February 1928; C. F. R. Harrison and ICI Ltd., British Patent 336,991 of 7 July 1929; T. G. Hunter and ICI Ltd., British Patent 330,025 of 16 March 1929.

63 Friedrich Bergius, German Patent 301,231 of 9 August 1913; British Patent 18,232 of 1 August 1914; US Patent 1,251,954 of 1 January 1918 (filed 3 August 1914).

64 H. A. Humphrey and ICI Ltd., British Patent 308,995 of 30 November 1927; Kenneth Gordon, "The Development of Coal Hydrogenation by Imperial Chemical Industries, Ltd.," *J. Inst. Fuel*, 9 (1935): 69–89, on pp. 73–74.

65 H. W. Strong and ICI Ltd., British Patent 335,215 of 19 June 1929. See also H. W. Strong and ICI Ltd., British Patent 345,305 of 20 December 1929; and ICI Ltd., R. Holroyd and C. Cockram, British Patent 344,975 of 28 December 1929.

66 P. J. Askey, R. Holroyd, C. Cockram, and ICI Ltd., British Patent 363,445 of 11 October 1930.

to improve the yield of gasoline and other light petroleum fractions. In the first stage, or liquid phase, the hydrogenation of solids or high-boiling liquids occurred. Hydrogen gas passed into the coal-oil paste where it dissolved readily at a reaction temperature and pressure of 450° C and 250 atm. The catalyst for the liquid phase, added as a fine powder, remained suspended during the reaction by the agitating effect of the hydrogen gas flow. The second stage, or vapor phase, resulted in hydrogenation of the lower-boiling oils obtained in the liquid phase and which vaporized at the reaction temperature. Vaporized oils flowed with hydrogen gas at 500° C over a fixed bed of catalyst where hydrogenation occurred. The plant converted 100 long tons of dry, ash-free coal into 62 long tons of gasoline, 28 long tons of gases, and a residue containing the ash and 6 long tons of solid carbonaceous substances. The gases, which consisted of methane and ethane (and excess hydrogen), reacted with steam in the water gas reaction to produce hydrogen, the residue burned as boiler fuel. Overall coal consumption, including the amounts required to generate steam for hydrogen manufacture and for power production, totaled 3.15 long tons of dry, ash-free coal or 3.65 long tons of actual coal per long ton of gasoline. The plant had a 43% thermal efficiency and a projected gasoline cost of 7d/Imperial gal.[67]

In the years 1929–31, the Billingham pilot plant carried out the first industrial-size hydrogenation of bituminous coal for an extended time. IG Farben's plant in Leuna did not begin industrial-size hydrogenation of bituminous coal until 1934. Billingham's successful operation enabled ICI to make several important decisions. Bergius's previous testing of British coals at Rheinau-Mannheim convinced ICI that it could make the hydrogenation process more economically attractive if it modified the process to give only one product, gasoline, rather than the usual variety of products, the tar acids, heavy oil, and pitch.[68] Also, ICI elected at this time to base the operation on hydrogenation of bituminous coal rather than coal tars, because of the tars' unavailability in quantities sufficient to aid significantly British petroleum demands. Both decisions proved most fortunate, resulting by 1930 in the Billingham pilot plant's producing gasoline of good quality suitable for use in most automobiles.

67 Imperial Chemical Industries, Ltd., "Coal Hydrogenation," *Fuel*, 10 (1931): 481–84; H. Harper, R. Scott, and ICI Ltd., British Patent 361, 135 of 2 December 1930; C. F. R. Harrison, E. D. Kamm, and ICI Ltd., British Patent. 338,576 of 16 July 1929; Gordon, "The Development of Coal Hydrogenation" (note 64), pp. 70–71.

68 Gordon, "The Development of Coal Hydrogenation" (note 64), p. 71.

As a result of Billingham's success, ICI decided to inform the British government of its findings. It would ask the government to support the development of an industrial-size synthetic petroleum industry, arguing that Britain, with no domestic source of natural petroleum, required an industrial-size synthetic petroleum industry for her national security. Also, ICI recognized that establishing a synthetic petroleum industry provided an excellent opportunity to reactivate its unused high-pressure equipment. The ammonia market had collapsed in late 1929 leaving ICI with an idle hydrogen production unit and several ammonia converters that it could use to synthesize petroleum. At a meeting in Billingham on 26 November 1930, attended by officials from ICI and the ruling Labour Party, Sir Harry McGowan (1874–1961), ICI's chairman, revealed in confidence that the Billingham plant had succeeded technologically, having achieved a 60% weight conversion of coal to oil. Economically, McGowan said, the conversion remained unpromising.[69]

The British government initially doubted that ICI had obtained such a high yield, but all doubt disappeared when Gordon demonstrated to members of the Fuel Research Board the correctness of ICI's yields. Later in 1930 McGowan informed the government of ICI's high production cost per Imperial gallon of synthetic gasoline, 8 d (3 pence, where 1 pence = £0.01), a figure more than double the cost of imported gasoline, 3 d (1.1 pence). He believed, however, that the low price of imported gasoline would probably increase after the current glut on the world market ended. In view of the present unfavorable economic situation, ICI could not develop a synthetic petroleum industry unless it received a government subsidy.[70]

But even while ICI was discussing its need for a subsidy, a new problem arose. By 1930, most of the original Bergius patents that ICI had obtained after gaining control of the British Bergius Syndicate had now lapsed. IG Farben and Standard Oil of New Jersey, which in 1930 had formed their own stock company, the Joint American Study Corporation (JASCO), to manufacture and market synthetic gasoline, controlled the remaining patents on high-pressure coal hydrogenation.[71] Consequently, ICI opened negotiations

69 Reader, *Imperial Chemical Industries* (note 61), pp. 162–75. Gordon, "The Development of Coal Hydrogenation" (note 64), p. 72.

70 Reader, *Imperial Chemical Industries* (note 61), pp. 170–75. Conversion factors for British currency (1930 value): 1p = 2.40 d; 1£ = 240 p; 1£ = 20 s; and 1£ = $4.86.

71 Joseph Borkin, *The Crime and Punishment of I.G. Farben* (New York: The Free Press, 1978), p. 51; Reader, *Imperial Chemical Industries* (note 61), pp. 166–70; Richard Sasuly, *I.G. Farben* (New York: Boni & Gaer,1947), pp. 144–47; Josiah E. DuBois Jr., *The Devil's Chemists* (Boston: The Beacon Press, 1952), p. 148.

FIGURE 4.3 Sir Harry McGowan (1874–1961)
SOURCE: BAIN NEWS SERVICE, LIBRARY OF CONGRESS, WASHINGTON, DC, CONTROL NO. 2014717648

for patent sharing and exchange of technical information with IG Farben and Standard Oil. Patent negotiations continued, and in April 1931 the three companies, joined by Royal Dutch-Shell, which already had £250,000 invested in Bergius's Rheinau-Mannheim plant, formed a holding company, the International Hydrogenation Patents Company (IHP) in Vaduz, Liechtenstein. IHP controlled the global rights to the Bergius process, with ICI's interests confined to the British Empire. In the same agreement, IHP arranged for distribution of its products through the Anglo-Iranian Oil Company (British Petroleum Company), Royal-Dutch Shell, and Standard's subsidiary, the Anglo-American Oil Company (Esso Petroleum Company).[72]

After shutting down the successful 10-ton pilot plant at Billingham in 1931, ICI continued research on coal hydrogenation in its laboratories. Improvement in the process came from these investigations as well as from its information exchange with IHP. In an important 1932 study, Holroyd and Cockram found that the addition of hydrogen chloride gas to the coal converters aided hydrogenation. Its acidic vapors neutralized the alkaline coal ash that remained

72 R. Holroyd, C. Cockram, and ICI Ltd., British Patent 427,883 of 30 December 1933.

after hydrogenation and usually weakened the catalyst's influence. The gas also seemed to catalyze the reaction. Unfortunately, hydrogen chloride's corroding of metallic heat exchangers and coolers and its clogging of reaction tubes by forming ammonium carbonate and chloride deposits offset its benefit.[73]

Holroyd and Cockram eliminated the corroding-clogging problem in 1933 upon developing a stable suspension of alkali or alkaline earth (calcium hydroxide) in oil. By scrubbing the hot vapors, which consisted of the hydrogenation products and hydrogen chloride, with the alkali-oil suspension, they effectively neutralized the hot vapors. Upon discovering the catalytic effect of tin and chloride compounds and successfully neutralizing any acid present in the vapor phase, ICI had in 1933 put the hydrogenation of bituminous coal on an absolutely sound scientific foundation.[74]

7 Politics and Economics

The politics and economics of ICI's coal hydrogenation program also had improved considerably by 1933. A sterling (British pound) crisis, the devaluation of the British pound resulting from a rise in unemployment in 1929–31 followed by an abandoning of the gold standard on 19 September 1931, brought down the Labour Party at the end of August 1931, and after the October General Election, Labour Prime Minister Ramsay MacDonald (1866–1937) returned to head the strongly conservative, pro-business National Government. By the fall of 1932 Britain's petroleum suppliers operating in Romania, which included Standard Oil of New Jersey and Royal Dutch-Shell, reached an agreement to limit production and increase prices.[75] The 1930s Depression had led to a global oil glut, but Romanian production, unlike that in other leading petroleum producing nations, the United States, Venezuela, and Iran, had continued to increase, making it in 1931 the world's fourth largest producer.[76] Standard Oil also bought out the independently operating Cleveland Oil Company (Ohio) in 1932 and thereby effectively ended price-cutting in Britain. The cost of

73 They added carbon tetrachloride to the reaction. It decomposed and reacted with hydrogen to form hydrogen chloride gas.

74 R. Holroyd, C. Cockram, and ICI Ltd., British Patents 432,958 and 433,020 of 5 January 1934; Gordon, "The Development of Coal Hydrogenation" (note 64), p. 74.

75 Reader, *Imperial Chemical Industries* (note 61), pp. 176–77; Payton-Smith, *Oil* (note 14), pp. 8–9.

76 R. Wigginton, "Notes on Recent Developments in Fuel Technology," *Fuel*, 11 (1932): 281.

imported gasoline quickly rose to 5–6 d (2–2.5 pence), nearly double the price in 1930.[77]

At this time, therefore, ICI renewed efforts to obtain a government subsidy in order to develop a British synthetic petroleum industry. It considered a subsidy crucial to its well-being. ICI had already invested £1 million (about $4.8 million), including the cost of constructing and operating the 10-long ton pilot plant, and had available at Billingham a water gas unit capable of generating enough hydrogen gas to produce 150,000 long tons/yr of gasoline.

In addition to its economic argument, ICI made an appeal to Britain's need for national security and stability. ICI pointed out that construction of an industrial-size coal hydrogenation plant at Billingham could initiate an entirely new British industry. As these were Depression years, its construction would provide jobs for thousands of coal miners and workers in coal-dependent industries, thus relieving Britain's unemployment problem. The extra coal consumption, ICI estimated, would give work to 2,000 miners, the manufacture of gasoline and related research would engage another 2,000 workers, and plant construction would employ about 7,000 people for 18 months. As its final argument, ICI called attention to Britain's total dependence on imported petroleum. In 1933 alone, Britain imported 8.5 million long tons of petroleum at a cost of £21 million. The Billingham plant, ICI acknowledged, while only providing about 4% of Britain's fuel needs, would be the first step in a movement toward total independence of imported petroleum.[78]

Economics provided the main opposition to granting ICI a subsidy. The government stood to lose the duty it received on each gallon of imported petroleum. In a plant producing annually 150,000 long tons or 45 million Imperial gallons of gasoline, the loss amounted to £750,000/yr with the existing duty of 4d/Imperial gal (1.47 pence/Imperial gal) and twice that amount for a duty of 8d/Imperial gal (2.94 pence/Imperial gal). The government also countered ICI's appeal to British national security arguing that petroleum imports brought by ship to different ports were much less vulnerable to enemy air attacks from the Continent than the Billingham site. Even with the limited range of aircraft available after World War I it seemed unwise to rely on Billingham as a major source of chemicals and gasoline in wartime. In fact, ICI's Kenneth Gordon in 1936 remarked that Billingham's location made it "so vul-

77 Reader, *Imperial Chemical Industries* (note 61), pp. 176–77.

78 W. A. Bone, "Economics and National Aspects of the Production of Oil from Coal," *J. Inst. Fuel*, 9 (1936): 277–82. See p. 278. "Hydrogenation of Coal," *Chemistry and Industry*, 54 (1935): 934–37. The average value for the pound in US dollars for 1933 was 1£ = $4.23.

nerable to air attack" that the government could neglect the Billingham plant "as a source of supply in the event of war."[79]

After much debate and discussion in spring 1933 between William Henry Coates (1882–1963), ICI's treasurer, and Sir Horace Wilson (1882–1972), the government's chief industrial adviser and close adviser of Neville Chamberlain (1869–1940), then chancellor of the exchequer, the government accepted the validity of ICI's arguments. With 2 million workers unemployed, serious morale problems in the coalfields, and no spending of public funds required, Wilson and Chamberlain decided to strike a deal with Coates and ICI. On 23 June 1933, the government agreed to grant a subsidy on gasoline manufactured from British coal. Prime Minister Ramsay MacDonald announced the government's decision in the House of Commons on 17 July 1933, had a bill ready on 21 December, and with little opposition passed the bill into law the next spring as the British Hydrocarbon Oils Production Act, 1934. Five years earlier, in 1928, the government had exempted from import duty gasoline obtained from British shale and light oil made from British coal. The passage of the Hydrocarbon Oils Production Act now exempted gasoline manufactured from British coal and heavy oils and provided a subsidy of 4d/Imperial gal (1.47 pence/Imperial gal) for a period of nine years beginning 1 April 1935 and subject to review if the price of imported gasoline varied.[80]

During the entire period of ICI-government negotiations the major oil companies operating in Britain offered little opposition. Under the terms of the 1931 IHP agreement, Royal Dutch-Shell and Standard, which had extensive global marketing organizations, completely controlled the distribution of ICI's synthetic gasoline. The agreement limited ICI's rights only to countries in the British Empire and to 25% of the petroleum products each country consumed. It also included a marketing commission for Royal Dutch-Shell and Standard. The major oil companies therefore had a potential profit-making arrangement with ICI and one that they could control.[81]

ICI never waited for passage of the Hydrocarbon Oils Production Act. Once McGowan had confidence in the government's intentions, ICI began constructing the industrial-size Billingham plant in August 1933, completing it

79 Payton-Smith, *Oil* (note 14), p. 22. Gordon's quote is from Reader, *Imperial Chemical Industries* (note 61), p. 263.

80 Reader, *Imperial Chemical Industries* (note 61), pp. 178–80; "Development of Coal Oils in Great Britain," *Petroleum Times*, 33 (1935): 255–58; Payton-Smith, *Oil* (note 14), pp. 21–22. In the Finance Act of 1938 the government increased the guaranteed preference to 8 d and extended its duration to 1960.

81 Reader, *Imperial Chemical Industries* (note 61), p. 170.

in November 1934. The plant had the capacity to produce annually 150,000 long tons of gasoline, 100,000 long tons to come from coal and 50,000 long tons from creosote, an oil obtained by distilling coal tar. It cost £3 million, although this increased to £5.5 million upon including the value of the hydrogenation units previously existing at Billingham. Hydrogenation of creosote oil took place for the first time on 7 February 1935. On 9 April 1935, a tanker of 1,000 tons capacity left Billingham and steamed down the River Tees with the first shipment of synthetic gasoline. Hydrogenation of coal began shortly after in June 1935.[82]

In constructing the Billingham plant, ICI decided to complete first the vapor-phase converters that composed the plant's stalls and to use only creosote oil. ICI believed that if it succeeded with the more difficult vapor-phase conversion little chance of encountering any problem with the liquid-phase operation would arise. The plant, when finally completed, contained three first-stage liquid-phase converters, two for coal paste, one for heavy oil; and two second-stage vapor phase converters. The water gas reaction provided hydrogen gas.

As in previous Billingham operations, the conversion required (1) cleaning the coal by froth flotation, usually to less than 2.5% ash, (2) grinding the clean coal, (3) mixing the clean ground coal with an oil obtained previously from the process, or with creosote oil or a low temperature tar, to make a 50% coal-oil paste, (4) injecting the paste into the liquid phase converters, (5) adding hydrogen gas to a pressure of 250 atm at 450° C, which converted most of the paste into middle oil (boiling point range 170–300° C). Further hydrogenation of the middle oil took place in the second stage vapor phase converters where middle oil vapors and hydrogen gas passed over a solid catalyst of tin (II) oxalate (SnC_2O_4). The vapor phase hydrogenation, after cooling and distilling produced mainly gasoline and some residual middle oil.

The first or liquid-phase hydrogenation also produced a coke residue that served as fuel, a small amount of crude oil containing unconverted coal (5% by weight), and ash. Recovery and distillation of the crude oil gave heavy oil, middle oil, and gasoline. The heavy oil passed to the heavy-oil converter for further reaction with hydrogen giving more gasoline and middle oil; the middle oil went to the vapor phase converters for additional hydrogenation. ICI's

82 "Petrol from Coal," *J. Inst. Fuel*, 9 (1935): 65–68; Gordon, "The Development of Coal Hydrogenation" (note 64), p. 75. To estimate the cost of the Billingham plant in US dollars: 1£ = $5.03 (1934 average) and 1£ = $4.90 (1935 average).

FIGURE 4.4 General view of the Billingham plant
SOURCE: KENNETH GORDON, "THE DEVELOPMENT OF COAL HYDROGENATION BY IMPERIAL CHEMICAL INDUSTRIES," *J. Inst. Fuel*, 9 (DECEMBER 1935): 69–89

two-phase process converted practically all of the coal into gasoline, gaseous hydrocarbons (methane, ethane, propane), and a small residue.[83]

To produce one long ton of gasoline the Billingham plant consumed five long tons of raw coal which included the coal used for fuel and for hydrogen production from coke in the water gas reaction. Gasoline yield per long ton of bituminous coal actually hydrogenated increased significantly, to about 60%,

83 Gordon, "The Development of Coal Hydrogenation" (note 64), pp. 77–84; and "Progress in the Hydrogenation of Coal and Tar," *J. Inst. Fuel*, 20 (1946): 42–58. See pp. 42–43; also "Process and Power Steam at Billingham," *Canadian Chemistry and Metallurgy*, 20 (1936): 10–12.

and as high as 80 or 90% with creosote oil as the raw material. Creosote oil produced an excellent quality gasoline requiring very little refining and had an 87-octane number.[84] ICI distributed its gasoline through the oil companies designated in the 1931 IHP agreement, Royal Dutch Shell, British Petroleum, and the Anglo-American Company.[85]

The 1933 arrangement that brought the Billingham plant into existence called for the gasoline produced to come from the hydrogenation of bituminous coal. By 1936 ICI had overcome most of the process's technical difficulties, running one of its conversion units for seven consecutive months, although the average run in 1937–39 was usually about two and one-half months. During this time, rising coal costs, and consequently the cost of producing hydrogen gas, had begun to make coal hydrogenation uneconomic. Profits for 1937 and 1938 totaled only £94,669 and £119,933, for net returns of 2.01 and 2.78%. ICI's research manager, the ever-optimistic Kenneth Gordon, expected no dramatic improvement for 1939: £247,059 or a 5.65% return. As a result, ICI turned increasingly to hydrogenating creosote oil. By autumn 1939 when war broke out in Europe, ICI ceased completely to hydrogenate coal, producing instead from creosote oil over 100,000 long tons of aviation fuel, 93.5–96 octane number with lead added.[86] Between 1940 and 1945, gasoline production ranged from 76,000 long tons to 148,000 long tons.

84 Gordon, "The Development of Coal Hydrogenation" (note 64), p. 83.

85 Gordon, "Progress in the Hydrogenation of Coal and Tar" (note 83), p. 42.

86 In addition to the Billingham plant, ICI, Shell Oil, and Trinidad Leaseholds, Ltd. operated a high-pressure oil hydrogenation plant and refinery on a 300-acre site at Heysham on the Lancashire coast, a site considered safe from German air attack. The three firms formed a holding company, Trimpell Limited, in September 1939. ICI and Shell jointly operated the plant using gas oil that Trinidad Leaseholds supplied from Trinidad. The British government paid for the plant's construction, £6,200,000. Heysham opened in August 1941 and had an annual capacity of about 300,000 long tons of aviation fuel, which after adding lead had a 93.5 octane number. This was 60% of Britain's total aviation fuel production of 500,000 long tons/yr. Shell also began operating in February 1939 a 32,000 long ton-per-year iso-octane plant at Stanlow. Later improvements in refining enabled all three plants to produce aviation fuel with octane numbers greater than 100 and were responsible for the British Air Force's success in intercepting and destroying the German V-1 flying bombs over Britain. See Gordon, "Progress in the Hydrogenation of Coal," (note 83), pp. 42–43, 56–58; Kenneth Gordon, the Melchett Lecture, 1947, "Hydrogenation in the Fuel and Chemical Industries," *J. Inst. Fuel*, 21 (1947): 53–61, see pp. 59–61; K. Gordon and J. A. Oriel, "Aviation Fuel Production," *Transactions of the Fuel Economy Conference*, The Hague, 1947, 2 Vols. (London: Lund Humphries and Company, Ltd., 1948), Vol. 2, pp. 509–19, see pp. 509–13; Payton-Smith, *Oil* (note 46), pp. 173–77, 386–87; Reader, *Imperial Chemical Industries* (note 61), pp. 266, 275.

FIGURE 4.5 Control room (top), high pressure converters (bottom)
SOURCE: KENNETH GORDON, "THE DEVELOPMENT OF COAL HYDROGENATION BY IMPERIAL CHEMICAL INDUSTRIES, LTD.," *J. Inst. Fuel*, 9 (1935): 69–89

TABLE 4.3 Output of gasoline at Billingham 1935–45

Year	Gasoline produced from coal (long tons)	Gasoline produced from creosote (long tons)
1935	11,000	58,000
1936	35,000	76,000
1937	44,000	74,000
1938	52,000	91,000
1939	28,000	120,000
1940		148,000
1941		111,000
1942		92,000
1943		76,000
1944		106,000
1945		97,000

SOURCE: KENNETH GORDON, "PROGRESS IN THE HYDROGENATION OF COAL AND TAR," *J. Inst. Fuel* 20 (1946): 42

The decision to stop hydrogenating coal was not only economic but a requirement of Britain's defense plan. During the wartime air raids the government ordered all high-pressure plants, the Haber synthetic ammonia plants and methanol plants, to release their pressure and shut down. This action eliminated any secondary damage, from fires and explosions that the escaping gases might cause if a bomb struck a high-pressure plant. When reacting only gases, as in the ammonia and methanol plants, it took only minutes for a shutdown, but a coal hydrogenation plant required a much longer time. In order to avoid heavy coke formation, plant workers had to displace the coal in the converters with oil, making a rapid shutdown exceedingly difficult if not impossible.[87] Britain's serious coal shortage during the war and her increasing need of hydrogen gas to synthesize the ammonia gas so essential for fertilizers also contributed significantly to ICI's decision to stop hydrogenating coal at Billingham.

87 Gordon, "Progress in the Hydrogenation of Coal and Tar" (note 83), p. 43.

8 Conclusion

In the years after World War II's ending, the economics of coal hydrogenation had not improved. The cost of coal, the increasing cost of producing hydrogen gas, about 40% of the total operation, prevented ICI's return to high-pressure coal hydrogenation. Creosote oil's price rise in relation to gasoline's market price, even with a government subsidy, also eliminated any profit margin on gasoline production from creosote. New petroleum discoveries in the Middle East and the United States led to an abundance of gasoline in the 1950s. In view of this situation, ICI turned more and more to the production of chemicals instead of gasoline from coal and its derivatives. In September 1958, the synthesis of gasoline at Billingham ended.[88]

Twelve years later, Britain's liquid fuel availability changed dramatically. Beginning in October 1970, British Petroleum (BP) discovered and developed a source of natural petroleum, the North Sea deposits which enabled Britain to achieve self-sufficiency in 1980, only five years after production began in June-November 1975. By 1982 Britain, with a daily yield of 1.5 million barrels was the fifth largest petroleum producer after the United States, Saudi Arabia, the Soviet Union, and Mexico.[89] But the North Sea fields, as all naturally occurring oil fields, face eventual depletion. Because Britain's coal reserves are much greater than its petroleum deposits and because of the 1973–74 and 1979–81 Middle East oil crises the government's National Coal Board once again looked at the conversion of coal to oil in the 1970s–80s, but no further research and development followed once the oil crises subsided.[90]

What Britain's scientists and engineers accomplished at Birmingham University, the government's Fuel Research Station, and ICI's Billingham coal

88 C. Cockram and E. W. Sawyer, "Hydrogenation at Billingham in Retrospect," *The Industrial Chemist*, 35 (1959): 221–29; E. A. Blench, "The Billingham Enterprise" (note 61), pp. 926–33, 960–68.

89 Roland Gribben, "Oil Self-Sufficiency is Just Around the Corner," *Gas World*, 184 (October 1979): 493–95; Philip Algar, "Though Only Formed in 1973, Offshore Supplies Office is Now A Vital Part of the North Sea Scene," *Petroleum Review*, 34 (January 1980): 26–27; Peter Rodgers, "What Will Happen When North Sea Oil Runs Out: The U.K. Government's Already Getting Rattled," *Petroleum Review*, 33 (November 1979): 8–10; Jane Charteris, "Britain's Other Oil: Onshore Production Makes an Important Contribution to National Wealth," *Petroleum Review*, 33 (December 1979): 40–45; "North Sea Gas," *The Lamp*, 66 (Summer 1984): 18–19.

90 Andrew Stratton, "Coal As a Raw Material For the Chemical Industry," *J. Soc. Chem. Ind.*, 18 (1978): 551–60; M. St. J. Arnold and P. F. M. Paul, "Aviation Fuels From Coal," *J. Inst. Energy*, 55 (1980): 55–62.

hydrogenation plant in the 1920s–50s, nevertheless remains a classic example of how the discovery, quantification, and application of the fundamental laws and principles of science advanced the coal-to-oil synthesis from the laboratory, to the pilot plant, and to its successful industrial establishment. Britain's three-stage advancement repeated the success of Germany's synthetic fuel program.

CHAPTER 5

Synthetic Fuel Production in Prewar and World War II Japan: A Study in Technological Failure

> It has been concluded by some investigators that the research and scientific ability of the Japanese is mediocre. This is believed to be a fallacy. It is rather concluded on the basis of research discussed herein, that their progress has been conspicuous and that within the next few decades Japan may well become one of the foremost technical nations of the world.
>
> Report of the US Naval Technical Mission to Japan, "Japanese Fuels and Lubricants," 1945. Quoted in *Petroleum Processing* 2 (1947): 819.

1 Introductory

Japan largely lacks supplies of many essential natural resources including petroleum, coal, and iron ore. As her industrial base and economy expanded in the 1920s and 1930s, Japan's dependence on imports of these resources became increasingly evident. The onset of the 1930s Depression further threatened Japan's lifeline, and in effort to become economically independent and self-sufficient in natural resources (autarky) Japan's militaristic government pursued a policy of territorial expansion. Beginning in 1937, Japan's military forces swept out of Manchuria into China and then into Southeast Asia in search of strategic materials such as petroleum, coal, copper, zinc, and rubber.

To achieve independence in petroleum, the Japanese developed a dual approach: they would acquire natural petroleum sources in Southeast Asia and at the same time establish a synthetic fuel industry for the conversion of coal to oil. Actually, the Japanese had begun research on synthetic fuel in the 1920s, only a few years after other countries like Germany and Britain that lacked sources of natural petroleum. They did excellent laboratory research on the coal hydrogenation and Fischer-Tropsch (F-T) conversion processes, but in their haste to construct large synthetic fuel plants they bypassed the intermediate pilot-plant stage and failed to make a successful transition from small

 DOI:10.1163/9789004690912_006

to industrial-size production. Unable to synthesize liquid fuels from coal, they instead derived significant quantities from the technologically simpler coal carbonization and oil shale distillation processes. In the last year of World War II, the Japanese attempted to revive their synthetic fuel industry and entered into an agreement with IG Farben for technical assistance. Germany's defeat ended this final effort. The Japanese synthetic fuel industry presents a good case study of technological failure. It shows that high-quality basic scientific research does not necessarily translate into industrial-size technological success.

The US Naval Technical Mission to Japan arrived at the Japanese naval base at Sasebo (Kyûshû) on 4 September 1945 and remained in Japan until 1 November 1946. It consisted of four members, three naval commanders plus a language officer who served as an interpreter and translator, with a technical representative from the US Bureau of Mines added later. The members traveled to the assigned locations in an ambulance especially fitted with litters for sleeping, extra racks for drinking water and gasoline, and a trailer for food and miscellaneous goods. They interrogated 14 leading Japanese fuel scientists, among them Gen-itsu Kita at Kyoto Imperial University, Yoshisada Ban at the Imperial Fuel Research Institute in Kawaguchi (Tokyo), Hikoshi Sogabe, a Mitsui technical director at the Miike Synthetic Oil plant in Ômuta, and several other fuel science personnel. The Naval Technical Mission compiled the interrogations, related fuel documents, some of which the Japanese scientists translated into English, and wrote comprehensive reports of their investigations. The Naval team inspected the fuel research institutes and fuel plants, such as the First Naval Fuel Depot at Ôfuna, Kyoto Imperial University, Imperial Fuel Development Company plant at Ube, and the Miike Synthetic Oil plant in Ômuta and included descriptions and photographs in their reports. The Naval Technical Mission's reports, its visits to the synthetic fuel and coal conversion facilities and plants, and the Japanese scientists' pre-war publications provided a comprehensive record of Japan's coal-to-oil conversion program.

2 Overview of Japan's Coal Conversion Processes, Seven Year Energy Plan

In the 1930s Depression years when economic problems threatened Japan's well-being, the Japanese government adopted a policy of autarky. Being a country with limited natural resources, the Japanese intended to achieve economic independence and self-sufficiency in natural resources through territorial expansion. But as Japan's military forces began their drive out of

Manchuria and into China's major cities and then into Southeast Asia, the need for additional petroleum to fuel the expansion become increasingly evident. Petroleum consumption had risen from 40,000 metric tons in 1912, to 666,000 metric tons in 1924, and tripled to 1,980,000 metric tons in 1932. Production had not kept pace. In 1932 Japan proper produced only 242,000 metric tons petroleum, Taiwan another 35,000 metric tons, a combined production that represented about one-seventh of Japan's total peacetime consumption. Imports from the United States, British Borneo, the Dutch East Indies, and the Soviet Union made up the balance. Although Japan contributed only 0.1% of the world's total petroleum production, the empire had comparatively large coal reserves estimated in 1935 at 22 billion metric tons.[1]

During this same period Germany and Britain attempted to eliminate their lack of naturally occurring petroleum by synthesizing liquid fuel from coal. The Japanese adopted a similar approach beginning in 1936. By that time they had calculated a minimum 400–500 year lifetime for their coal reserves if converted to liquid fuel at an annual rate of 15 million tons. The Japanese investigated the three known coal-to-oil conversion processes: low-temperature carbonization (LTC), high-pressure coal hydrogenation (direct coal liquefaction), and Fischer-Tropsch synthesis (F-T, synthesis gas process or indirect coal liquefaction). Of the three processes, LTC, the oldest and simplest technologically, required neither catalysts nor expensive and complicated high-pressure apparatus. No synthesis occurs in LTC, only a thermal decomposition of coal in the absence of air (destructive distillation) at 500–700° C giving semicoke (a more active fuel than coke), coal tar, and gases. Distilling the tar derives the oil rather than synthesizing it. Japanese industrialists had introduced LTC for the production of semicoke not liquid fuel, hoping to substitute semicoke for the more expensive charcoal commonly used in household heating. Only in the early 1930s after the rise of Japan's militaristic government did Japanese fuel scientists and industrialists recognize LTC as a good liquid fuel source and include it in a comprehensive synthetic fuel plan. Because of its low technological requirements LTC became the most successful of Japan's three coal conversion processes, producing more liquid fuel than hydrogenation and F-T combined.

1 Yoshio Kodama, "The Coal-Hydrogenating Industry in Japan," *J. Fuel Soc. Japan*, 14 (1935): 71–2; Kodama, "Hydrogenation of Coals," *J. Fuel Soc. Japan*, 14 (1935): 691–701; Kodama, "The Coal-Hydrogenating Industry in Japan," *Chemical Abstracts*, 29 (1935): 7043; Kodama, "Hydrogenation of Coals," *Chemical Abstracts*, 29 (1935): 5629. Throughout the chapter, tons means metric tons; articles cited in Japanese journals are in Japanese.

Both synthetic processes, the reaction of hydrogen gas (hydrogenation) with coal or tar to give petroleum liquids, and F-T, the conversion of coal and steam to synthesis gas (a carbon monoxide-hydrogen mixture) and then to petroleum liquids, give considerably more liquid fuel per ton of coal than LTC. In the 1930s coal hydrogenation was farther advanced than F-T, and it became the focus of Japan's synthetic fuel program. The Japanese carried out extensive and first-rate laboratory research and placed great faith in hydrogenation's rapid and successful development. They also conducted high quality laboratory research on the F-T synthesis. But in both cases the Japanese failed to make a successful transition out of the laboratory because they leaped to industrial-size synthetic fuel production with insufficient pilot plant experience. Curtailing or by-passing the intermediate pilot plant stage of development led to serious operating problems and doomed their synthetic fuel program to technological failure. The failure also illustrates the generally lower level of pre-World War II Japanese engineering compared to the West's. A wartime atmosphere clearly compounded the urgency, and consequently the two synthetic processes contributed little to Japan's liquid fuel sources.

In addition to large coal reserves the Japanese possessed a rather significant potential source of liquid fuel in the 5 billion metric tons of oil shale located at Fushun in southern Manchuria. Like LTC, oil shale distillation is a low-yield, low-technology process that requires heating the shale to separate the oil. The Japanese constructed only one plant, at Fushun, but during the World War II years it outproduced all the coal conversion plants.

To ensure industry's participation in a synthetic fuel program the Japanese government offered several financial incentives beginning in July 1936 when the seventy-first Special Imperial Diet introduced its petroleum substitutes policy. This war measure included subsidies and laws to encourage liquid fuel production, indemnities for losses incurred in synthetic fuel production, and a ¥ 4.5 million grant from the Ministry of Finance to the Ministry of Commerce and Industry. The following May the government established the Fuel Department within the Ministry of Commerce and Industry to promote synthetic fuel production. Four naval and three army officers served as department directors and section chiefs. In August 1937 the government proposed a Seven Year Plan that contained two major provisions for its implementation: the Imperial Fuel Development Company Law enforced from September 1937 and the Synthetic Oil Production Industry Law from January 1938. At the same time (January 1938) the government established Teikoku Nenryô Kôgyô KK (Imperial Fuel Development Company) in Tokyo to administer the Imperial Fuel Development Company Law. Imperial Fuel Development Company also manufactured and sold synthetic fuel and financed its production by Japanese industry. The Seven Year Plan called for the construction of 87 synthetic fuel plants by 1944, producing annually 6.3 million barrels (one million kL) each of synthetic gasoline and fuel oil. But shortages of plant equipment and strategic metals including nickel and iron, the high purchase and shipping cost of equipment and material resulting from the yen's decreasing value, and the serious technological scale-up problems severely handicapped the plan. When the war ended in 1945 the Japanese had constructed only 15 plants that in 1944 achieved a peak production of 717,000 barrels (114,000 kL).[2]

Unlike the German and British governments which did not take an active role in their industries' early work on synthetic fuel, the Japanese government, acting through the navy, participated from the beginning. The navy did not depend on private industry for petroleum research and development or for the production of refined products. It recognized the severity of Japan's problem and led the way in trying to make the empire self-sufficient in liquid fuel. But at the same time the navy's participation generated an unnecessary rivalry with non-military scientists over research decisions and promoted undue secrecy and security concerns, which contributed nothing to the synthetic fuel program.[3] By 1942 its synthetic fuel program, despite claims to the contrary, had failed. Later, in the war's last year and after the navy's program collapsed, the Japanese army tried to revive the production of synthetic liquid fuel. Japan's defeat ended the army's short-lived effort.

During the war years the Japanese government also increased petroleum imports and their storage. In 1941 Japan required 30 million barrels of petro-

2 "Synthetic Oil in Japan: An Attempt to Gain Self-Sufficiency," *Industrial Chemist*, 23 (1947): 333–40; "Coal Hydrogenation in Japan: A Seven Year Plan," *Iron and Coal Trade Review*, 135 (1937): 716; "Hydrogenation Industry," *J. Fuel Soc. Japan*, 16 (1937): 8; "Coal Oil in Japan," *J. Fuel Soc. Japan*, 18 (1939): a47; Shingo Ando, "Artificial Petroleum Industry in Japan and Manchukuo," *J. Soc. Chem. Ind. Japan*, 43, no. 9 (1940): 265B-66B; "Hydrogenation of Coal-Japan," *Chemistry and Industry*, 55 (1936): 131; "Japan Seeks NEI Oil For Co-Prosperity Sphere," *Oil and Gas Journal*, 39, no. 33 (1940): 82–83, 108, 110, 112–13; "Japan's Synthetic Output Virtually at Standstill," *Oil and Gas Journal*, 44, no. 51 (1946): 96; C. S. Goddin and D. P. Thornton, Jr., "Low-Temperature Carbonization of Coal Produced Most of Japs' Synthetic Oil," *Petroleum Processing*, 3 (February 1948): 121–31, on 122–23. The designation KK, kabushiki kaisha, means a joint stock limited company. For units conversions use below.

1 kiloliter (kL) = 1.03 metric tons
1 kL = 6.11 barrels (bbl)
1 kL = 264 gal
1 metric ton = 256 gal

3 Barbara Molony, *Technology and Investment: The Prewar Japanese Chemical Industry* (Council on East Asian Studies, Harvard University, 1990), pp. 226–33.

leum for wartime military and civilian needs and had the capacity to refine 35 million bbl/yr. It produced only 2.5 million bbl/yr of crude petroleum, but at the time of Pearl Harbor (7 December 1941) the Japanese had stored 60 million barrels of crude and refined products. They had acquired these stockpiles mainly from the United States prior to the Japanese invasion of China and United States' imposing moral embargoes on petroleum.[4] The Japanese intended to use these stockpiles until synthetic fuel production and crude imports from Borneo and Sumatra in the Dutch East Indies increased enough to meet their liquid fuel requirements. By mid-1943 Borneo's and Sumatra's petroleum production equaled expectations, but the Japanese seriously miscalculated the effect of the United States' submarine blockade and the submarines' increasing success in sinking Japanese tankers. The losses reached such heights that in 1944–45 Japanese petroleum supplies declined rapidly and neared exhaustion. The seriousness of Japan's petroleum shortage greatly reduced the time-schedule for synthetic fuel research and development. It resulted in the army's final desperate attempt to develop a synthetic fuel industry.[5]

Japan's unsuccessful prewar and World War II synthetic fuel program has provided a case study in technological failure. It failed, despite highly promising laboratory research and the government's considerable economic incentives, because Japan's impatient engineers tried to make the transition from laboratory to industrial development without pausing to acquire sufficient pilot plant experience. Japanese engineers needed pilot plant experience to test plant engineering design, apparatus, chemical reactions, and economics of the synthetic fuel processes on a larger-than-laboratory scale before committing to the construction of an industrial-size plant. The inherent complexity of the synthetic fuel processes made the transition more difficult but not unattainable.

The United States' prewar embargoes on petroleum and machinery coupled with the wartime exigencies further compounded the urgency, but they did not create it. Such a pattern of impatient behavior in Japanese engineering appeared much earlier, immediately after the Meiji restoration of 1868.

---

4 Michael A. Barnhart, *Japan Prepares for Total War* (Ithaca, New York: Cornell University Press, 1987), pp. 130, 156, 166, 180, 190, 196, 255–56. Embargoed materials included aircraft, machine tools, scrap iron, steel, aluminum, nickel, copper, zinc, vanadium, manganese, and tungsten. The embargo on lead tetraethyl, crude oil, and aviation gasoline (87-octane and higher) began in late July 1940.

5 D. P. Thornton, Jr., "Desperate Japs Tried to Make 'Avgas' Even from Pine Tree Roots, Needles," *Petroleum Processing*, 2 (November 1947): 815–20.

The Japanese were anxious to industrialize, and to acquire Western know-how quickly they brought large numbers of Western scientists and engineers to staff their newly established universities, technical institutes, and industries. Ten years later, in the 1880s, their scientists and engineers traveled to the West to acquire the know-how.[6] For the most part the Japanese strategy succeeded, although in their haste to establish a soda ammonia industry in the 1920s, Japanese engineers also tried to eliminate the pilot plant stage and suffered equally disastrous results. A precedent for technological impatience and haste therefore existed, and it helps explain the synthetic fuel program's failure in the 1930s–40s.[7]

3 Development of Japan's Oil Shale, Derived and Synthetic Liquid Fuel Sources to 1930

Japanese research on producing liquid fuel from coal and shale began in the 1920s first with oil shale distillation and LTC and then with coal hydrogenation

6 Eikoh Shimao, "Some Aspects of Japanese Science, 1868–1945," *Annals of Science*, 46 (1989): 69–91; see also Ian Inkster, "Science, Technology, and Economic Development – Japanese Historical Experience in Context," *Annals of Science*, 48 (1991): 545–63.

7 Yuji Jido, "The Establishment of Ammonia Soda in Japan and its Technology," *Ritsumeikan Business Management*, Ritsumeikan Business Management Society, 29, no. 3 (September 1990): 1–82 (in Japanese).
Haber's 1908 high-pressure synthetic ammonia process, the hydrogenation of nitrogen gas with hydrogen gas, $N_2 + 3 H_2 \rightarrow 2 NH_3$, advanced through the well-established three stages of development, from laboratory to pilot plant to industrial-size plant at Oppau in September 1913, with fewer problems than the Bergius and F-T coal conversion processes. Major reasons for the successful ammonia synthesis was the near purity of the reactant gases, N_2 and H_2, which greatly reduced the possibility of side reactions and increased the yield of the main product, ammonia; its economics, catalysts, and high-pressure apparatus. The hydrogenation of coal (Bergius or F-T) was a much more difficult hydrogenation because of the varying percentage composition of carbon and hydrogen in the different coal types, or ranks, (lignite, subbituminous, bituminous, anthracite coals); the complex and not fully known structures of the different coal types, or ranks; and the presence of other elements (oxygen, sulfur, nitrogen, mercury, and others) in coals. All of these factors influenced the effectiveness of reaction conditions, catalysts, temperature, and pressure, and required extensive testing in a plant of sufficient size (pilot plant) to give significantly predictable reaction results that Bergius, Pier, and other synthetic fuel scientists could reproduce in a larger, costlier, industrial-size plant. Coal hydrogenation's complexity and high industrial plant construction and operating costs made pilot plant testing a necessity. In the ammonia synthesis, the absence of such complicating factors permitted reduced pilot plant testing and even a possible bypass of the pilot plant stage without disastrous consequences in a crisis, such as the Japanese World War II fuel crisis.

and F-T synthesis. With their victory in the 1904–05 Russo-Japanese war, the Japanese extended their influence in Manchuria, and by 1920 they had recognized the possibility of distilling oil from the huge shale deposits at Fushun. The semigovernment South Manchuria Railway Company (SMR), or Minami Manshû Tetsudô KK, in which by law only the Japanese and Chinese governments and their citizens held shares, directed the research aimed at providing Japan's Imperial Navy with a source of fuel oil.[8] Of the major powers, only the British and United States navies had begun their conversion from coal to oil-fired turbine ships, Britain from 1906 to 1914, the United States from 1909 to the 1920s.[9] The saving in fuel storage space enabled a ship to increase its firepower by introducing bigger guns and to reduce its total weight without sacrificing armor by eliminating the dead weight of coal. Refueling at sea, greater speed and range (40–50%) provided equally important advantages of the conversion to oil-fired ships.[10] Japan's conversion began in 1909 with the navy's liquid fuel investigations at the Naval Research Laboratory in Tokuyama (Honshû). Three years later the navy submitted to the government a program to start construction in 1913 on eight dreadnoughts and eight armored cruisers.[11] Initially, the navy's turbine ships consumed both coal and oil in their furnaces although its faster ships increasingly burned oil. The navy completed its first two exclusively oil-fired cruisers, the Tatsuta and Tenryû, in 1919.[12]

Fushun's oil shale deposit, 150 meters thick, lay right above a coal seam. It contained about 5 billion metric tons shale with 2.3 billion metric tons at a depth of 300 meters. Because the miners used an open-cut method to remove the coal, they first had to strip the shale, a strong-smelling, dark brown semi-solid at normal temperatures. This easy accessibility rather than quality led SMR to investigate Fushun's shale as a potential liquid fuel source.

The process consisted of crushing the shale to 150 mm particle size and separating the oil particles from the smaller dust particles with a Cottrell electrostatic precipitator. SMR used the dust in cement manufacture and distilled the oil particles in a retort at 550° C. Distillation produced 75% spent shale

8 "Synthetic Oil in Japan" (note 2), pp. 333–40. Oil shale is the rock containing the oil. Shale oil is the oil distilled from the rock.

9 E. B. Potter, ed., *Sea Power: A Naval History* (Englewood Cliffs, New Jersey: Prentice Hall, 1960), p. 392.

10 Joseph Borkin and Charles A. Welsh, *Germany's Master Plan: The Story of Industrial Offensive* (New York: Duell, Sloan and Pearce, 1943), pp. 186–87.

11 William L. Langer, *An Encyclopedia of World History* (5th ed.; Boston: Houghton Mifflin Company, 1972), p. 923.

12 Hector C. Bywater, *Sea-Power in the Pacific* (reprint of 1921 ed.; New York: Arno Press and the New York Times, 1970), pp. 211–12, 215.

which the miners used as packing material; 21% gases which provided fuel for the boilers, heaters, and stills; and only 6% crude oil which on further distillation gave a mixture of 1% gasoline, 60% heavy oil, 23% crude paraffin, and 6% coke. SMR conducted additional distillation research in Scotland, where it tested a Pumpherston retort, and with the Japanese navy in 1923.[13] By 1927 SMR had improved the shale process economically by designing and constructing 40 larger, 50 metric ton/d, internally heated retorts that distilled more shale and at the same time permitted large-scale recovery of by-product ammonium sulfate for fertilizer.

According to the terms of a contract negotiated with SMR in 1928, the Japanese government agreed to purchase all the oil produced at Fushun. So despite the low yield, SMR continued its shale treatment, doubling each retort's capacity to 100 metric ton/d in 1928, although actual throughput was 20 to 30% less. Constructed at an estimated cost of ¥ 8.5 million ($1.96 million), and in full operation at the end of 1929, the plant had an annual production capacity of 53,000 tons fuel oil, 20,000 tons ammonium sulfate, and 15,000 tons paraffin.[14]

Research on LTC began in 1921 as part of a general program at the Imperial Fuel Research Institute, a 50% government-owned institute in Kawaguchi (Saitama Prefecture, Honshû), a northern suburb of Tokyo. The government had established the institute that same year to investigate the utilization of Japanese coals, and at the time the Naval Technical Mission arrived in January 1946 the institute had expanded to 40 buildings, 121 personnel, and occupied 483,624 ft^2. Yoshisada Ban (b. 1891) directed the institute, T. Shimura was the assistant director and supervised the F-T group, M. Kurokawa supervised the coal hydrogenation group, and N. Sato the LTC group. The institute had an annual budget of ¥ 500.

In 1921 LTC's primary purpose was to produce semicoke as a substitute for charcoal which cost twice as much but enjoyed widespread use in household heating and cooking because it lacked semicoke's slightly objectionable odor.[15] Yoshikiyo Ôshima (1882–1957), at Tokyo Imperial University and the institute's

13 Koki Ishibashi, "Present Status of Properties and Refining of Fushun Shale Oil," *Proceedings of the Second World Petroleum Congress*, 2 Vols. (Paris: Journal des Economistes, 1937), Vol. 2, pp. 183–96, on pp. 183–84.

14 "Synthetic Oil in Japan" (note 2), p. 340; "Shale Industry Authorized," *J. Fuel Soc. Japan*, 7 (1928): 39; R. W. Rutherford, "Oil from Coal in Japan," *Coke and Smokeless Fuel Age*, 6 (April 1944): 68–70. Conversion of yen to dollars: ¥ 1 = $0.4641. The yen varied in the 1930s: 1929 ($0.4610); 1931 ($0.4885); 1933 ($0.2565); 1935 ($0.2871); 1936 ($0.2902); 1937 ($0.2879); 1938 ($0.2845); 1939 ($0.2596). See *Statistical Abstract of the United States* (Washington, DC: Government Printing Office).

15 "Synthetic Oil in Japan" (note 2), p. 339.

director, and the chemist Ban who became the institute's director in 1932, conducted the initial experiments in both the laboratory and "first stage plant." In their 1925 study, *Technical Paper No. 1*, the institute's first published paper, Ban reported that a typical Fushun coal, carbonized in a 6-metric ton/d cast steel retort for 8 hours, gave 66% semicoke and 12% tar. Although semicoke production was the objective, Ban found the tar, when free from pitch and acid, comparable to a heavy petroleum oil.[16]

Only two other Japanese centers of LTC research existed in the 1920s: the Ôsaka Gasu Kôgyô Kaisha KK plant at Seimi (near Ôsaka, Honshû) and the navy's Tokuyama Research Laboratory. Each had a different objective. Ôsaka Gasu used a retort that its president, the American-educated engineer Kôtarô Shimomura (1861–1937), designed in 1923 to produce a semicoke for blending (mixing) with domestic, high-volatile bituminous coal.[17] The company, which had been producing coke since 1898, operated a trial plant at Iwasaki (Honshû) in April 1923 before constructing a semicommercial plant of similar design at Seimi in September 1923. The Seimi plant underwent expansion and modernization, giving it an annual rated input capacity of 3,000 metric tons, but from 1923 to 1926 it operated sporadically and carbonized only 700 metric tons of coal to 500 metric tons of semicoke. Mechanical operating problems and financial difficulties led to its dismantling in 1930 and to the end of Ôsaka Gasu's LTC program.[18]

---

16 Yoshisada Ban, "The Low-Temperature Carbonization Plant at the Imperial Fuel Research Institute," *Proceedings of the Second International Conference on Bituminous Coal*, 2 Vols. (Pittsburgh: Carnegie Institute of Technology, 1928), Vol. 2, pp. 303–11. To test the quality of tar, Ban used a single cylinder, four cycle, vertical type Niigata diesel engine with cylinder dimensions 280 mm by 416 mm and a rating of 33 bhp at 230 rpm. Neither a pitch-free oil nor one high in tar acid performed satisfactorily, although a partial removal of tar acid provided good quality diesel oil.

17 Kôtarô Shimomura, Japanese Patent 50,532 of 21 August 1923. Shimomura studied organic chemistry at Worcester Polytechnic Institute where he received the Bachelor of Science Degree in 1888 and did research under Ira Remsen at Johns Hopkins for one year (1888–89). Shimomura became the president of Doshisha University in Kyoto (1904–07). He received the doctor of engineering degree from the Japanese government in 1915. At that time no Japanese University had the authority to award this degree. See "Personal History of Kôtarô Shimomura," Doshisha University Archives, 3 pages. See also Genji Jimbo, Noriaki Wakao, and Masahiro Yorizane, "The History of Chemical Engineering in Japan," in *History of Chemical Engineering*, ed. William F. Furter (Washington, DC: American Chemical Society, 1980), pp. 273–81.

18 William T. Reid, "Low-Temperature Carbonization of Coal in Japan," *US Bureau of Mines Information Circular 7430* (February 1948), pp. 1–82, on pp. 17, 20, 53–6. Reid was a member of the US Bureau of Mines (pp. 32–3). His report is the most comprehensive and accu-

The navy's program ran from 1925 to 1932 and had as its objective the production of liquid fuel for the fleet. Its Tokuyama laboratory tested coals for their tar content in a horizontal-rotational Thyssen retort before abandoning LTC research because of insufficient tar production and difficulty in marketing the by-product semicoke. LTC's rapid expansion did not occur until 1936 when Japan's militaristic government established the petroleum substitutes policy as a war measure to encourage liquid fuel production. It became Japan's most productive and economical method of obtaining oil from coal in the 1930s–40s.[19]

Four years after the start of LTC in Japan, in 1925, studies on the high-pressure hydrogenation of coal and tar had begun at the Imperial Fuel Research Institute. In 1925 the Japanese navy also began a coal hydrogenation program at Tokuyama, and in 1928 SMR initiated a third hydrogenation program at its Central Research Laboratory in Dairen, near Port Arthur, in south Manchuria.[20] Indeed, in the 1920s–30s Japanese scientists and engineers regarded coal hydrogenation as the best way to eliminate their almost total dependence on imported oil.[21] Shale distillation and LTC gave only a 6–10% yield or about 13 gal/metric ton. In addition, no market existed for the semicoke, which accounted for 65% by weight of the coal carbonized, whereas coal hydrogenation offered the possibility of completely converting coal to oil. With the exception of Germany, coal hydrogenation probably received more

rate English language source of information on Japan's LTC industry. Some of the same information appears in the reports of the US Naval Technical Mission to Japan (note 19).

19 Goddin and Thornton (note 2), p. 122; US Naval Technical Mission to Japan (Nav. Tech. Miss. Japan), Microfilm Reel JM-200L, Report X-38 (N)-7, Japanese Fuels and Lubricants, Article 7: "Progress in the Synthesis of Liquid Fuels from Coal," p. 21, German Document Retrieval Project, Synthetic Fuels Collection, Texas A&M University Archives. The reports contained in this microfilm are the best source on Japan's synthetic fuel industry. The US Naval Technical Mission was in Japan from 4 September 1945 to 1 November 1946. It published a few of its 185 reports in technical journals (1946–48). The Operational Archives Branch, US Naval History Division, Washington, DC, published a complete microfilm edition of all the reports in 1974. Information on Japan's synthetic fuel industry also is available in *Materials about Five Year Plan*, Part 2, Volume 3 (Liquid Fuel), edited by the Commerce and Industry Section of SMR, microfilm YD/324, section MOJ 275, National Diet Library, Tokyo; and in GHQ/SCAP, Record Group 331 (Records of General Headquarters/Supreme Commander for the Allied Forces), National Archives and Records Administration, Washington, DC. According to members of the US Naval Technical Mission, very little information on the Japanese navy's LTC program existed.

20 Nav. Tech. Miss. Japan (note 19), p. 9.

21 "Synthetic Oil in Japan" (note 2), p. 334.

TABLE 5.1 Japan's petroleum production (metric tons) during the 1920s

Year	Mainland	Formosa	Total
1923	1,576,353	13,5701	1,589,923
1924	1,579,212	19,882	1,599,034
1925	1,637,453	23,032	1,660,485
1926	1,496,477	80,333	1,576,810
1927	1,421,602	126,553	1,548,155

SOURCE: COURTESY OF AUTHOR

serious study in Japan than anywhere else.[22] Japan's total petroleum production had fallen in the 1920s despite the tenfold development of Formosa's oil fields, and coal hydrogenation appeared to offer the most promising and reliable source of liquid fuel.[23]

The Imperial Fuel Research Institute conducted its early hydrogenation research at the laboratory batch scale and used a small autoclave of 600 cm^3 capacity. The autoclave of nickel-chromium steel, externally heated by an electric heater fitted around it, withstood pressures of 250 atm. Shingo Ando, Saburô Tashirô, and Ôshima conducted much of the research, studying the effect of catalysts, temperature, and pressure on low-temperature tar obtained from Fushun sub-bituminous coal. Of the several metallic oxide catalysts tested, nickel (II) oxide (NiO) and iron (III) oxide (Fe_2O_3) proved most effective. Hydrogenation occurred below the experimentally determined reaction temperature of 400–450° C, but above 400–500° C the coal decomposed to give coke. No hydrogenation occurred below 75 atm, an increased pressure reduced costs. They cautioned, however, that establishing conditions of maximum yield at the laboratory batch scale had little value without considering industrial-size economics.[24]

Both the navy and SMR had taken a strong interest in hydrogenating coal, tar, and oil shale because units of the fleet had begun their conversion from

22 Nav. Tech. Miss. Japan (note 19), p. 11; see also note 1. Neither did the Japanese seriously consider industrializing the methanol synthesis from water gas (a carbon monoxide-hydrogen gaseous mixture produced from the reaction of steam with coal, coke, or methane) similar to synthesis gas, because of little peacetime demand for methanol.

23 "Production of Petroleum in Japan," *J. Fuel Soc. Japan*, 7 (1928): 73.

24 Saburô Tashirô, "Hydrogenation of Neutral Oil of Low-Temperature Tar," *J. Fuel Soc. Japan*, 67–70; Yoshikiyo Ôshima and Saburô Tashirô, "Hydrogenation of Japanese Coals," *J. Fuel Soc. Japan*, 70–73. See Ando's numerous papers in *J. Soc. Chem. Ind. Japan*, 1931–40.

coal to liquid bunker fuel. From 1921 to 1925 Toru Ogawa and his coworkers at the navy's Tokuyama Research Laboratory conducted tests on the theory and application of high-pressure hydrogenation. Experiments treating bituminous and sub-bituminous coals from Fushun, Sakito, and Takashima in small, batch scale, autoclaves, such as a 2,000 cm^3 rotating autoclave and a 600 cm^3 shaking autoclave, and with an iron (III) chloride ($FeCl_3$) or sodium formate (HCOONa) catalyst, showed that hydrogenation varied with coal type and required conditions above 400° C and 200 atm.[25]

The navy had other coals tested in Germany. In 1926–27 it sent samples of Shimabara coal to Bergius at his Rheinau-Mannheim plant. Bergius conducted nine experiments in a 1 metric ton/d continuous unit, hydrogenating the coal at 455° C, 200 atm, with an iron (III) oxide catalyst. Compared to Fushun and German coals, Bergius found Shimabara coal difficult to liquefy, giving only a 30–35% yield of an oil rich in asphaltic components, and resembled crude petroleum.[26]

Despite the poor results obtained with Shimabara coal the navy decided to expand its hydrogenation program, and in 1927 it constructed a 3-metric ton/d continuous unit. It tested mainly sub-bituminous Oyama coal from Fushun and obtained yields as high as 75.4% at 400° C and 200 atm. Three years later, in 1930, the navy began construction on Japan's first semicommercial coal hydrogenation plant, a pilot plant of 7 metric ton/d coal paste capacity that became operational in 1932.[27] Laboratory investigations continued during this period (1928–32), with Ogawa's research team in Tokuyama, Dr. Shigeru Komatsu (1883–1947) at Kyoto Imperial University, and Ryounosuke Abe (b. 1898) at SMR's Central Research Laboratory in Dairen conducting about 2,000 experiments. Specific investigations dealt with catalysts and their x-ray analyses; properties of coal pastes; reaction mechanisms, temperatures, times, and pressures.[28] Hydrogenation tests included Fushun oil shale, Japanese and

25 Toshia Yokota, 'Coal Liquefaction Research in the Imperial Japanese Navy," *J. Fuel Soc. Japan*, 16 (1937: 469–80; 51–5 (in English). The process required (1) crushing the coal to a powder and (2) mixing it with sufficient heavy oil to form a paste before reacting with hydrogen gas.

26 Yokota (note 25), pp. 51–55; Nav. Tech. Miss. Japan (note 19), p. 9. According to members of the Technical Mission, Ogawa visited IG Farben's Leuna plant in 1926. BASF's (IG Farben's) archives in Ludwigshafen contain no record of his visit.

27 Tôru Ogawa, "Hydrogenation of Fushun Coal," *J. Fuel Soc. Japan*, 11 (1932): 118–9; Shigeru Komatsu, "Liquefaction of Coal," *J. Fuel Soc. Japan*, 7 (1928): 1240–50. See also Ogawa's and Abe's numerous papers in *J. Soc. Chem. Ind. Japan*, 1931–40.

28 Yokota (note 25), pp. 51–55; Nav. Tech. Miss. Japan (note 19), pp. 9, 11. A typical sub-bituminous coal such as Oyama contained 42% volatile matter and 10–12% ash.

Asian mainland coals, especially the catalytic effect of the iron oxides, FeO, Fe_2O_3, and Fe_3O_4, on Fushun coal. Ogawa and Komatsu also showed that of 50 possible catalysts, zinc chloride ($ZnCl_2$) proved most effective. Other experiments with Fushun oil shale, gave a highly saturated gasoline, kerosene, light and heavy oils.[29]

By 1927, however, after Matthias Pier found that metallic sulfide catalysts resisted sulfur, corrosion, and cost less, SMR's iron oxide – sulfur-tin hydroxide catalyst [Fe_2O_3-S-$Sn(OH)_2$] became one of the standard Japanese hydrogenating catalysts. Laboratory reactions at 420–480° C and 300 atm for 1 hour gave complete liquefaction and prevented excessive coke and gas formation. The tests included 50 coals from northern China, forty from Manchuria, and a few from Japan, Korea, and Formosa.[30]

In 1926 Franz Fischer and Hans Tropsch at the Kaiser Wilhelm Institute for Coal Research in Mülheim, Germany, published their first paper on the synthesis gas process (F-T synthesis) using newly developed cobalt-thorium catalysts.[31] The following year Dr. Gen-itsu Kita (1883–1952) at the Institute of Physical and Chemical Research, Kyoto Imperial University, began the first Japanese study of the F-T synthesis. Because Japan had no domestic source of expensive cobalt or thorium metal, Kita initially focused on preparing cheaper iron and nickel catalysts. Working in the laboratory and with a small pilot plant of 100 m^3/hr synthesis gas throughput, Kita in 1937 developed an iron catalyst whose yield approached those obtained with cobalt-thorium catalysts at atmospheric pressure. But, like all iron catalysts, it had high sensitivity to any sulfur present in coal. Outside of Kita's research, and that of his colleague and successor Dr. S. Kodama, no other Japanese study of the F-T synthesis occurred until practically the mid-1930s.[32]

During the 1920s–30s the Japanese made efforts to keep informed of international developments in fuel research. The list of Japanese delegates to

29 "Hydrogenation," *J. Fuel Soc. Japan*, 12 (1933): 6.

30 Ogawa (note 28), pp. 118–19.

31 Franz Fischer and Hans Tropsch, "Die Erdölsynthese bei gewöhnlichen Druck aus den Vergasungsproducten der Kohlen," *Brennstoff-Chemie*, 7 (1926): 97–104; Franz Fischer, "Liquid Fuels from Water Gas," *Industrial and Engineering Chemistry*, 17 (1925): 574–76. Synthesis gas is a variable mixture of carbon monoxide (CO) and hydrogen (H_2) usually prepared by reacting steam with coal. Treating the gas with catalysts converts it to liquid hydrocarbon fuels such as gasoline and diesel oil.

32 Nav. Tech. Miss. Japan (note 19), pp. 10, 16. See also Gen-itsu Kita, *Über die Benzinesynthese aus Kohlenoxyd und Wasserstoff unter gewöhnlichen Druck* (Tôkyô: Institute of Physical and Chemical Research, 1936). Prior to the outbreak of World War II the Japanese imported cobalt from Katanga, Belgian Congo (Zaire), and thorium from the United States.

the London World Power Conference of September-October 1928 contains twenty-eight names and includes representatives from the military, government, industry, and universities. The conference proceedings do not indicate whether all actually came to London, but of those in attendance nine presented papers. Ôshima presented his paper on LTC, the others dealt with the Japanese coal, petroleum, and gas industries. About 1,500 delegates representing 48 countries heard 170 papers at the thirteen-day conference.[33] At least one Japanese delegate attended each of the first three International Conferences on Bituminous Coal held at the Carnegie Institute in Pittsburgh in November 1926, 1928, and 1931. In 1926 Ôshima and in 1928 Ban spoke on LTC research. At the 1931 meeting Ôshima and Yashitami Fukuda, his colleague from the Imperial Fuel Research Institute, discussed the combustibility of carbonaceous compounds.[34]

Japanese delegates attended the first and second World Petroleum Congresses held in the 1930s. The navy sent two officer-engineers to the London meeting of July 1933. Bergius who two years earlier in 1931 received the Nobel Prize in chemistry for his invention of high-pressure coal hydrogenation gave a short address at the Congress's banquet in addition to a paper on recent developments in coal hydrogenation. Pier, and Gordon of Imperial Chemical Industries (ICI), both synthetic fuel experts, also gave papers. Four Japanese delegates attended the 1937 meeting in Paris. The navy's two delegates described their research on the hydrogenation of cracked gasoline, Shingo Ando of the Imperial Fuel Research Institute discussed hydrogenating low-temperature tar, and Kôki Ishibashi of SMR spoke on shale oil refining at Fushun.[35]

33 *The Transactions of the Fuel Conference: World Power Conference*, 2 Vols. (London: Percy Lund, Humphries & Co., 1928). See Vol. 1, pp. cvii-cvix for the list of Japanese delegates, and pp. xlii-cxiv for the complete list of delegates and other conference representatives. The table of contents in each volume lists the papers presented by Japanese delegates. Ôshima's paper, "Coal Carbonisation in Japan," is in Vol. 2, pp. 426–36.

34 Ban (note 16), pp. 301–11; Y. Kosaka and Y. Ôshima, "The Formation of Naphthalene During High-Temperature Carbonization," *Proceedings of the International Conference on Bituminous Coal* (Pittsburgh: Carnegie Institute of Technology, 1926), pp. 463–71; Yoshikiyo Ôshima and Yashitami Fukuda, "The Effect of Ash on Reactivity and Combustibility of Carbon Materials," *Proceedings of the Third International Conference on Bituminous Coal*, 2 Vols. (Pittsburgh: Carnegie Institute of Technology, 1931), Vol. 2, pp. 448–84. At the 1931 meeting Bernard Lewis of the US Mines Branch read Ôshima and Fukuda's paper. Because there is no list of delegates, there is no way of knowing whether they actually attended.

35 A third Japanese delegate, Dr. J. Somiya attended the 1933 congress, but the delegate list gave him no affiliation; *Proceedings of the World Petroleum Congress*, 2 Vols. (London: Offices of the Congress, 1933), Vol. 1. Japanese delegates published the following papers at

4 Government Support of Japan's Synthetic Fuel Industry

The development of Japan's synthetic fuel industry ran parallel to the military's increasing and eventual control of government in the 1930s–40s. Prior to the 1930s the government had made little progress in establishing a synthetic fuel industry other than a 1923 discussion of a liquid fuel policy designed to make the country independent of foreign petroleum. But in 1933, after annexing Manchuria and its huge coal and oil shale deposits, Japan's militaristic government decided to aid the LTC industry and to encourage greater oil production from SMR's Fushun shale mine. Beginning in 1934 it offered subsidies (paying none that year) for the construction of additional LTC plants. Japan proper had produced a mere 330,000 metric tons of natural crude oil in 1936, the last reliable prewar statistic, another 175,000 metric tons came from Sakhalin, the 505,000 metric tons representing the empire's total production.

Two sources of derived liquid fuel contributed smaller amounts: 125,000 metric tons from the Fushun shale plant and 4,000 metric tons total from the four operational LTC plants. For comparison, in 1936 Japan imported 2.14 million metric tons of crude oil valued at $37.7 million and 2.5 million metric tons of petroleum products.[36] The wide gap between domestic sources and imports showed clearly the great need for an energy policy designed to give self-sufficiency in liquid fuel. In view of this alarming situation, especially with the military about to launch its Asian expansion program, the Japanese government in 1937 prepared its Seven Year Plan for synthetic fuel production.

The first sign of military control had appeared much earlier in 1928–32 and resulted directly from dissatisfaction with foreign policy established at the Washington Conference in autumn 1921 and from the ensuing Depression. In accepting the conference's limitation on American, British, and Japanese battleships and aircraft carriers, Kijurô Shidehara (1872–1951), Japanese ambassador to Washington, endorsed the American objective of a peaceful, politically harmonious, and economically interdependent capitalist world order. Because the United States was Japan's biggest supplier of capital and her best trading partner, accounting for 40% of Japan's exports in the 1920s, cooperation with the Americans made sense. Shidehara argued that Japan adopt

---

the 1937 congress: Shingo Ando, "Catalytic Hydrogenation of Low-Temperature Tar Under High Pressure," *Proceedings of the Second World Petroleum Congress*, 2 Vols. (Paris: Journal des Economistes, 1937), Vol. 2, pp. 237–48; Kosuke Kûdô and Haruki Fujimoto (an engineering commander in the Japanese Navy), "On the Hydrogenation of Cracked Gasoline," *Proceedings Second World Petroleum Congress*, 271–75; Ishibashi (note 13).

36 Morio Otaki, "Japanese National Petroleum Policy," *World Petroleum*, 8 (September 1937): 64; "Synthetic Oil in Japan" (note 2), p. 333.

an internationalist policy and expand its overseas trade peacefully; expansion and promotion of Japan's economic base and interests in China and Manchuria should occur without resorting to aggression.

Opposition to Shidehara's internationalist policy emerged in the government and especially within the leadership of the Kwantung army, those military divisions organized in 1917 and stationed in Manchuria to protect Japanese economic interests. The Japanese had been in Manchuria since 1905, the result of the Treaty of Portsmouth that concluded their successful 1904–05 war against Russia. Russia had obtained a twenty-five-year lease on Manchuria's Kwantung province from China in 1898, but the 1905 treaty transferred control to Japan where it remained until the end of World War II. By 1928 Shidehara's opponents feared that any further advance of Chiang Kai-Shek's Koumintang (Nationalist) army and expansion of his authority in North China would lead to loss of Japanese interests in Manchuria. While Manchuria's seizure never occurred at this time, the government's failure to punish the Kwantung extremists demonstrated its ineffectiveness and invited possible later insubordination.

Global economic collapse added to the government's problems. Japanese exports dropped by 50% from 1929 to 1931 causing a sharp rise in urban unemployment and a drastic fall in commodity prices that reduced real income to about two-thirds its 1925 level. The ongoing Manchurian situation provided further erosion of confidence in the government's ability to control events and reached a climax on the evening of 18–19 September 1931. A small explosion on SMR's rail line just north of Mukden (Shenyang) triggered Japanese aggression and became the impatient Kwantung army's excuse to mobilize its troops, conquer Manchuria's three eastern provinces, and on 18 February 1932, declare the provinces' independence as the puppet state of Manchukuo. In Japan proper, ultranationalism or jingoism took hold. The following year, on 24 February 1933, when the League of Nations at its General Assembly in Geneva unanimously condemned Japan's aggression in Manchuria, the Japanese delegation walked out of the League headquarters.

The Manchurian Incident represented a turning point in Japan's foreign policy. Denounced internationally for its actions, Japan abandoned the policy of cooperation with the major powers, which it had followed since the 1868 Meiji Restoration, and pursued a course designed to further its own interests. During the years 1932–36 cabinets headed by admirals twice ruled the country, Manchuria (Manchukuo) became an integral part of Japanese industrial might, and the economy made a quick and successful recovery from the Depression. Because of its new international outlook, Japan underwent a significant military and industrial build-up that amounted to a two-year, 26% rise in government expenditures. The government also abandoned the gold standard in December 1932, a move that led to a large increase of exports. As a result of such stimulation, economic growth in the 1930s doubled that of the 1920s. Manchuria's chemical industry alone expanded 400% from mid-1937 to 1939, Japanese investments there in 1938 totaled ¥ 47 million although they fell the next year because of overproduction and government restrictions on investments and profits.

While the economy grew steadily in the 1930s so did ultranationalism and violence. The government increasingly found itself under the military's influence and more willing to resort to military solutions. In the summer of 1937 the military got its war but not with the Soviet Union in Manchuria as the Japanese general staff had believed since 1931. Manchuria's industrialization, including synthetic fuel and shale distillation development, was part of Japan's overall plan to strengthen its military presence there. The strategy risked another Russo-Japanese conflict. On 7 July 1937, when a small clash occurred between Chinese and Japanese troops stationed at the Marco Polo Bridge outside of Peking, the Japanese, therefore, tried to contain it but could not. Instead, Japan stumbled into a full-scale war with China that became a prelude to World War II in the Pacific.[37]

One month later the Japanese government drafted its Seven Year Plan and passed the two laws aimed at its implementation, the Synthetic Oil Production Industry Law and the Imperial Fuel Development Company Law. It expected the plan, which called for the construction of 87 liquid fuel plants in Japan and Manchuria and the production by 1944 of one million kL/yr of gasoline and one million kL/yr of fuel oil, to meet 50% of civilian demands. Of the 87 proposed plants: each of the 10 high-pressure coal hydrogenation plants would produce 100,000 kL/yr or 1 million kL/yr total; 11 F-T plants would produce a total of 500,000 kL/yr; and each of the 66 LTC plants would treat 100,000 metric tons of coal yearly to give the remaining 500,000 kL. The Manchurian plants supplied 25% of the production. The 87 plants' total cost of ¥ 750 million included ¥ 360 million for hydrogenation, ¥ 121 million for F-T, ¥ 115.5 million for LTC, ¥ 50 million for the amount already spent on synthetic fuel plants, and ¥ 15/metric ton of output for coal field development. The government expected to pay for the plan by increasing the tax on gasoline.

37 Kenneth B. Pyle, *The Making of Modern Japan* (Lexington, MA: D. C. Heath and Co., 1978), pp. 137–50; "Chemical Industry in The Far East," *The Chemical Trade Journal and Chemical Engineer*, 104 (31 March 1939): 316; G. C. Allen, *A Short Economic History of Modern Japan 1867–1937* (London: George Allen & Unwin Ltd., 1946), pp. 98–99, 146–49.

TABLE 5.2 Seven year plan

Plant	Gasoline and fuel oil production	Cost
10 high-pressure hydrogenation	100,000 kL/yr per plant	¥ 360 million total for 10 plants
11 F-T	500,000 kL/yr total	¥ 121 million total for 11 plants
66 LTC	100,000 metric tons coal/yr treated to give 500,000 kL/yr	¥ 15.5 million for 66 plants ¥ 50 million already spent on synthetic fuel plants ¥ 15/ton for coal field development
87 total	2 million kL/yr gasoline and fuel oil	¥ 750 million

SOURCE: COURTESY OF AUTHOR

At first the Synthetic Oil Production Industry Law provided subsidies and payments for losses only to the larger liquid fuel plants in Japan, Korea, Sakhalin, and Formosa. The F-T and hydrogenation plants had a minimum oil production of 10,000 kL/yr. LTC plants using coal or lignite had to carbonize 100,000 metric ton/yr. By 1941 with the military's expansionist program well underway, the government offered support to every liquid fuel plant that derived crude oil, fuel oil, or lubricants from any raw material except natural petroleum and to liquid fuel plants whose annual production exceeded 1,000 kL of aviation gasoline blending stock. In addition to subsidies, the law exempted participating companies from income and local taxes, import duty on equipment, and allowed them to expropriate land for plant construction.

The Imperial Fuel Development Company Law, the other half of the Seven Year Plan, underwent revisions in 1940, 1941, and 1942, but its major articles summarized here remained the same. (1) The Imperial Fuel Development Company shall transact all business that promotes synthetic oil production. (2) It shall have a capitalization of ¥ 100 million, one half supplied by the government, which it can increase with government approval and did increase to ¥ 250 million by the end of the war. (12) It shall finance synthetic oil enterprises, and they may manufacture or sell synthetic oil or manage other businesses necessary for promoting synthetic oil production, when permitted by the government. (13) It can issue debentures up to an amount equal

TABLE 5.3 Government subsidies for construction of LTC plants

Year	Estimated subsidy ¥	Amount paid ¥
1934	449,500	0
1935	505,300	165,556
1936	552,000	287,092
1937	517,300	455,102

SOURCE: COURTESY OF AUTHOR

to five times its paid-up capital. (30) It is not obliged to pay dividends on government-held shares unless the amount of annual profit reaches 6% of the value of paid-up shares not held by the government. (32) It shall pay no corporation and business taxation in its first year of existence and in the following ten years.

The manner of subsidy payment also changed during the Depression and war years. From 1934, when the government offered subsidies only to encourage construction of LTC plants, to 1937, before the Synthetic Oil Production and Imperial Fuel Development Company Laws went into effect, the government paid a small subsidy of ¥ 10/metric ton (9¢/gal) on tar produced in Japan proper and ¥ 2/metric ton on coal carbonized in Sakhalin. Table 5.3 gives the total subsidies paid and the amount the government expected to pay.

Beginning in January 1938, when the government granted a subsidy on all synthetic fuel, the amount it paid for gasoline and other oil produced by hydrogenation, F-T, and LTC appears in Table 5.4.

In 1942 the government introduced its third and final policy. It decided not to subsidize production but instead offered a fixed price including suitable profit for the synthetic fuel. A government company, Sekiyu Kyôhan KK, bought synthetic fuel from the producers, pooled it with natural petroleum products obtained from petroleum refining companies, and then sold the pooled product at a fixed price. The government's purchase price depended on synthetic fuel's production cost and determined the pooled product's retail price as shown in Table 5.5.[38]

Table 5.6 lists the subsidized amounts Imperial Fuel Development Company received from 1937 to 1944. The total subsidy during this period exceeded

38 Nav. Tech. Miss. Japan (note 19), 321–6. See note 14 for conversion rates of yen to dollars.

TABLE 5.4 Government subsidies for all synthetic fuel produced

Year	Process	Subsidy (¥/metric ton)	
1938	H	0	0
	F-T	0	0
	LTC	43.88	17.20
1939	H	77.00	34.00
	F-T	66.00	47.00
	LTC	47.00	18.00
1940	H	72.00	24.00
	F-T	60.00	50.00
	LTC	39.00	13.00
1941	H	72.00	24.00
	F-T	60.00	50.00
	LTC	39.00	13.00

H ≡ Hydrogenation, F-T ≡ Fischer-Tropsch, LTC ≡ Low Temperature Carbonization
SOURCE: COURTESY OF AUTHOR

TABLE 5.5 Purchase and retail price of synthetic oil

Product	Process	Purchase price ¥/kL	Purchase price ¢/gal	Official retail price ¥/kL	Retail price ¢/gal
Aviation Gas.	H	600	52.2	444	38.7
Motor Gas.	H	400	34.8	307	26.7
Gas Oil A-1	H/LTC	230		152	
Gas Oil A-2	H/LTC	220		141	
Gas Oil B-1	F-T	340	29.6	177	15.4
Gas Oil B-2	F-T	330		167	
Gas Oil B-3	F-T	320		157	
Semi-Diesel Oil	LTC	180	15.7	88	7.7
Bunker Fuel Oil	LTC	100	8.7	84	7.3

SOURCE: COURTESY OF AUTHOR

TABLE 5.6 Subsidies to Imperial Fuel Development Company

Year	Subsidy amount ¥
1937	148,094
1938	878,387
1939	1,938,000
1940	2,625,057
1941	4,086,052
1942	4,050,650
1943	7,497,335
1944	8,519,288

SOURCE: COURTESY OF AUTHOR

¥ 29 million ($6.7 million). From 1940 to 1944 Japan produced 430,116 kL (113.6 million gal) of liquid fuel giving an average subsidy of ¥ 69/kL or 6¢/gal.[39]

The Japanese did not intend to rely solely on the development of a synthetic fuel industry to provide them with petroleum and had established several alternative or supplementary oil sources. Before passage of the Seven Year Plan and the outbreak of war with the United States in December 1941 they had imported large quantities of lubricants and fuels from the United States. The navy used California crude as its source of bunker fuel, diesel oil, kerosene, aviation and motor gasoline. Fushun oil shale also contributed an important component to the bunker fuel and diesel oil for the navy's submarines. After its aggression in Asia Japan no longer had American crude oil available, but by March 1942 it already had completed occupation of the entire Dutch East Indies, gaining access to the crude oil of Borneo and Sumatra. While production from the East Indies equaled expectations by mid-1943, the success of American submarines in sinking Japanese tankers had increased so much that by 1944 oil supplies fell rapidly and significantly shortened the time for research and development on synthetic fuel.

The navy still had access to Fushun oil shale, but the fuel situation became so desperate in the war's closing days that the navy resorted to such unlikely

39 Reid (note 16), 7–12. Reid used ¥ 1 = $0.23. This is approximately the average for the years 1940 ($0.2346) and 1941 ($0.2349). These were the only figures available during the war years. The new rate established in April 1949 was set at ¥ 360 = $1.00, in coordination with American occupation forces.

fuel sources as pine roots, soy beans, and other vegetables available in Japan. Most of Japan's crude oil refineries had long since shut down for lack of stock or because of American air raids that destroyed two-thirds of their capacity. Ships and aircraft remained idle, and pilots went into battle with little training because of fuel shortages. 100-octane gasoline remained available only in laboratory quantities. A battleship sunk at the Battle of Okinawa in May-June 1945 had edible soy bean oil in its bunkers.

In the early years of the war petroleum storage had reached 60 million barrels of crude and refined products, much of it in large reinforced concrete tanks buried in hills and camouflaged with heavy vegetation. At the Sasebo (Kyûshû) Naval Base, which had a total storage capacity of 4 million barrels, one large reservoir held a half-million barrels. Total storage capacity was 1 million barrels at Kure (Honshû) and 9 million barrels at Tokuyama. The navy initiated fuel research in 1909 at Tokuyama and erected its first refinery there in 1920. By wartime, refining capacity had increased to 9,500 bbl/d.

Construction on a 17,000 bbl/d refinery at Yokkaichi (Honshû) began in 1940. The two refineries carried out crude distillation, thermal- and hydrocracking, and accounted for 25% of Japan's total refining capacity. The following year, the navy established its Ôfuna (Honshû) research center. Prior to that time, the navy had conducted research at Tokuyama and other locations, but once the war began it centralized activities at Ôfuna. In 1941 the navy also named its different research centers, Fuel Depots, with Ôfuna the First Naval Fuel Depot. Its 100 acres, containing 37 steel and concrete laboratories, made Ôfuna one of the world's largest petroleum research centers. At the time of its establishment in 1941, Ôfuna had 1,000 workers and a staff of 4 chemists, chemical engineers, and technicians. When it ceased operation on 15 August 1945, Ôfuna's director Vice-Admiral Nobusuki Yamaguchi supervised 3,210 workers including 450 staff members among them Komatsu at Kyoto Imperial University who was a civilian adviser to the Department of Fuel Research; Japanese navy engineering commander and Ôfuna's foremost chemist Dr. Haruki Fujimoto; Dr. Keisaku Mitsui engineering commander Japanese navy and key researcher on coal hydrogenation; and A. Nakei engineering lieutenant-commander Japanese navy and key researcher on the F-T synthesis. Ôfuna which consisted of 70 modern buildings, experienced no bomb attacks during the war. When the four-member Naval Technical Mission visited the Ôfuna Fuel Depot on 4 September 1945 and interrogated its chief scientists they learned that Yamaguchi on 1 August 1945 had ordered the burning of all research files. The Naval Mission then had 100 Japanese personnel return to the Depot for three months to reassemble the information in English using laboratory notebooks, personal files, and other available records.[40]

5 Oil Shale and LTC Emerge as Major Sources of Liquid Fuel

In 1937 when the Japanese government drafted its Seven Year Plan, oil shale, distillation and LTC were the only industrial processes available for the production of liquid fuel other than the refining of crude petroleum. SMR's Fushun oil shale plant started operation in 1927 with 40 retorts, each of 50 metric ton/d capacity. SMR increased their capacity to 100 metric ton/d in 1928, and in 1935 it doubled the number of retorts to 80. The plant became the Japanese empire's largest producer of oil from coal or shale and the world's largest oil shale plant before production fell off during the last three war years because of equipment shortages, technical difficulties, and wartime stringencies. Prior to 1935 SMR's engineers distilled the crude oil shale, obtaining 70–75% fuel oil and small quantities of low-octane gasoline. But after constructing a continuous-operating, 110 metric ton/d Dubbs cracking plant in 1935 they improved gasoline yield and quality. Cracking at 480° C gave kerosene and a 50–55% gasoline yield with 62–69-octane number.[41]

Japan in the 1920s had four operating LTC plants with a total production of 4,000 metric ton/yr. The industry never developed in the 1920s because sales revenue from coal tar, one of LTC's two main products, could not support the cost of carbonization, and semicoke, the other product, had not found a suitable market. In the 1930s when the Japanese began their expansion into Asia the government recognized the strategic value of the tar and in 1934 introduced subsidies for LTC.[42] From 1934 to 1945 Japanese industrialists and Imperial Fuel Development Company constructed ten LTC plants in Japan proper,

---

40 "Synthetic Oil in Japan" (note 2), pp. 336, 340; Thornton (note 2), pp. 815–22.

41 Ishibashi (note 13), pp. 183–96. Nippon Oil had operated a Dubbs cracking plant at its Tsurumi refinery since 1924; Thornton (note 5), pp. 815–20; "Japan's Oil Industry Under Private Operation," *World Petroleum*, 20 (August 1949): 36–37.

42 Goddin and Thornton (note 2), p. 130; Nav. Tech. Miss. Japan (note 19), p. 21; Even the unwanted semicoke found use during the war years. Because of gasoline shortages the government prohibited civilian gasoline consumption. The Japanese therefore built portable semicoke-burning gas producers for their vehicles, as charcoal, their more expensive main fuel, also was in short supply. The vehicles performed satisfactorily despite a 22–30% ash content that made necessary frequent removal of clinker from the gas producers. A. C. Fieldner and P. M. Ambrose, "Annual Report of Research and Technologic Work on Coal," *US Bureau of Mines Information Circular 7446* (January 1948), pp. 1–113, on p. 89. In order to reduce gasoline consumption the Japanese government

TABLE 5.7 Fushun oil shale production

Year	Production
1936	123,320 metric tons
1938	145,000 metric tons est.
1939	133,000 metric tons est.
1940	150,000 metric tons est.
1941	138,000 kL
1942	163,000 kL
1943	160,000 kL
1944	83,000 kL
1945	17,000 kL

SOURCE: COURTESY OF AUTHOR

Sakhalin, and Korea. Five of them, at Ube (Honshû), Kawasaki (Honshû), Wakamatsu (Kyûshû), Agochi (Korea), and Nagoya (Honshû), produced tar for a never-developed hydrocracking program. Three others remained under construction when the war ended. The army and navy conducted little LTC research and contributed nothing to its industrial development.[43] Of the three coal conversion processes, LTC cost the least. At a time when structural steel sold for ¥ 15/ton, a LTC plant annually producing 10,00 metric tons of liquid fuel cost ¥ 2.2–¥ 5 million, one-half to one-seventh the construction cost of an industrial-size F-T or coal hydrogenation plant.[44]

German LTC retorts, especially the Lurgi Spülgas retort, performed best, and Japanese industrialists widely adopted them. They installed Lurgi retorts in some of the largest plants, at Naihoro (Sakhalin), Agochi, Wakamatsu, and the Imperial Fuel Development Company plant in Ube. The Ube Kôsan KK (Ube Mining Co. Ltd.) plant at Ube, contained German-designed Koppers retorts. The Japanese also used the American-designed Knowles oven in the Nihon Yuka Kôgyô KK (Nihon Liquefaction Industry Co. Ltd.) plant at Kawasaki, a retort that M. Kawaguchi designed in 1936 for the Nippon Seitetsu KK (Nippon Iron Manufacturing Co. Ltd.) plant in Wanishi (Hokkaidô), and one that

had enforced since 1 July 1939 a regulation that required all gasoline sold to have a 10% alcohol content, "Chemical Industry in Japan," *Chem. Trade J.*, 104 (17 March 1939): 270.

43 Goddin and Thornton (note 2), p. 130. The Japanese intended to construct six additional LTC plants, but they began construction on only three of them; Reid (note 18), p. 21.

44 Reid (note 18), p. 34.

S. Mimura, an engineer with the Tôkyô Gasu KK, designed in 1938 and first used in 1940 in the Tôkyô Gasu plant in Tsurumi near Yokohama (Honshû) and the Tôhô Kagaku Kôgyô KK (Tôhô Chemical Industry Co. Ltd.) plant in Nagoya.[45] The Knowles ovens were duplicates of those found in the United States and Japanese-constructed from American-furnished plans. Kawaguchi's retort replicated the Thyssen retort introduced in Germany prior to 1920. The retorts varied considerably in size from the large Lurgi (64 ft long) to the smaller (12–25 ft long) Koppers and Mimura retorts.

All retorts operated around 500° C and provided a continuous output of tar, except for the intermittently operating Knowles oven. They differed in that the German-designed retorts (other than the Thyssen) were vertically arranged and stationary, the two Japanese retorts were horizontal and rotating, and the Knowles oven was horizontal and stationary. But the reason for the greater success of the German retorts, particularly the Lurgi, resulted more from their design and construction than from their orientation and motion. No Japanese-designed retort operated successfully.[46]

The Mimura retorts found in the Tôkyô Gasu and Tôhô Kagaku Kôgyô plants had a cylindrical shape and constructed by forming and welding mild-steel plates. Each measured 25.5 ft long, 25 in inside diameter, and had an internal agitator. They sat horizontally, three high in banks of 24, and externally heated by a gas-fired refractory furnace. Despite numerous attempts to control the temperature, a difficulty aggravated by the strongly caking (agglomerating) properties of certain Japanese coals (high-volatile bituminous B coal), overheating excessively damaged Mimura's retorts. Their unnecessarily complicated operation also caused problems. The arrangement of the driving units required shutting down an entire bank to repair one of them, and the small amount of space between retorts made repair difficult. Tôhô Kagaku Kôgyô, which had a capitalization of ¥ 50 million, completed the Nagoya plant in October 1940, but no more than 80 of the 120 retorts installed ever operated at any one time. At best they carbonized 82 metric ton/d of an expected daily maximum of 300 tons. Wartime haste prevented any adequate design testing and contributed to the Mimura retort's very low 3.43% efficiency.[47]

45 Goddin and Thornton (note 2), pp. 130–31; Nav. Tech. Miss. Japan (note 19), pp. 19–22.

46 Reid (note 18), pp. 16–18.

47 Reid (note 18), pp. 16–18. Caking coals stick together or agglomerate and cause clogging of apparatus. Overheating occurred whenever air leaked into the combustion space surrounding the retorts. This disturbed the air-fuel ratio and at times raised the retort's temperature to more than 700° C. Because of their mild steel composition the retorts, agitators, and other metallic components had lifetimes that ranged from 15 to 150 days

If the Nagoya plant was merely a failure, the Tsurumi plant was clearly a disaster. Construction on the plant, located midway between Tokyo and Yokohama in the center of one of Japan's most industrialized areas, began in 1938 with a design of 240 Mimura retorts laid out in 10 banks, each with 24 retorts. The plant experienced the same difficulties as the Nagoya plant and gave so much trouble that Tôkyô Gasu dismantled it by the end of 1943 and sold most of the equipment including the retorts to Tôhô Kagaku Kôgyô for use at Nagoya. The Tsurumi plant cost ¥ 4 million (est.) excluding the tar processing equipment. It had a maximum coal capacity of 4,300 metric ton/yr for an annual investment cost of ¥ 930/ton. This was nineteen times greater than the cost of operating Lurgi retorts at the Nissan Ekitai Nenryo Kôgyô KK (Nissan Liquid Fuel Co. Ltd.) plant in Wakamatsu and showed the utter futility of trying to carbonize strongly caking coals.

William T. Reid, a supervising engineer with the US Bureau of Mines' Combustion Research Section in Pittsburgh, inspected the Japanese LTC plants after the war. Reid's inspection, which lasted from April through August 1946, occurred because Arno C. Fieldner, chief of the Bureau's Fuels and Explosives Branch, long had stressed the importance of investigating foreign developments in coal conversion for their value in assessing the United States' future energy needs. In the early 1920s Fieldner had visited several LTC plants in Germany and England, the Bergius coal liquefaction plant in Rhineau-Mannheim, and wrote papers discussing European coal research programs.

Reid had a similar mission, and he spent five months in Japan as a consultant assigned to the Mining and Geological Division, Natural Resources Section, General Headquarters of the Supreme Commander for the Allied Powers in Tokyo.[48] He wrote in his report that Tsurumi's engineers found "that the entire operation of the carbonizers had been a 'nightmare' and that they were relieved when the company finally decided to abandon the plant." The engineers even destroyed the operating data because they found it too humiliating to retain.[49] Nagoya and Tsurumi, respectively, produced only 826,000 and 148,000 gallons of liquid fuel, mostly bunker fuel oil, and 71,925 and 4,108 metric tons of semicoke.[50]

---

and usually failed at the welded joints. Uneven heating also warped the retort and caused by-product gas leakage through the end seals. Mimura redesigned the seals but could not stop the leaks because the retorts were not rigid enough to prevent warping at the reaction temperature.

48 Reid (note 18), 1, 7.

49 Reid (note 18), 56–63, 67–9, on 69.

50 Reid (note 18), 28.

In comparison with the poorly performing Nagoya and Tsurumi plants, the eight other LTC plants equipped with German-designed retorts experienced much greater success. Ube Kôsan installed Koppers continuous, vertical retorts because they carbonized best the low-rank sub-bituminous A coal available at the nearby Ube-Onada coal field. Plant construction took place in three stages, from May 1934 to December 1937. Cost of its 68 retorts, which Ube Kôsan built entirely in Japan without German assistance, all accessory equipment, and buildings totaled ¥ 5,887,997, increasing to a nominal capitalization of ¥ 81,637,500 at the war's end. The 68 retorts had a 600 metric ton/d combined coal capacity, but plant operators considered satisfactory a maximum daily input of 500 metric tons. Records show that from 1940 to 1945, the Ube plant carbonized 568,693 metric tons of coal, representing a continuous operating rate of 438 metric ton/d or a 73% capacity. Ube contributed 19.61 million gallons of liquid fuel from 1938 to 1945 making it Japan proper's biggest liquid fuel producer. The most serious damage the Koppers retorts experienced in their eleven-years' operation was clinker (slag) formation on the retorts' walls.[51]

The Naihoro and Wakamatsu plants received assistance in installing their retorts from Lurgi Gesellschaft in Frankfurt. Lurgi engineers also may have assisted at the Agochi plant in northern Korea, but because of this plant's location, above the 38th parallel and under Soviet occupation when the war ended, Nippon Chisso Hiryô KK (Nippon Nitrogen Fertilizer Co. Ltd.) in Tokyo, which shared ownership with Chôsen Jinzô Sekiyu KK (Chosen Artificial Oil Co. Ltd.), lost contact with the Agochi operation. The Naihoro plant, on the west coast of Sakhalin Island about fifty miles north of its peninsula's southern tip, also fell under Soviet occupation when the war ended. Sakhalin has remained within the Soviet Union, but Imperial Fuel Development Company, which acquired the plant in 1937, provided operating information to the Naval Technical Mission.

Mitsubishi Sekitan Yuka Kôgyô KK (Mitsubishi Coal Liquefaction Industry Co. Ltd.) started construction on the Naihoro plant in 1934 after Mitsubishi obtained patent rights from Lurgi. On completion in 1935 Naihoro was the first Japanese plant to operate with Lurgi retorts. Initially the plant had two retorts, each of 150 metric ton/d capacity; it added another two, each of 200 metric ton/d capacity, in 1937, for a total capacity of 700 metric ton/d. Despite being the first Lurgi-equipped Japanese plant it experienced few serious problems other than sulfur present in coal corroding the gas mains, and coal dust

51 Reid (note 18), 46, 52, 28. Records of fuel production are available only from 1938–45; see p. 48.

clogging the gas distributors and reducing each retort's capacity. Mitsubishi invested ¥ 28,116,942 in the Naihoro plant which had a conversion rate of 231,000 metric ton/yr, giving an annual investment cost of ¥ 122/metric ton. Its much higher investment cost compared to other plants with Lurgi retorts, such as Wakamatsu, explains why Mitsubishi sold the plant to the Japanese government shortly after passage of the Seven Year Plan. Imperial Fuel Development Company continued its operation, and from 1940 to 1945 the Naihoro plant produced 18.79 million gallons of liquid fuel. Although one of the Japanese empire's largest liquid fuel producers, it had only a 5.85% efficiency.

The Nippon Chisso Hiryô-Chôsen Jinzô Sekiyu in Agochi, Korea, operated the second largest LTC plant after Ube Kôsan's LTC plant. Chôsen Jinzô Sekiyu also operated a successful synthetic methanol plant that during the war years, 1940–45, produced 7.1 million gallons methanol, producing 3.4 million gallons methanol in the peak year 1944. The Army Fuel Board and the Naval Office of Accounts and Supplies received about 90% of the methanol which, according to Chôsen Jinzô Sekiyu, kamikaze airplanes burned in their internal combustion engines.

The LTC plant at Agochi started operation in 1940 with the construction of four Lurgi retorts each having a coal capacity of 150 metric ton/d. Much of the brown coal came from the nearby Agochi mine, but no records of its analysis exist. The more complete plant production figures show that during the plant's operating years of 1940–45, it consumed about 600,000 metric tons of coal and produced 19.53 million gallons of liquid fuel, most of it, 16.49 million gallons of bunker fuel oil, and 140,000 metric tons of semicoke. Toward the end of the war Nippon Chisso Hiryô-Chôsen Jinzô Sekiyu began construction on three additional 150 metric ton/d retorts but never finished their construction. Its LTC plant had a capitalization of ¥ 90 million at that time.[52]

Nissan Ekitai's plant in Wakamatsu and the Imperial Fuel Development Company plant in Ube were the last of the plants to make a major contribution to Japan's LTC industry. Two companies, Nissan Kagaku Kôgyô KK (Nissan Chemical Industry Co. Ltd.) and Imperial Fuel Development Company, shared equally in establishing Nissan Ekitai in February 1939 with a ¥ 10 million capitalization that increased to ¥ 25 million (¥ 20 million paid) in November 1944.

The Wakamatsu LTC plant employed 700 workers. It lay on the Chikuho coal field's northern edge close to Japan's largest coal producing area where in 1936 seven mines produced 1.72 million metric tons of coal, far more than the plant's expected 400,000 metric ton/yr consumption. The coal, a low-caking,

52 Reid (note 18), 77–80.

high-volatile bituminous C coal, appeared similar to American coals from Colorado, Utah, Illinois, and other midwestern fields.

Nissan Ekitai contracted with Lurgi in February 1939 to construct four 300 metric ton/d retorts each costing ¥ 5 million, with operation to begin between March 1941 and November 1944. Lurgi provided technical information and sent a three-man engineering team who remained in Wakamatsu from September 1940–June 1941 to assist in constructing the first retort. But except for the burners and some special refractory equipment, Japanese engineers built the first retort and then built and operated the remaining three without German assistance.[53]

Nissan Ekitai expected each retort to carbonize annually 100,000 metric tons of coal, yielding 10,000 metric tons of liquid fuel and 65,000 metric tons of semicoke at an annual investment cost of ¥ 500/metric ton for liquid fuel and ¥ 76/metric ton for semicoke. Actual yields produced much less. During its lifetime, 16 April 1941–8 August 1945 when American bombing shut it down, the Wakamatsu plant carbonized 626,240 metric tons of coal to give 16.87 million gallons of liquid fuel, a 9.64% yield that Nissan Ekitai considered satisfactory, and 396,529 metric tons of semicoke, a 63.3% yield. The liquid fuel consisted mainly of bunker fuel oil (82.8%) that contained excessive carbon residue but satisfactorily powered nonmilitary craft such as fishing boats. Civilians, in fact, consumed the Wakamatsu plant's entire liquid fuel output.[54]

The Naval Technical Mission inspected the Wakamatsu plant on 24–25 October 1945 and interrogated N. Yamada, the plant's director, describing him as "unusually well-informed on the technical details of the low-temperature process employed at this plant;" S. Kunii head of LTC, T. Fujii head of mechanical engineering, and T. Monden office manager. Reid of the US Bureau of Mines, who made a follow-up inspection of Japan's LTC plants in April-August 1946, noted that at the time of his inspection Nissan Ekitai's Wakamatsu LTC plant remained the only LTC plant in operation.

The Imperial Fuel Development Company's Ube LTC plant and the Wakamatsu LTC plants had similarities. Ube had four of the same 300 metric ton/d

53 Reid (note 18), 22. Capitalization increased to ¥ 25 million during the period March 1941–November 1944. The original plan included a research pilot plant for hydrocracking tar with equipment supplied by Friedrich Krupp AG. Nissan Ekitai also contacted IG Farben about constructing a larger hydrocracking plant with a gasoline yield of 10,000 kL/yr, but neither negotiation led to anything. Nav. Tech. Miss. Japan (note 19), p. 22; "Nissan liquid Fuel Co.," *J. Fuel Soc. Japan*, 18 (1939): 29.

54 Nav. Tech. Miss. Japan (note 19). Despite the damage Wakamatsu plant officials expected to resume operation in November 1945. Reid (note 18), pp. 24–25, 77, 27, 30.

Lurgi retorts plus two banks of 72 Koppers retorts (36 each) that had a 600-metric ton/d total coal capacity. Imperial Fuel Development Company's joint ownership of the Wakamatsu plant and the experience gained there made German assistance unnecessary in the construction and operation of its Ube plant. A partnership between Imperial Fuel Development Company and Ube Kôsan, called Ube Yuka Kôgyô KK, had incorporated the plant in August 1939 with a capitalization of ¥ 50 million. Later, in 1944, Imperial Fuel Development Company acquired complete control and ran the plant until the war's end. Ube had a design annual production of 40,000 kL (10.6 million gallons) of gasoline, obtained from carbonizing the non-caking sub-bituminous A coal found in Ube-Onada's fourteen major mines and then hydrocracking the light tar oil. A plan to add six more Lurgi carbonizers in 1945 increased projected annual gasoline production to 90,000 kL.

From April 1944 to June 1945 the Ube plant carbonized 401,437 metric tons of coal, but had a low 5.54% liquid yield: 23,092 kL of bunker fuel oil and a trace of gasoline. The low yield, less than the 9.1% predicted for Lurgi retorts, resulted in part because the Koppers retorts, which had a low inherent tar yield, carbonized more than half the coal. Water gas formation and the coal's low rank also contributed to the poor yield.

The USAAF bombed the Ube plant four times, on 2, 15, 23 July, and 5 August 1945, the last attack permanently ending its operation. At that time Ube had a staff of 350 and 1,200 workers. The Naval Technical Mission arrived at the plant on 28 October 1945 and interrogated its top five officials: K. Iwamoto plant manager; Mr. Ito head of general affairs department; Mr. Uwaba head of production department; Mr. Iki gas engineer; and Mr. Tarwada business manager.[55]

In addition to the ten (of the proposed 66) operational LTC plants, three other plants had reached various stages of construction when the war ended in August 1945: two in Hokkaidô and one in Chielin, (Chilin, Jilin, Kilin), Manchuria. While only the Kawasaki plant used American-designed Knowles ovens, the two plants under construction in Hokkaidô, at Hiragashi and Takikawa, had Knowles ovens. Construction at Hiragashi reached 90% completion but remained largely incomplete (90%) at Takikawa where construction did not start until March 1945. Increasingly serious liquid fuel shortages in 1945, the cheaper construction cost and more successful operation of LTC

55 Reid (note 18), 37–42. The Imperial Fuel Development Company's plant at Ube also was producing 150,000 tons of ammonia annually. Construction began on a plant to produce lubricating oil from rubber in February 1943 and on a synthetic methanol plant later that year.

plants compared to either F-T or coal hydrogenation plants encouraged companies such as Hokkaidô Tanso Kôgyô KK and Nippon Jinzô Sekiyu to construct additional LTC plants. But in the end these plants remained unfinished because of inferior conception, manpower shortages, and the increasing difficulty of obtaining plant equipment and material that followed the United States' moral embargoes and the war's outbreak.[56]

Most of the LTC plants received severe damage from American wartime bombing. The Japanese also lost four LTC plants in Korea and Sakhalin because they fell under Soviet control after the war. From 1940 to August 1945 these four plants produced 56 million gallons of liquid fuel versus 53.7 million gallons from the six plants in Japan proper. Of the six, the Wakamatsu plant operated until 8 August 1945 when incendiary bombing caused ¥ 1 million damage and forced its temporary closing. Wakamatsu's four Lurgi retorts suffered only slight damage, and by August 1946 they resumed producing liquid fuel and semicoke. Because of Ube's importance as an industrial center, the United States bombed it extensively and during four raids in July and August 1945 almost totally destroyed the Imperial Fuel Development Company plant. On the night raid of 5–6 August 1945, 106 B-29s dropped 938 tons of bombs on Ube's petroleum refining units, completely destroying them and about 80% of other plant facilities. The USAAF suffered no losses during the attack.

With the war over and the end of the government's synthetic fuel subsidies, liquid fuel from LTC cost more than imported petroleum. Imperial Fuel Development Company decided that the semicoke's poor quality could not justify the Ube LTC plant's future operation. As a result, it tried to recover its investment by converting the LTC plant to the production of badly needed ammonium sulfate fertilizer. Because this required large quantities of ammonia gas, a LTC by-product, Imperial Fuel Development Company rebuilt the severely-damaged Koppers retorts and not the moderately damaged but inferior gas-producing Lurgi retorts.[57]

The Ube Kôsan plant also shut down because of bombing that destroyed or burned parts of it. The Koppers retorts suffered only slight damage, and in May 1946 Ube Kôsan already had converted one of the three banks of retorts to fertilizer production. Construction began on the other retorts. Bombing in July 1945 destroyed 90% of the 15 Knowles ovens installed at the Nihon Yuka Kôgyô plant in Kawasaki, but because the ovens performed well and produced

56 Reid (note 18), 80–2. Barnhart (note 4), pp. 130, 166, 196, 255–56. Chielin was also the site of an incomplete Fischer-Tropsch plant intended for methanol production.

57 Reid (note 18), pp. 12, 45–46.

TABLE 5.8 LTC plants, not complete list

Plant, capitalization	Location	Years of operation	Converter	Production capacity
Nihon Yuka Kôgyô KK (Nihon Liquefaction Industry Co. Ltd.)	Kawasaki (Honshû)		Knowles	
Ube Kôsan KK (Ube Mining Co. Ltd.), ¥ 81.6 million	Ube (Honshû)	1940–45	Knowles	68 retorts, 600 metric ton/d total
Imperial Fuel Development Company and Ube Kôsan partnership called Ube Yuka Kôgyô KK. In 1944 Imperial Fuel Development Company, ¥ 50 million	Ube (Honshû)	1939–45	Lurgi, Koppers	40,000 kL/yr 90,000 kL/yr
Nissan Ekitai Nenryo Kôgyô KK (Nissan Liquid Fuel Co. Ltd.) and Imperial Fuel Development Company, ¥ 10 million	Wakamatsu (Kyûshû)	1941–45	Lurgi	16.87 million gal. liquid fuel 1941–45
Nippon Chisso Hiryô KK (Nippon Nitrogen Fertilizer Co. Ltd.) and Chôsen Jinzô Sekiyu KK (Chosen Artificial Oil Co. Ltd.), ¥ 90 million	Agochi (Korea)	1940–45	Lurgi	150 metric ton/d
Mitsubishi, Imperial Fuel Development Company, ¥ 28.1 million	Naihoro (Sakhalin)	1935 Mit. 1940–45 Imperial Fuel	Lurgi	150 metric ton/d and 200 metric ton/d
	Hiragashi (Hokkaidô)	1945	Knowles	
	Takikawa (Hokkaidô)	1945	Knowles	
Mitsui Bussan KK – Imperial Fuel Development Company	Chielin (Manchuria)			unfinished
Tôhô Kagaku Kôgyô KK (Tôhô Chemical Industry Co. Ltd.), ¥ 50 million	Nagoya (Honshû)	1940–45	Mimura	
Tôkyô Gasu KK, ¥ 4 million	Tsurumi	1943	Mimura	
Nippon Seitetsu KK (Nippon Iron Manufacturing Co. Ltd.)	Wanishi (Hokkaidô)	1946	Kawaguchi	

SOURCE: COURTESY OF AUTHOR

a high-quality coke Nihon Yuka Kôgyô rebuilt them. The Nippon Seitetsu plant in Wanishi, located at the site of the largest steel plant in Hokkaidô, suffered no damage during the war. Its Kawaguchi retorts ceased operation in June 1946 mainly because of poor wartime maintenance and coal shortages.

The two dismantled plants, the Tôhô Kagaku Kôgyô plant in Nagoya and the Tôkyô Gasu plant in Tsurumi, had identical troublesome Mimura retorts. The Nagoya plant experienced severe difficulties from its opening in October

1940 but remained in operation until December 1944 when a severe earthquake partially destroyed it. The Navy Fuel Board took over the plant in June 1945 and operated it as the Second Fuel Depot. Its low production and incendiary bombing later that year completely ended its operation. After the war in November 1945 the Nippon Hiryo Kôgyô KK (Nippon Fertilizer Industry Co. Ltd.), a fertilizer distribution agency half-owned by the Japanese government, absorbed Tôhô Kagaku Kôgyô. But the plant's poor performance precluded any rebuilding and resulted in total dismantling of the Mimura retorts. The Tsurumi plant had undergone complete dismantling by the end of 1943, selling some of the retorts and carbonizing equipment to Tôhô Kagaku Kôgyô. Tsurumi ranked as the least efficient and poorest performing Japanese LTC plant.[58]

6 Research and Development of the Fischer-Tropsch Synthesis: A Promising Beginning

Kita by far carried out Japan's most important research on the F-T synthesis. What started at Kyoto in 1927 as a purely scientific laboratory study using small-scale glass apparatus reached the semicommercial pilot plant stage in 1937. Kita aimed to produce a catalyst superior to cobalt-thorium catalysts. By 1936 his best catalysts, those of nickel not of cheaper and readily available iron, equaled the activity of Ruhrchemie's cobalt-thorium catalysts.

Shortly after Kita began his pioneering work, centers of F-T research emerged at the Imperial Fuel Research Institute in Kawaguchi (Tokyo), Tôkyô Imperial University, and Mitsui Bussan KK in Rumoi (Hokkaidô). Shiro Watanabe, head of Mitsui's coal department in Tokyo, and Toyohisa Ayai, manager of Mitsui's German products branch, Doitsu Bussan KK, Mitsui, on 16 February 1937 as a result of almost two years of negotiating, also secured exclusive Japanese commercial rights to the F-T patents from Ruhrchemie AG which now controlled their licensing. Mitsui paid ¥ 7.2 million (RM 4.5 million) for the license which expired on 31 December 1945. The Japanese military played an active role in developing the F-T synthesis, but unlike oil shale extraction and high-pressure hydrogenation, the army, not the navy, promoted the research. The navy lacked interest, claiming the F-T process could not produce the heavy fuel oil needed for its ships, and conducted only limited catalytic research at the First Naval Fuel Depot in Ôfuna.

58 Reid (note 18), 53, 63, 67–69, 73.

The navy made coal hydrogenation, not F-T, its first priority. The navy wanted sole control of coal hydrogenation in Japan and strongly encouraged Mitsui to pursue rights to the F-T synthesis. Mitsui already had tried unsuccessfully to obtain from ICI a license to establish a Japanese coal hydrogenation industry and consequently expressed considerable interest in reaching an agreement with Ruhrchemie. Mitsui's reputation had suffered at that time because the government believed it had engaged in excessive profit taking and not followed Japan's national policy of developing heavy industry. Mitsui focused mainly on commercial trading and had coal mining as its only heavy industry. To improve its public image Mitsui decided to develop the F-T synthesis, and with the establishment of the Imperial Fuel Development Company in January 1938 Mitsui agreed to nationalize its license so that any company in the Japanese empire (Korea, Manchuria, Taiwan) had use of the F-T patents.[59]

At the Imperial Fuel Research Institute, its scientists and engineers carried out extensive F-T investigations in both the laboratory and semicommercial plant during the years 1933–40. Instead of cheaper iron catalysts, the institute worked with nickel-cobalt catalysts that had a one-month maximum life. Laboratory tests gave 100 g of liquid fuel for each m^3 of synthesis gas (CO: H_2) consumed. The semicommercial plant, which operated at 5 atm, contained a small vertical cylindrical converter (reactor), a product recovery system with air and water-cooled condensers, and a pressure-type scrubber. The plant had a liquid fuel capacity of 200 kg/d with a 100 m^3/hr synthesis gas throughput, but actual production never exceeded 30% of capacity. Plant engineers blamed the catalysts and the converter's inadequate temperature control for the poor performance. The plant never operated for more than a few days at a time, and when Mitsui's industrial plant at Miike came on stream in 1940 they shut down the troublesome Kawaguchi (Tokyo) plant.[60]

After Mitsui acquired the rights to prepare and use Ruhrchemie's catalysts, Soji Kono and other researchers at Rumoi, Koichi Matsubara at Tokyo Imperial University, and Kita concentrated almost exclusively on developing iron substitutes for standard (1 atm) and middle (10–15 atm) pressure F-T syntheses. Japanese scientists made significant contributions to understanding the catalysis of the F-T synthesis. Their investigations represented the kind of close

59 Ryoichi Ishida, *Story of Coal Liquefaction* (Tokyo: Chûô Shuppan Insatsu, 1990), pp. 27–42 (in Japanese). Mitsui signed an optional contract with Ruhrchemie on 11 February 1936, and during the next year it kept two or three researchers in Germany, studying the F-T synthesis. See also the US Naval Historical Center's *ATIS Report 4575*, document numbers ND 26-0004.1 through ND 26-0004.31 and Allen (note 37), pp. 148–49.

60 Nav. Tech. Miss. Japan (note 19), pp. 16–17.

cooperative industrial and university research that organizations like Mitsui and Mitsubishi Shoji strongly encouraged.

Mitsui's laboratory investigations at 275–280° C, 15 atm, and a C: H_2 synthesis gas mixture gave a 44% conversion to gaseous hydrocarbons (50% propane and 20% butanes) used to prepare isooctane. Investigations on iron catalysts continued in its small semicommercial pilot plant of 10 m^3/hr synthesis gas throughput and from 1942 in its industrial-size plant at Takikawa (Hokkaidô).[61] Matsubara tested a natural iron catalyst in 1943–44, but he obtained a maximum liquid hydrocarbon yield of only 86 g/m^3. Kita's best iron catalyst, when reacted in the laboratory with 40 cm^3/hr of a CO: 2 H_2 synthesis gas mixture at 240° C and 1 atm, gave maximum liquid yields of 98–102 g/m^3, considerably lower than the 151–158 g/m^3 obtained with the best cobalt-thorium catalyst. His iron catalyst had about a one-month lifetime before carbon poisoning of its surface destroyed its effectiveness.

Kita operated Kyoto's F-T semicommercial pilot plant for four years, 1937–41, carrying out reactions at 250° C, 1 atm, and a synthesis gas throughput of 100 m^3/hr. The Japanese army, hoping to expand Kita's program to industrial-size production, transported the plant to its Fuel Research Institute in Fuchu (Honshû) in 1942 where testing continued with natural iron catalysts, giving low liquid yields of only 73 g/m^3.[62] Although Japanese research failed to produce a superior iron catalyst, Kita established that pressures of 10–15 atm greatly favored iron catalysts. As a result, from 1938 and continuing throughout the war, Kita tested 40 cm^3 of different iron catalysts in a small, single-tube converter, but none performed as well as cobalt-thorium catalysts. He obtained maximum liquid yields of 115–120 g/$m^3$with an iron catalyst at 210–240° C, 10–15 atm, and a 40 m^3/hr throughput of a CO: H_2 synthesis gas mixture.[63] Kita and Kodama had published and patented their F-T research

61 "Synthetic Oil in Japan" (note 2), p. 339. Other pressure units used in the Japanese reports were kg/cm^2 and psi. 1 atm = 1 kg/cm^2 = 15 psi.

62 Nav. Tech. Miss. Japan (note 19), p. 17. Matsubara's catalyst (100 ochre – 1 Cu-1 H_3BO_3-6 K_2CO_3) consisted chiefly of ochre, a natural yellow earth 64.8% ferric oxide, obtained from Niwasaka near Fukushima. The composition of Kita's best iron catalyst was 100 Fe-25 Cu-2 Mn-125 kieselguhr-15 H_3BO_3 and of the best cobalt-thorium, 100 Co-5 Cu-12 U_3O_6-4 Mn-100 kieselguhr. After the war ended, the petroleum section of the US Naval Technical Mission inspected the Fuchu plant on 12 January 1946.

63 Composition of the iron catalyst was 100 Fe-25 Cu-125 kieselguhr-6 K_2CO_3. Later the Japanese developed an iron catalyst identical to that used in the Kyoto semicommercial pilot plant but also containing five parts of magnesium. It supposedly increased yields and was the best middle pressure catalyst. But in the one result repeated, its yield was 111 g/m^3. Goddin and Thornton (note 2), pp. 125, 127, 130. See also "Chemical Industry in the Far East," *Chem. Trade J.*, 108 (23 January 1941): 8.

in the pre-war years, and when the Naval Technical Mission arrived in Japan, it inspected the Fuchu F-T plant at the Fuel Research Institute on 12 January 1946. It visited Kita's Kyoto's research laboratories and interrogated Kita and Kodama a few days later, on 15–17 January 1946.

Construction of the first industrial-size F-T plants began in 1936, one year before Mitsui acquired an option on the rights to Ruhrchemie's catalysts and one year before the Japanese government prepared its Seven Year synthetic oil plan. The plan proposed an annual production of 1,000,000 kL each of gasoline and fuel oil, the 11 F-T plants to contribute 50% of the total, or 500,000 kL/yr. The Seven Year Plan included considerable subsidies for plant construction, yet by 1940 the government faced serious difficulty financing it and deliberately delayed new construction by limiting support to plants already constructed or scheduled for construction. The ruling affected five plants, three of which, at Miike (near Ômuta, Kyûshû), Amagaski (Honshû), and Takikawa (Hokkaido), the Japanese completed by the end of World War II. Mitsui-Imperial Fuel Development Company jointly operated the three plants, all of which had the same unit production costs and received subsidies. Mitsui licensed the two unfinished plants, in Manchuria at Chielin and Chinchu.

The three completed plants produced good quality liquid fuel, but production never approached capacity, the best being the Miike plant's nearly 50% output in 1943. Synthesis gas shortages, low catalytic activity, and leaks in the converter's cooling-water tubes making temperature control difficult remained major problems that Mitsui's engineers failed to eliminate at the pilot plant stage before beginning industrial-size F-T operation. These problems contributed to F-T's high production costs. Gas oil that had an official retail price of ¥ 152–177/kL cost ¥ 320–340/kL to produce.[64] The Naval Technical Mission interrogated Hikoshi Sogabe, a Mitsui technical director, and Naosuke Shônô, an assistant technical director of Imperial Fuel Development Company, and obtained information on the three jointly owned Mitsui-Imperial Fuel Development Company plants. M. Takei, head of the chemical department at the Miike plant, provided additional details on the Miike plant's operation

The Miike plant went on stream first, and of the three F-T plants the only plant the US Naval Technical Mission inspected in the months from 4 September 1945 to 21 February 1946. Construction on the ¥ 12 million Miike plant

64 Nav. Tech. Miss. Japan (note 19), pp. 18–19, 326; "Synthetic Oil in Japan" (note 2), pp. 335, 337–39; Goddin and Thornton (note 2), pp. 125, 129–30; Ando (note 2); "Hydrogenation of Coal-Japan" (note 2).

started in 1936, production followed in May 1940. The plant operated at 1 atm, 180–200° C, with a standard cobalt-thorium catalyst (21.2 cobalt-1.2 thorium oxide-2.4 magnesium oxide-75.2 kieselguhr) and a synthesis gas mixture of 2 H_2: CO, prepared by adding water gas to reformed coke oven (coal) gas. Five Koppers water gas producers, each consuming 100 metric tons of coke daily, manufactured the water gas, and a single Koppers gas reforming unit of capacity 456,000 m^3/d reformed the coke oven gas. The plant contained 56, steel "fin and tube" type horizontal Ruhrchemie converters in which pressurized cooling water circulated through horizontal tubes interlaced by vertical sheets. Each converter held about 4.5 metric tons of catalyst packed in the spaces between the tubes and sheets and had a synthesis gas throughput of 1,000 m^3/hr with an expected yield of 116 g/m^3. The plant achieved an actual throughput of 750 m^3/hr, and yields of only 70–85 g/m^3. It operated in two stages, using 30 converters in the first stage, 18 in the second, and had 8 in reserve.

The Miike plant had a total production capacity of 30,000 metric ton/yr (29,721 kL/yr), but during its peak year of 1943 the Miike plant produced 14,000 metric tons (13,870 kL) of liquid fuel including 40-octane number gasoline and 90-cetane number diesel oil, as well as propane, butanes, and paraffin waxes. Plants that used Ruhrchemie's standard cobalt-thorium catalyst typically produced low-octane gasoline and high-cetane diesel oil. Failure to meet production capacity resulted from the plant's temperature control problems, reduced activity of Japanese-manufactured catalysts, and from using coals that gave insufficient synthesis gas because of their low melting point ash and high sulfur content. Nevertheless, the Miike plant was Japan's most successful F-T plant before bombing in August 1945 damaged its gas generating capacity and ended its operation.

The largest of the three operating F-T plants, the Takikawa plant, had an initial capitalization of ¥ 20 million, later increased to ¥ 50 million. It had a 50,000 metric ton/yr (49,535 kL/yr) capacity and unlike the other F-T plants operated at a middle pressure of 10 atm and used iron and cobalt-thorium catalysts. Production started in December 1942 with 36 Ruhrchemie converters and 4 of Japanese design, including one developed at Kyoto that used an iron-magnesium catalyst. The Kyoto converter produced fuels with a high olefin (unsaturated) content characteristic of an iron catalyst, and although iron catalysts were sulfur sensitive they had a six month lifetime versus two months for cobalt-thorium. Like the other plants, the Takikawa plant never approached design capacity. In its four years of operation, 1942–45, production totaled 1,550 kL, 1,600 kL, 6,610 kL, and 2,480 kL.

The last of the three operating F-T plants, the ¥ 12 million, 1 atm Amagasaki plant, had a production capacity of 30,000 ton/yr. It contained 24 Ruhrchemie

TABLE 5.9 Fischer-Tropsch plants

Company	Location	Designed capacity	Type of operation	Catalyst	Output kL (actual)
Mitsui-Imperial Fuel Development, Nippon Jinzô Sekiyu Co., Miike Works ¥ 12 million, increased to ¥ 50 million in 1943, to ¥ 150 million in 1944	Miike (Kyûshû)	30,000 metric ton/yr oil output (29,721 kL/yr); 1,000 m^3/hr converter gas throughput	actual prod. 16,000 tons; Ruhrchemie license, 56 converters; 180°–200° C, atm pressure, two stage; actual throughput gas 750 m^3	$Co:ThO_2:MgO$: kieselguhr: 21.2:1.2:2.4: 75.2 catalyst life 120 days	1940 = little 1941 = 7,200 1942 = 9,290 1943 = 13,870 1944 = 11,070 1945 = 1,720
Takikawa Works	Takikawa (Hokkaidô)	50,000 metric ton/yr (49,535 kL/yr)	10 atm press.; 36 double tube, 32 single Ruhrchemie, 4 Japanese	Co and precipitated Fe catalyst and natural Fe catalyst	1941 = 0 1942 = 1,550 1943 = 1,600 1944 = 6,610 1945 = 2,480
Amagasaki Works ¥ 12 million	Amagasaki (Honshû)	30,000 metric ton/yr	atm press., 24 double tube Ruhrchemie converters	Same as Miike	1941–42 = 0 1943 = 200 1944 = 130 1945 = 0
Manchû Jinzô, Sekiyu Co., Chielin Works	Chielin (Manchuria)	1,000 kL/yr	unfinished		1941–45 = 0
Chinchu Works	Chinchu (Manchuria)	40,000 kL/yr	unfinished		1941–45 = 0
Total					55,710

SOURCE: COURTESY OF AUTHOR

converters and used a standard cobalt-thorium catalyst. Production began in September 1943, but in its two years of operation, 1943 and 1944, it produced liquid fuel yields of only 200 kL and 130 kL. Amagasaki experienced mainly the same problems as the Miike plant.[65]

65 "Synthetic Oil in Japan" (note 2), pp. 335, 337–39; Nav. Tech. Miss. Japan (note 19), pp. 18–19; Goddin and Thornton (note 2), pp. 122, 125, 127,130. Coke oven gas is mainly a mixture of hydrogen (53%) and methane (32%) plus carbon monoxide, nitrogen, other gases, and illuminants. The US Naval Technical Mission also visited the semicommercial Fuchu plant (see note 60). The capitalization of the Miike plant was ¥ 50 million in October 1943, and after Mitsui brought in two other partners to form Nippon Jinzô Sekiyu KK in October 1944, final capitalization was ¥ 150 million. See Ishida (note 59), p. 87, and Nav. Tech. Miss. Japan (note 19).

The two unfinished Mitsui-licensed F-T plants in Manchuria, Manchû Jinzô Sekiyu KK (Manchu Artifical Oil Co. Ltd.) in Chielin and Manchû Gosei Nenyro KK (Manchu Synthetic Fuel Co. Ltd.) in Chinchu, had capacities of 1,000 kL/yr and 40,000 kL/yr. This gave the five F-T plants a total capacity of 178,500 kL/yr, but for the duration of the war they produced only 55,720 kL.[66] The plants produced mainly diesel oil of 80–100-cetane number, most of which came from the Miike plant. The navy used small quantities for blending with 90% Borneo crude or for submarine fuel. The army, Japan's chief diesel oil consumer, burned about 50% of all Japanese-produced diesel oil in its tanks.[67]

7 Coal Hydrogenation: High Hopes but Little Success

Of the three coal conversion processes that the Japanese investigated and expanded to the industrial stage, high-pressure coal hydrogenation (liquefaction) proved by far the biggest disappointment. The 1937 Seven Year Plan called for the construction of 10 plants in Japan and Manchuria, each producing 100,000 kL/yr, although the plants proposed and the two actually constructed were much smaller. Plant proposals included a ¥12.5 million Ube Chisso Kôgyô KK (Ube Nitrogen Industry Co. Ltd.) plant and a ¥ 10 million Chôsen Sekitan Kôgyô KK (Chosen Coal Industry Co. Ltd.) plant. The two plants constructed were at Fushun, Manchuria, and Agochi, Korea, and, according to the Naval Technical Mission, only Fushun produced any synthetic petroleum, a meager 1,460 kL. In 1944, when synthetic and derived liquid fuel production in the Japanese empire, excluding Fushun oil shale, totaled 113,166 kL (750,000 barrels), hydrogenation contributed 690 kL. The remainder came from F-T, 18,000 kL, and LTC, 95,000 kL. Yet, the Japanese regarded hydrogenation as the most promising process for converting coal into oil.[68]

The navy had begun research at its Tokuyama laboratory in 1925 and in 1932 completed Japan's first pilot plant there. The navy in fact played the dominant role in developing coal hydrogenation because it wanted a reliable source of liquid fuel for the fleet, although at times its policies delayed and obstructed progress. Studies also began at the Imperial Fuel Research Institute in Kawaguchi (Tokyo) in 1925, with the construction of a small pilot plant in 1934, and by 1928 with SMR's experiments at its Central Research Laboratory in Dairen, Manchuria. But these were only preliminary investigations. The

66 Goddin and Thornton (note 2), p. 122; "Synthetic Oil in Japan" (note 2), p. 335.

67 Thornton (note 5), p. 818.

68 Goddin and Thornton (note 2), p. 122; "Synthetic Oil in Japan" (note 2), p. 334.

Naval Technical Mission, which spent 16–28 January 1946 at the Imperial Fuel Research Institute examining its LTC, F-T, and coal hydrogenation facilities and interrogating its scientists such as Ban, reported that after considerable laboratory research the Japanese plunged into industrial hydrogenation with insufficient pilot plant experience. Lacking this experience their program suffered from technical difficulties and general inefficiency made even worse by the United States' moral embargoes and the wartime unavailability of steel alloys suitable for constructing high-pressure equipment. The Naval Technical Mission believed that while the Japanese excelled as copyists ("excellent copyists") they demonstrated insufficient understanding or "not enough in the development of complicated engineering techniques, such as high-pressure hydrogenation."[69]

Because of the similarity between Japanese coals and British bituminous coals, the Japanese at first tried to obtain assistance from ICI which had considerable experience in hydrogenating bituminous coals. ICI had operated a pilot plant at Billingham and in 1935 completed construction of the first industrial plant to liquefy bituminous coal. Between 1931 and 1933 representatives from Mitsui and SMR contacted ICI. Mitsui approached ICI in May 1931. In March 1933 Yôsuke Matsuoka (1880–1946), an SMR vice-president and its president in August 1935, interviewed ICI's chairman, Harry McGowan, and in November 1933 Jôtarô Yamamoto (1866–1935), another SMR vice-president, accompanied by Mr. Buckney, a former employee of ICI's Brunner Mond (Japan) division, met with McGowan in Japan.

In each case ICI explained that it could not share information or participate directly in any Japanese hydrogenation project because of its arrangement with IG Farben, Standard Oil (NJ), and Royal Dutch-Shell. This arrangement of April 1931 established International Hydrogenation Patents Company (IHP) and International Hydrogenation Engineering and Chemical Company (IHEC) in The Hague. IHP controlled international licensing and access to its members' patents and technical information, IHEC handled plant engineering and design. The arrangement also limited ICI's direct participation in hydrogenation projects to the British empire. IHP intended to deal jointly with SMR, Mitsui, and Mitsubishi.

In 1937, following the ICI-SMR-Mitsui hydrogenation discussions and Mitsui's obtaining the F-T patent rights from Ruhrchemie, other than some continuing negotiations and invitations for Matsuoka to visit German and British plants and Standard's hydrogenation plants in Bayway, New Jersey, and

69 Goddin and Thornton (note 2), pp. 124–5; Nav. Tech. Miss. Japan (note 19), p. 15.

Baton Rouge, Louisiana, little more developed until 1938–39. At that time the army acting through Ogura Sekiyu (Ogura Oil Company) and Mitsubishi tried to reach an agreement with IG Farben. But the navy and its supporters like Ôshima, now a director of Imperial Fuel Development Company, claimed prematurely and mistakenly the imminent successful development of its own hydrogenation process, and until 1944–45 flatly blocked any army-IG Farben agreement. Discussion also went nowhere because of Japanese unwillingness to pay IHP's licensing fee which, depending on the amount of assistance and size of plant, required payment as high as RM 8 million ($3.2 million).[70]

The navy had continued basic and applied hydrogenation research at its Tokuyama laboratory and pilot plant during the 1920s and 1930s. The plant, which had a 7 ton/d coal-oil paste input (43–50% coal), consisted of a single converter and operated only in the liquid phase at 415° C and 200 atm. A Knowles generator of capacity 200 m^3/hr provided hydrogen by electrolyzing water. The Naval Technical Mission had no access to records from the pilot plant's early operation in 1932–38, but interviews with Tokuyama's personnel revealed that it operated one-third of the time. Tests in 1938–39 with subbituminous coals from Ôyama (Fushun) and Naibuchi (Sakhalin) reported a 77% liquid yield: 3–4% gasoline (bp under 180° C), 40% middle oil (bp 180–280° C), and 52–53% heavy oil (bp over 280° C). The plant's difficulties included uneven distribution of hydrogen and coal paste in the converter, severe erosion of the heavy oil discharge mechanism, and a lack of flow control instrumentation that hindered uniform operating conditions. Nevertheless, the navy valued the work at Tokuyama, particularly the investigations on Ôyama coal which SMR tested at its research laboratory in Dairen and its industrial-size plant in Fushun.[71]

The hydrogenation program at the Imperial Fuel Research Institute operated independently of the navy and of any industrial organization. It ran from 1931 to 1940, with expansion planned to the full demonstration stage.

---

70 Japan Hydrogenation, Box 184, Imperial Chemical Industries (ICI) Archives, London; Matthias Pier Collection, 1104/69 (Japan), 'Aktennotizen' of 3, 4, 5 July 1939, BASF Archives, Ludwigshafen; "Synthetic Fuels in Japan," *Chem. Trade J.*, 104 (17 February 1939): 162, Akira Kudo, "Strategy of IG Farben Toward Japan-Case of Synthetic Oil," *Japan Business History Review*, 22 (1987): 1–28 (in Japanese). Brunner Mond was one of four companies that merged to form ICI in 1926. Brunner Mond (Japan) in 1939 was a ¥ 4 million company that acted as a distributing organization for ICI's constituent and subsidiary companies, "Chemical Industry in the Far East," *Chem. Trade J.*, 104 (26 May 1939): 496.

71 Nav. Tech. Miss. Japan (note 19), pp. 12–13; "Synthetic Oil in Japan" (note 2), pp. 335–36. When the US Naval Technical Mission inspected the laboratory after the war they found relatively few documents on this research.

Researchers at the institute tested about one hundred Japanese coals and coal tars in a small 5-liter autoclave, showing the suitability of most coals, including Fushun, Hokuku, and Bibai bituminous, for hydrogenation. In 1934 they constructed a pilot plant to test the three best coals. The plant, like the navy's, contained one liquid phase converter that had a 3–5 metric ton/d coal-oil paste input (50% coal) and ran for 5,000 hours. They simplified its design features to give maximum operability while the institute studied the engineering and economics of an industrial-size plant.[72] Japanese coals tested well, but because of the institute's independent operation its Kawaguchi (Tokyo) plant contributed little to the industrial development of coal hydrogenation before its shutdown in 1940.[73]

The navy and SMR took the lead in designing and constructing the first industrial-size plant at Fushun. Construction began in 1936 after nearly ten years of cooperation between the Naval Research Laboratory and SMR's Central Research Laboratory. T. Miyama served as Fushun's technical director from 1937 to 1942, R. Abe was the assistant director. Preliminary operation started early in 1939 hydrogenating Ôyama coal at 430° C and 200 atm in a single, liquid phase converter. Addition of a second converter later that year brought the plant to its 10,000 metric ton/yr coal capacity (40% coal-60% oil paste input of 4 metric tons/hr). SMR invested ¥ 16 million in the plant.

SMR and the navy initially disagreed on the choice of catalyst. The navy wanted zinc chloride (1% by weight of coal), but SMR objected because of its cost and corroding effect, and sulfide catalysts that SMR had tested for several years became the preferred catalysts. Water gas obtained from coke and steam supplied hydrogen gas.[74] The navy insisted on using only Japanese-made machinery and equipment, arranging to manufacture the high-pressure converters at its Kure Naval Yard. They measured 7.3 m long, 95 cm inner diameter, and were the largest Japanese steel companies could manufacture at the time. The navy's insistence on domestic manufacture proved unfortunate, because two of the problems that plagued Japan's synthetic fuel program were the lack of good heat-resistant alloy steels and facilities for constructing converters of sufficient size. The navy also intended to design converters with propeller-type stirring devices, but SMR argued that they would leak badly. The

72 Goddin and Thornton (note 2), p. 124.

73 Nav. Tech. Miss. Japan (note 19), pp. 13–14. The papers are those cited in note 24.

74 Goddin and Thornton (note 2), p. 125. The water gas reaction requires a 95% iron-5% chromium catalyst and a temperature of 540° C. A coal hydrogenation plant with a 10,000 metric ton input should yield about 5,000 kL of liquid fuel.

TABLE 5.10 Fushun production of gasoline by hydrogenation

Product	Weight (%)
Aviation gasoline	17.4
Motor gasoline	17.7
Cresols and phenols	3.2
Gasols	3.5
Pitch coke	20.0
Gas and losses	38.2

SOURCE: COURTESY OF AUTHOR

converters finally installed contained no stirrers and consisted of two 7.3 m sections bolted together.[75]

Fushun operated with only liquid phase hydrogenation until the addition of two vapor phase converters in February 1940. The two converters, arranged in series processed daily 40 kL of liquid phase heavy oil and low-temperature tar at 200 atm, 430–440° C, with a fixed-bed vapor phase catalyst of 40% molybdenum sulfide and 60% acid clay. Fushun never met expectations. The combined liquid-vapor phase gasoline yield, 50% of it aviation quality 91–92-octane number, resulted from the hydrogenation of about 35% by weight of coal charged. In view of the Fushun plant's poor performance coal hydrogenation ceased in July 1942, and from that time until the war's end Fushun manufactured aviation gasoline by hydrocracking gas oil and kerosene.[76]

Because of the Fushun plant's location in Soviet-occupied Manchuria, US military authorities rejected the Naval Technical Mission's plan to inspect the plant and obtain operating data and production figures. The Naval Technical Mission's inability to visit Fushun resulted in its obtaining incomplete and sometimes contradictory information. Miyama, Fushun's technical director 1937–42 and from 1942 a technical director at Imperial Fuel Development

75 Goddin and Thornton (note 2), p. 5; Nav. Tech. Miss. Japan (note 19), p. 15. The Ishii Iron works in Kamata, near Tokyo was one of the firms that manufactured equipment for the coal liquefaction and Fischer-Tropsch synthesis, "A New Japanese Chemical Plant Maker," *Chem. Trade J.*, 104 (7 April 1939): 340.

76 "Synthetic Oil in Japan" (note 2), pp. 336–37; Munehiro Miwa, "The Naval Fuel Depot and Coal Liquefaction," *Kaqakushi Kenkyû* (*Journal of the History of Chemistry, Japan*), 4, no. 41 (1987): 164–75 (in Japanese).

Company in Tokyo, when interrogated on 11 and 17 January 1946 said the Fushun plant had experienced no particular mechanical operating difficulties. He provided operating details from his personal notes, explaining that a bombing attack on Tokyo had destroyed his home and the official plant documents he had in his possession. Naosuke Shônô, assistant to the chief engineer at Imperial Fuel Development Company and a hydrocracking specialist, disagreed with Miyama's assessment, but gave no specifics. Fujimoto at the Ôfuna First Naval Fuel Depot reported that from 1939 to the cessation of coal hydrogenation in July 1942 Fushun operated irregularly, but by that time a reasonably successful process had emerged.

According to Miyama, Fushun from 1939 to 1945 produced 15,000 kL of mostly heavy oil. Yet, the Naval Technical Mission reported oil production of only 770 kL in 1942 and 690 kL in 1944, revealing a huge discrepancy between the two figures. Fushun also had no clear purpose. Miyama and Shônô called Fushun an experimental plant and not industrial-size, despite its large size. But the Naval Technical Mission believed that because of Japan's extreme need for oil the Fushun plant if successful would have operated continuously and at its rated capacity.

The remaining coal hydrogenation plant at Agochi, in northern Korea, was also in Soviet-occupied territory, and the few details on its operation came from interrogating Shônô and Mitsui at the navy's Ôfuna Fuel Depot. Chôsen Sekitan Kôgyô KK (Chosen Coal Industry Co. Ltd.) built the plant beginning in 1937 at a cost of ¥ 30 million, and it went on stream in 1939 with an annual coal capacity of 20,000 metric tons (5 metric tons/hr of a 40% coal-60% oil paste). According to Mitsui, the navy and Chôsen Jinzô Sekiyu operated the plant, but during the war it came increasingly under the navy's control. The plant had an expected annuai production of 10,000 kL of heavy oil from the liquid phase hydrogenation of Agochi brown coal, but Shônô indicated that the plant's two converters (each 13 m long and 1.60 m inner diameter) failed to liquefy coal successfully and produced excessive coking because of severe overheating in the reaction zone. The inability of the converter's thin walls to withstand operating pressures of 200 and 250 atm also contributed to Agochi's failure, and by autumn 1942 the navy ordered its conversion to synthetic methanol production.

Agochi's engineers gave incomplete and contradictory production figures, like those given for Fushun. The Naval Technical Mission reported that the Agochi plant failed to produce any synthetic fuel before its conversion to synthetic methanol production, whereas reports from plant personnel such as the engineer Eiji Munekata, indicated that for one month in 1942 the plant produced daily 103 kL of synthetic fuel. Agochi had an expected annual pro-

TABLE 5.11 Coal hydrogenation plants

Company	Location	Designed capacity	Type of operation	Catalyst	Output kL (actual)
Manshû Jinzô Sekiyu Co. (SMR-Japanese Imperial Navy) ¥ 16 million	Fushun, Manchuria	2.5 metric ton/hr 10,000 ton/yr Ôyama coal	liquid phase: 200 atm, 430° C; vapor phase: 200 atm, 430° C 1st stage, 440° C, 2nd stage	liquid phase: precipitated FeS; vapor phase: 40% MoS_3 + 60% acid clay; Fe_2O_3; Fe_2O_3:S: $Sn(OH)_2$ 100:17:1; $ZnCl_2.2H_2O$; MoO_3, $Sn(OH)_2$	1941 = 0 1942 = 770 1943 = 0 1944 = 690 1945 = 0
Chôsen Sekitan Kôgyô KK (Chosen Coal Industry Co. Ltd.) Chôsen Jinzô Sekiyu Co. ¥ 30 million	Agochi, Korea	5 metric ton/hr paste, 50,000 ton/yr Agochi brown coal, 10,000 kL/yr heavy oil	200–250 atm; 1939–42 unable to function, 1942–45 converted to methanol synthesis	See above	0
Ube Chisso Kôgyô KK (Ube Nitrogen Industry Co. Ltd.) ¥ 12.5 million	Ube proposed				
Chôsen Sekitan Kôgyô KK (Chosen Coal Industry Co. Ltd.) ¥ 10 million	no location proposed				
Total					1,460

SOURCE: COURTESY OF AUTHOR

duction of 10,000 kL, which makes the daily production figure seem high, but the difference could have resulted from the method of reporting production figures. The Naval Technical Mission in reporting synthetic fuel production at both Fushun and Agochi included only the lighter, lower boiling fractions, specifically gasoline, kerosene, diesel oil, and fuel oil. Adding the very heavy, high boiling fractions of the first-stage, liquid phase hydrogenation such as pitch, lubricating oils, cresols, and gas oils would give a higher yield, and thereby account for the difference in production.

In addition to the serious technological problems, Agochi's engineers also had to endure the navy's interference in both the administration and operation of the plant. Naval personnel assigned to the plant had no training for their duties and often were a nuisance to the civilian engineers. They promoted secrecy and hindered the dissemination of research ideas, even to the extreme of distrusting Ôshima, the Agochi plant's director. The navy insisted on using its own process and tried to dictate the choice of catalysts and heating coils for the reactors, both of which proved inferior and which civilian engineers replaced.[77] That the Japanese abandoned coal hydrogenation at both the Fushun and Agochi plants so early in the war, convinced the Naval Technical Mission, despite having inspected neither plant, that the Japanese program never reached the same degree of advancement as the German and British programs.[78]

8 Japan's IG Farben Connection

At a meeting in Berlin on 27 September 1940, Japan, Germany, and Italy signed the Tripartite Pact. To Britain and the United States, which had not yet entered the war, the Tripartite Pact "gave the impression of a united fascist force rolling toward global domination." But many Japanese leaders did not trust Germany nor did they want their "Greater East Asia" policy linked to Hitler's "New Order" in Europe.[79] This same element of distrust permeated the Japanese-IG Farben contract discussions on high-pressure coal hydrogenation.

Nor did harmony exist between the Japanese navy and army on developing coal hydrogenation. The army had grown impatient with the navy's slow progress and joined by Ogura Sekiyu (Oil) in Yokohama opened discussions with IG Farben in 1938–39. An IG Farben document of 3 May 1939 described the construction of a 200-metric ton/d gas oil hydrogenation plant producing 90-octane number leaded gasoline (0.9% lead tetraethyl). The plant's total cost, RM 8.1 million ($3.24 million, ¥ 14.1 million) included IHEC's license fee, which varied according to the plant's size and complexity, but the navy

77 Nav. Tech. Miss. Japan (note 19), pp. 14–15; "Synthetic Oil in Japan" (note 2), p. 337; "Hydrogenation Industry" (note 2), p. 8.

78 Nav. Tech. Miss. Japan (note 19), p. 15; Goddin and Thornton (note 2), pp. 122, 125; Molony (note 3), pp. 228–31. According to Molony, Chôsen Jinzô Sekiyu's sister company, Kilin Artificial Oil, planned to build a second synthetic fuel plant in Kilin (Jilin) Manchuria but never completed its construction. This location was the same as that of an incomplete F-T plant and the two proposed plants may be the same. Coal hydrogenation and F-T are both synthetic processes, and sometimes both are called coal liquefaction.

79 Bernard Bailyn et al., *The Great Republic: A History of the American People*, Vol. 1 (Boston: D. C. Heath and Co., 1977), p. 1169.

claimed imminent development of its hydrogenation process and successfully blocked construction.[80]

The same document revealed discussions with Japanese firms on eleven other hydrogenation projects, but once again none resulted in any plant construction. One of the discussions, with Dr. S. Endo, a chemical engineer with the navy and a Nissan Ekitai director, considered adding a hydrogenation plant to Nissan's LTC plant in Wakamatsu. Karl Ruhl, IG Farben's representative in Japan, estimated that a plant hydrogenating 13,500 metric tons of tar with an expected annual gasoline yield of 10,000 metric tons would give IG Farben a profit of ¥ 1.5 million ($0.35 million). Another proposal submitted to Ube Chisso in June 1937 called for a 20,000-metric ton/yr coal and tar hydrogenation plant, followed by a discussion in December 1938 for a plant hydrogenating annually 40,000 metric tons of tar to 28,400 metric tons of aviation and automobile gasoline. IHEC suggested charging Ube Chisso a RM 2.5 million ($1.0 million) license fee. A third project considered the construction of a Manshû Yuka Kôgyô KK (Manchurian Liquefaction Industry Co. Ltd.) plant in Hsinking, Manchuria. Manshû Yuka, which had reorganized with a capitalization of ¥ 20 million, planned to hydrogenate tar obtained from LTC using its own Kuroi process. Its representative, Mr. Jun Hama, while in Berlin on 31 October 1938 outlined the company's plans to produce 1 million metric tons of tar and to convert one-third of it (350,000 metric tons) to gasoline using both the F-T synthesis and SMR-Navy hydrogenation process. The proposed plant would increase Manshû Yuka's gasoline production from 10,000 to 50,000 metric ton/yr.

Other enquiries mentioned in the document came from Nippon Yuka KK (Nippon Liquefaction Co. Ltd.) and Sumitomo Kagaku KK (Sumitomo Chemical Co. Ltd.) A Mr. Haga who represented Nippon Yuka in New York City visited Ludwigshafen on 1 September 1938 to discuss constructing a plant with an annual production of 10,000 metric tons of gasoline from bituminous coal. But Nippon Yuka showed no interest in paying a license fee. Sumitomo Kagaku's three representatives, Shinroku Nakao, Shinjirô Kodoma, and Jirô Shioya, met

80 Pier Collection, 1104/69 (Japan), 'Zusammenfassung' of 3 May 1939 and 'Aktennotizen' of 3, 4, 5 July 1939 (note 70); E. B. Peck and Irvin H. Jones, "Technical Assistance on Synthetic Oils Rendered the Japanese by the IG Farbenindustrie AG," *Combined Intelligence Objectives Sub-Committee*, APO 413 (July 1945), pp. 1–13. At the end of May 1939 Ogura Sekiyu merged with Mitsubishi and Nihon Sekiyu to form a new industrial organization; Nav. Tech. Miss. Japan (note 19), p. 9.

with IG officials in Ludwigshafen and Leuna on 7 December 1937 and later in London with Leonard Patrick of ICI of and Mr. Tillmann of IHEC.[81]

Throughout the 1930s, IG Farben maintained continuous though limited contact with the Japanese navy, army, and industry but had no later discussions with any of the above-mentioned companies. The navy most actively pursued negotiations, but declined to pay IG Farben's license fee, which was the same as IHEC's prewar fee, generally RM 5 million for a plant hydrogenating 100,000 metric tons of coal. The navy objected mainly because it considered the prospect of having to pay for failing to develop its own process embarrassing and instead offered to purchase plant equipment at inflated prices. Negotiations with the navy resumed early in 1941, when Matsuoka, SMR's president and now Japan's foreign minister, met with IG Farben officials in Ludwigshafen to discuss acquisition of a coal hydrogenation license.

Another meeting took place in Ludwigshafen on 3 November 1941, at which time Japanese naval officials informed Pier of the many difficulties they encountered with their small hydrogenation plant at Tokuyama. Operating problems continued to plague Japan's two industrial plants at Fushun and Agochi, and by the end of 1942, when these plants ceased to operate, negotiations between the Japanese and IG Farben went very slowly.

Fritz Ringer, an agent in IG Farben's Petroleum Department who handled negotiations with the Japanese, believed two factors contributed to the delay. The first arose from jurisdictional disputes between Mitsubishi-Army and SMR-Navy, each of whom was developing its own process, and the second because the Japanese while negotiating were attempting to circumvent IG Farben and obtain hydrogenation equipment from Friedrich Krupp AG in Germany.[82]

IG Farben also delayed negotiations. It tried to avoid giving either the army or navy exclusive rights to the process and succeeded until late November 1944. IG Farben considered the Japanese potential rivals in the production of oil from coal and wished to stall as long as possible any sharing of valuable

---

81 Pier Collection, 1104/69 (Japan), 'Besprechung' of 9 June and 1 July 1939 (note 70); Peck and Jones (note 80), pp. 2–3, 10. The documents also included arrangements with Friedrich Krupp AG, ICI, and Bethlehem Steel in the United States. R. Holroyd, ed., "Report on Investigations by Fuels and Lubricants Teams at the IG Farbenindustrie AG," *US Bureau of Mines Information Circular 7375* (August 1946), pp. 1–75, (p. 60). In a brief note "Chemicals in the Far East," *Chem. Trade J.*, 108 (9 May 1941): 284, Kyôwa Kagaku Kenkyûsho KK made the remarkable claim that it had produced isooctane from the direct fermentation of starch and that Japanese aviation authorities had tested and approved it.

82 Akira Kudo, "The Tripartite Pact and Synthetic Oil," unpublished manuscript, courtesy of author.

FIGURE 5.1 Guests at luncheon honoring General Hiroshi Ôshima, Ludwigshafen, 17 May 1943. From left to right facing: Matthias Pier BASF-IG Farben; Hermann Waibel, IG director, business manager, and trade export manager, IG Farben; Martin Müller Cunradi, director of Oppau Works, General Hiroshi Ôshima, ambassador to Germany; Hermann Schmitz, head of IG Farben; General Mitsuhiko Komatsu, military attaché to the Japanese embassy in Berlin
SOURCE: PIER COLLECTION, BASF ARCHIVES, LUDWIGSHAFEN

information with them. By this time the navy had abandoned hydrogenation. When the German government instructed IG Farben to conclude an agreement with its ally, IG Farben finally signed a general licensing contract with the Japanese army on 11 January 1945.[83]

The German government's instructions called for IG Farben to provide the Japanese with "hydrogenation know-how" related to the preparation of gasoline and diesel oil, and IG Farben limited its disclosure precisely to that information. The Japanese obtained little or nothing on the preparation of isooctane and aviation gasoline, on the DHD hydrocracking process (Druck-Wasserstoff-Dehydrierung), on the preliminary hydrogenation process (Vorhydrierungs Verfahren), on the complete operations with catalyst 6434, that gave a gasoline of 91-octane number (leaded), and on the manufacture of synthetic lubricants.

83 Vertrag zwischen dem Dainippon, Teikoku Rikugun Daijin [Kaiserlich Japanischen Heeresminister (The War Office of Japan)] und der IG Farbenindustrie Aktiengesellschaft, Frankfurt am Main, 11 January 1945, Pier Collection, 1104/69 (Japan) (note 70).

FIGURE 5.2 Cover of Luncheon announcement honoring General Ôshima
SOURCE: PIER COLLECTION, BASF ARCHIVES, LUDWIGSHAFEN

The IG Farben-Army contract signed by the War Office of Japan (Dainippon Teikoku Rikugun Daijin) acting for the Imperial Japanese Army and written in both languages, consisted of fifteen pages in German, eighteen pages in Japanese. Each party received the two translations. Major General Mitsuhiko Komatsu, Military Attaché to the Japanese Embassy in Berlin, represented the War Office. Other participants included General Osamu Otani, the army officer supposedly responsible for successfully concluding the negotiations before his capture by American forces in south Germany; Colonel Yoshida, who returned to Japan in late 1944; Major Tosishada Kinoshita, a German-speaking army engineer who carried on most of the discussions; and Drs. Kani and Fukao of Mitsubishi, who had experience in petroleum refining. Ringer, Pier, and Heinrich Bütefisch represented IG Farben.

According to the contract, IG Farben granted the army rights to the hydrogenation process in Japan proper, Manchukuo, and China for the production of gasoline, diesel and jet fuel, kerosene, oils, and waxes from crude petroleum, bituminous coal, brown coal (lignite), peat, wood, shale, and other raw materials. It permitted access to IG Farben's operating techniques, patents, and catalysts as long as the information did not violate the 1931 IG-IHP-IHEC contractual arrangement that controlled the international rights to the high-pressure coal hydrogenation process.

The Japanese army agreed to accept all responsibility "for the legality of any compulsory licensing of IHP patents" and to pay IG Farben's price of RM 18 million ($7.2 million) with no future royalty for the information. Payment schedule required 20% on closing the contract, 30% within six weeks after delivering to the army's delegate in Germany the information essential for constructing the first industrial-size plant, and 50% within five years after closing the contract (10% annually). The army and all persons in charge of construction or connected to the program pledged to keep secret the process and to develop it exclusively in the contract's areas of Japan, Manchukuo, and China. Should any dispute arise regarding the contract's terms or language the two parties should solve it in a friendly manner, giving preference to the German-language version. The contract remained in effect for ten years, after which time the army could use the patents without cost.

The contract also contained a clause (Number 5, Technical Support) requiring payments to IG Farben for designing each plant, supplying equipment and personnel, and training Japanese operators. IG Farben set up a design department in Heidelberg, which it later transferred to Unterlockwitz, and agreed to a RM 800,000 payment for designing the first plant. But the Japanese built no industrial-size hydrogenation plants in 1945, and this part of the contract never went into effect.

Because IG Farben and the Japanese army reached an agreement so late in the war two questions require an answer. Why did the signing take place at all and what did the Japanese expect to benefit from the signing? One reason for the signing was the German government's instruction that IG Farben conclude an agreement with the Japanese.[84] IG Farben cooperated because it, and not the German government, decided how much and what information to give the Japanese. Nor did contracts with the Japanese represent any great change in IG Farben's Far East policy. It had been a principal sponsor of Japan's synthetic ammonia, fertilizer, and pharmaceutical industries since the 1920s and had business dealings with the Japanese well before Germany and Japan established a political axis. At the time of World War I, goods sold in Australia that carried the label "Made in Germany" also had a Japanese trademark.[85] To some extent the 1945 contract represented only a continuation of IG Farben's longtime business arrangements with Japan.

84 Holroyd (note 81), pp. 58–59.
85 Borkin and Welsh (note 10), p. 48.

While the Japanese delegation remained in Germany negotiating the terms of the hydrogenation contract, the German army made its final and futile grasp for victory at the Battle of the Bulge (16 December 1944–16 January 1945). The negotiations continued despite Germany's certain defeat. Because the SMR-Navy program had failed, Japan's desperate need for petroleum forced Mitsubishi-Army to look once again to IG Farben for assistance at a time when the Japanese military situation also approached disaster. American victories at Saipan in the Marianna Islands (15 June-19 July 1944) and in the Philippines (19 June 1944–23 February 1945) resulted in the loss of heavy oil from the South Seas, making a petroleum shortage inevitable. With the United States controlling sea and sky, preventing petroleum imports, Japan did not have many options other than to try once again to develop its synthetic fuel industry.

Preliminary negotiations began in Heidelberg in the spring of 1944 and turned serious by November. IG Farben often held meetings in Heidelberg because many of its officials, including Pier, lived in Heidelberg located a short distance from Ludwigshafen and relatively untouched by Allied bombing. During the negotiations the Japanese implied that they had constructed three, 300 atm, liquid and vapor phase plants, each producing 50,000 metric tons (68,800 kL) of gasoline from the hydrogenation of low-temperature tar. The US Naval Technical Mission, however, uncovered no evidence of any plant construction, and the Japanese mentioned only one location, at Kinsei (Kiensi or Chinsi) in Manchuria, southwest of Mukden and opposite Koro Island.

In December Japanese army officials met again with IG Farben's representatives in Heidelberg and informed them that they had abandoned the Kinsei site and planned to construct a plant on Sakhalin Island. The Sakhalin proposal called for a considerably larger 300 atm plant, producing 93,000 metric tons (128,000 kL) of aviation gasoline from 600,000 tons of Utibuti (Sakhalin) coal. The most economical plan put forward at the meeting called for hydrogenating coal and low-temperature tar. After signing the contract on 11 January 1945, the Japanese received substantial operating details on the Scholven plant in the Ruhr. Scholven was the first German plant to hydrogenate bituminous coals similar to Japanese coals, and the Japanese planned to follow IG Farben's procedure at Sakhalin. IG Farben also provided limited information on three hydrogenating catalysts, 3510, 5058, and 6434. One month later, on 8 February 1945, Major Kinoshita identified the two other plant sites as Mukden (Manchuria) and Hokkaidô on the Japanese mainland.

During the entire period of negotiations, Ruhl remained in Japan as IG Farben's representative. Ruhl specialized in nitrogen fixation and did not have the most recent information on coal and tar hydrogenation or have close contact with developments in Germany. Ruhl sent no reports and communicated

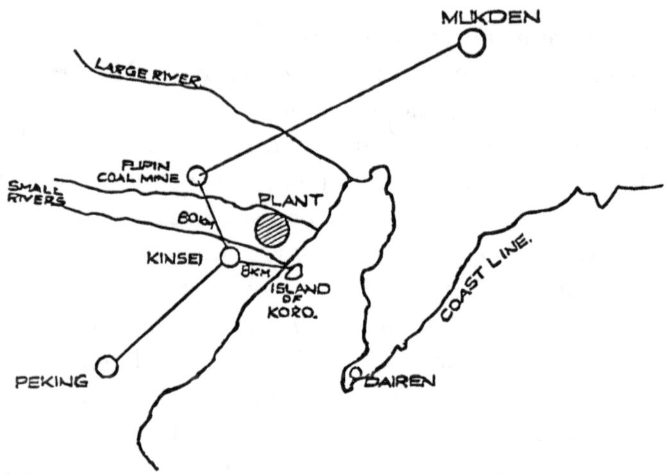

FIGURE 5.3 Proposed coal hydrogenation plant at Kinsei
SOURCE: R. HOLROYD, ED., "REPORT ON INVESTIGATIONS BY FUELS AND LUBRICANTS TEAMS AT THE IG FARBENINDUSTRIE, AG," *US Bureau of Mines Information Circular 7375* (AUGUST 1946), PP. 1–75 (FOLLOWING P. 58)

with IG Farben only by wireless telephone (radiotelephone) every seven to ten days. Instead, the frequent visits the Japanese made to IG Farben's Ludwigshafen and Leuna plants, which IG Farben reluctantly permitted, provided most of the information. In fact, a Japanese delegation traveled to Heidelberg only eight days before American troops occupied the city in late March 1945.

In interviews with members of the Technical Oil Mission (CIOS) and other allied investigators who went to Germany in May 1945, both Pier and Ringer revealed that IG Farben withheld information on lead tetraethyl and prevented inspection of plants that manufactured it. IG also provided no information on jet fuel, which it obtained as middle oil upon hydrogenating brown coal. But Pier and Ringer pointed out that General Otani, who had a major role in the negotiations, spent much time in Berlin on aviation affairs. They thought he could have received information on jet fuel from the Reichsministerium and on jet engines from the Junkers or Messerschmidt firms.

Ringer said that the Japanese had a particular interest in catalyst manufacture, and in order to help them improve their catalysts IG Farben began sending selected information on plant designs and catalysts in December 1944, one month before signing the contract. The information went by submarine, an eight-week trip, but Ringer acknowledged not knowing whether the information reached Japan. IG Farben also agreed reluctantly, after repeated Japanese urging, to send two or three of its catalyst experts to Japan. They planned to leave by submarine in May 1945, but with the war in Europe ending at that time no trip took place. In fact throughout the entire negotiation period the Japanese clearly conveyed the urgency of their situation while IG Farben always responded slowly.

Pier, Ringer, and Donath, Pier's chief assistant, had little complimentary to say about the capabilities of Japanese technicians who visited the German plants after the contract's signing in 1945. Donath got the impression "that they were either not experts or dissembled well." Pier and Ringer believed "they were not well-informed or clever technologists" and doubted "whether they could have accumulated sufficient information to be of practical advantage to Japan." The German scientists agreed unanimously that any synthetic fuel obtained from coal or its derivatives would have only a minor influence on the war's outcome.[86]

9 Conclusion

At the time IG Farben entered into negotiations with the Japanese army, Germany had practically twenty years' experience in the industrial production of synthetic liquid fuel. Twelve coal hydrogenation and nine F-T plants converted coal and tar into liquid fuel and lubricants. Japan, too, had considerable experience in synthetic liquid fuel production. But in 1945, after experimenting for twenty years, Japan had only three F-T and no coal hydrogenation plants in operation. Unlike the German plants, the Japanese plants had no effect on the outcome of World War II. The older and technologically simpler LTC and distillation process produced most of the liquid fuel the Japanese obtained from coal and shale oil.

The Seven-Year Plan of 1937 had called for the annual production of 12.6 million barrels of synthetic gasoline and fuel oil. Yet in 1944 Japan produced only 750,000 barrels, all of which came from F-T and LTC plants. Oil shale distillation contributed 582,000 barrels. The Japanese carried out considerable laboratory research on coal hydrogenation, more than any other country except Germany, and developed their own standard iron oxide-sulfur-tin hydroxide [Fe_2O_3–S–$Sn(OH)_2$] catalyst that performed reasonably well. They also obtained rights to the F-T process from Ruhrchemie for ¥ 7.2 million, and chemists such as Kita spent years developing cheaper iron catalysts to replace

86 Pier Collection, 1104/69 (Japan), reports of 20 February, 8 March, 25 March 1945 (note 70); Peck and Jones (note 80), 1–6. Donath told me the same thing when I interviewed him in August 1982. According to *US Bureau of Mines Information Circular 7375*, Kinsei was to produce 70,000–100,000 metric ton/yr of aviation gasoline from Fupin coal.

Ruhrchemie's more expensive cobalt-thorium catalysts. Yet their iron catalysts always gave lower yields.

In their rush to synthesize liquid fuel from coal, the Japanese leaped into industrial-size production lacking sufficient pilot plant experience, and consequently neither hydrogenation nor F-T proved particularly productive. The Japanese enjoyed their greatest success with LTC, the simplest of the three coal conversion processes. It required no expensive and elaborate high-pressure equipment or extensive catalyst research, but even in LTC plants the German designed Lurgi or Koppers and American-designed Knowles retorts performed best. In fact, the Japanese synthetic fuel experience seems to support the often-made charge that the Japanese excelled as adapters or modifiers of existing technologies but lagged in creativity and innovation. Such an assessment, however, ignores or insufficiently acknowledges their successful laboratory research. Japanese scientists practiced sound laboratory science, they lagged in their technological scale-up.

The Japanese synthetic fuel experience therefore requires scientists, engineers, and historians of science and technology to examine why some scientific ideas, technologies, or inventions succeed while others fail. Is it the innovation's intrinsic quality or social-economic events that determine success or failure? For the synthetic fuel industry both answers are correct. In Germany, coal hydrogenation succeeded technologically six years before the Nazi government rose to power in 1933 and proceeded to subsidize industrial-size development. Industrialization may not have occurred otherwise. ICI's technologically successful coal hydrogenation process received similar economic incentives from the British government in the 1930s. The F-T synthesis, though not as advanced as coal hydrogenation in Germany and non-existent in Britain, nevertheless succeeded technologically but failed economically. The United States Bureau of Mines' post-World War II synthetic fuel program succeeded technologically, but because of unfavorable economics and the abundance of cheap petroleum, neither coal hydrogenation nor F-T achieved industrialization in the United States.

In Japan the opposite happened. The Japanese government provided every possible economic and social incentive in the 1930s–40s. Japanese investigators understood the science of synthesizing liquid fuel from coal and tar and carried out a successful laboratory research program. Technologically the production of synthetic fuel failed because the impatient Japanese tried to make the transition from a small laboratory process to industrial-size production without first developing a successful intermediate pilot plant operation. The pre-World War II embargoes and the war clearly compounded Japan's urgency. In addition, the navy's interference at the two coal hydrogenation plants hindered synthetic fuel production. But the leap from the laboratory to industrial development proved too much for the Japanese to overcome. This same impatience and rush to industrial production largely contributed to the failure the Japanese initially experienced when they tried to establish a soda ammonia industry in the 1920s, although both cases also clearly indicate the generally lower level of Japan's technological development in the pre-World War II period.

By the 1930s the Japanese had recognized that the lack of petroleum seriously would hinder their long-contemplated Asian expansion. The very real possibility of a major Pacific war led to their establishing a large reserve of crude petroleum and refined products and to developing a synthetic fuel industry within their empire to guarantee independence from foreign petroleum supplies. But in the end the Japanese empire collapsed and thereby eliminated the sources of crude petroleum for Japan's refineries. The failure of the Japanese synthetic fuel industry completed the collapse.

CHAPTER 6

Canada's Mines Branch and Its Synthetic Fuel Program for Energy Independence

1 Introductory

In the mid-1970s many of the industrialized nations started major synthetic fuel programs aimed at increasing their domestic petroleum production. The 1973–74 Arab oil embargo that resulted in a fivefold jump in petroleum prices from about $2/bbl to $10/bbl sparked their concern. The embargo's shock hardly had waned before another oil shock aggravated the situation. The 1979–81 Iranian hostage crisis tripled prices to $30/bbl, a skyrocketing global petroleum demand soon pushed prices into the $40/bbl range.

The 1970s were not the first time that industrialized nations had tried to lessen or even eliminate their dependence on costly and sometimes erratic supplies of imported petroleum and move towards energy independence. By the 1920s the mass production of automobiles, the conversion of ships from coal to oil-fired, the technological transformation occurring in the Western world, and the introduction of the airplane showed clearly that petroleum reigned as the fuel of the future. Recognition of this increasing dependence on liquid fuels led to the construction in the late 1920s and early 1930s of the first coal hydrogenation (liquefaction) plants in petroleum-poor Germany and Britain.

A similar concern emerged in the United States after World War II. The war showed how quickly the United States could consume huge quantities of petroleum and led the Bureau of Mines to establish a coal-to-oil demonstration program that ran from 1949 to 1953. In neighboring Canada, officials at the federal government's Mines Branch had worried about Canada's lack of significant petroleum deposits from the first decade of the 1900s. They too believed that production of synthetic liquid fuel might eliminate possible energy shortages. Initially, the Mines Branch, at its Fuel Testing Station in Ottawa, Ontario, concentrated on high-pressure coal hydrogenation, but by the 1940s the emphasis had shifted to hydrogenating bitumen, a very heavy slow flowing low-grade oil, extracted from Alberta's vast oil (tar) sands, to give a high-quality oil.

The Mines Branch assumed a leadership role in Canadian synthetic liquid fuel research. The government actively supported the research, and in this

 | DOI:10.1163/9789004690912_007

respect the Mines Branch program resembled those that emerged in Germany, Britain, Japan, and the United States. But while synthetic fuel research outside of Canada reached the industrial stage, Canada's Mines Branch program expanded to the pilot-demonstration plant stage. A move to industrial development did not occur because the Mines Branch lacked science-engineering expertise, but because the Mines Branch committed to basic research rather than industrial development of Canada's low-grade oil reserves, such as Alberta's oil sands, Canada's most abundant natural resource.

The Mines Branch began research on a hydrocracking process it called canmet in 1970 and in the ensuing decades continued its basic and pilot plant research before entering into a commercial licensing agreement with Universal Oil Products (UOP) of Chicago in 2007.[1] An examination of the technological evolution and economics of the Mines Branch's twentieth-century synthetic fuel program follows and provides a synthesis and analysis of the prominent role that the Mines Branch, renamed Canada Centre for Mineral and Energy Technology (CANMET) in 1975, has played in Canada's energy development.

1 Only a few countries such as Germany, Britain, the United States, Japan, and Canada attempted to establish synthetic fuel industries. The literature on their development is rather limited, much of it deals with the German synthetic fuel industry. See Harald Beck, "Zur Geschichte der Kohleverflussigung," *Kultur and Technik*, 7 (1983): 164–73; Manfred Rasch, "On the Prehistory of the Liquefaction of Coal up to 1945," *Energy in History*, Hans-Joachim Braun and Wolfgang König, eds. (Düsseldorf, 1984), pp. 478–86; Anthony N. Stranges, "Friedrich Bergius and the Rise of the German Synthetic Fuel Industry," *Isis*, 75 (1984): 643–67; Arnold Krammer, "Fueling The Third Reich," *Technology and Culture*, 19 (1978): 394–422; Wolfgang Birkenfeld, *Der synthetische Treibstoff 1933–1945* (Gottingen: Muster-Schmidt Verlag, 1964). For developments in the United States see Kenneth Mernitz, "Progress at a Price" (unpublished PhD dissertation, University of Missouri, 1983); Joseph Marchese, "Government and Energy: The Demonstration Program for Synthetic Liquid Fuels, 1944–1955" (unpublished PhD dissertation, Princeton University, 1983); Anthony N. Stranges, "Synthetic Petroleum from Coal hydrogenation: Its History and Present State of Development in the United States," *Journal of Chemical Education*, 60 (1983): 617–25. The only article on the history of the British synthetic fuel industry is Stranges, "From Birmingham to Billingham: High-Pressure Coal Hydrogenation in Great Britain," *Technology and Culture*, 26 (1985): 726–57. Anthony N. Stranges, "Synthetic Fuel Production in Prewar and World War II Japan: A Case Study in Technological Failure." *Annals of Science* 50 (1993): 229–65. The reports of the 1945–46 US Navy's Technical Mission to Japan, Microfilm x-38(N)-7, German Document Retrieval Project, Synthetic Fuel Collection, Texas A&M University Archives, are the main source on Japan's synthetic fuel industry. The Naval Technical Mission remained in Japan from 4 September 1945 to 1 November 1946. It published some of its reports in technical journals from 1946 to 1948 and a complete microfilm set in 1974.

2 Early Concern for Fuel Supply

The Geology and Mines Act of 27 April 1907, officially established the Department of Mines with its two divisions, the Geological Survey and the Mines Branch, under the Minister of Mines' control and management. Neither division was entirely new. The Geological Survey had existed since 1842, becoming a branch of the Interior Department in 1877, while an organization with functions and duties similar to the Mines Branch had existed within the Department of Interior since 1901.[2]

Eugene Haanel (1841–1927) served as Superintendent of Mines in the Department of Interior with headquarters in Ottawa since 5 June 1901, and he continued as superintendent after establishment of the twenty-one member Mines Branch in 1907.[3] Haanel, a German-born mineralogist, had been professor of science at Victoria University in Cobourg, Ontario, for fifteen years and spent the last twelve years at Syracuse University before returning to Canada. During his nineteen years as Mines Branch superintendent, fuels received priority, at first solid fuels, especially Ontario's and Quebec's vast peat bogs, and by 1910 Canada's oil reserves.[4]

In the early years of the twentieth century a global emphasis on peat as a solid fuel had emerged. Erik Nyström (b. 1877) of the Mines Branch visited plants in the United States and in Europe where in 1907 he made a complete survey of the peat industry in Sweden, Norway, Finland, Germany, Holland, and Ireland.[5] That same year, Haanel's son, Benjamin Franklin Haanel (1877–1958), inspected peat-burning gas producers and gas engines in several New York City factories and at the University of Illinois' fuel testing laboratory.

---

2 Establishment of the Interior Department occurred in 1873.

3 Alexander Ignatieff, *A Canadian Research Heritage* (Ottawa: Canadian Government Publication Centre, Supply and Service, 1981), pp. 3–5; "Annual Report of the Department of the Interior for the Year 1900," *Report 25*, (Ottawa: Government Printing Bureau, 1901), p. ii.

4 B. F. Haanel, "The Fuel Resources of Canada and Their Utilization for the Production of Power and Other Purposes," *Transactions of the First World Power Conference* 1924, 5 Vols. (London: Percy Lund Humphries & Co., Ltd., 1925), Vol. 1, pp. 256–324. Page 261 has figures on peat supply.

5 E. Haanel, "Summary Report of the Mines Branch of the Department of Mines for the Year Ending December 31, 1913," *Report 285*, Mines Branch (Ottawa: Government Printing Bureau, 1914), p. 4; E. Haanel, "Summary Report of the Mines Branch of the Department of Mines for the Fiscal Year 1907–8," *Report 21*, Mines Branch (Ottawa: Government Printing Bureau, 1908), p. 5; Eric Nyström, "Peat and Lignite: Their Manufacture and Uses in Europe," *Report 19*, Mines Branch (Ottawa: Government Printing Bureau, 1908), pp. 1–247; B. F. Haanel, "Final Report of the Peat Committee," *Report 641*, Mines Branch (Ottawa: Government Printing Bureau, 1925).

In 1908 he studied gas producers in Berlin and at the Körting Brothers company in Hanover.[6] The Mines Branch conducted experimental tests in 1912–13 at the recently opened Fuel Testing Station in Ottawa and planned to begin industrial-size testing the next year. Superintendent Haanel hoped to reduce the nearly $21 million that Canada spent on imported coke and coal, but the outbreak of World War I on 1 August 1914 ended the testing.[7]

In 1907, the year of the Mines Branch's founding, only a small Canadian oil and natural gas industry existed. Industry workers searched and drilled mainly in Western Ontario where in 1857 James Miller Williams (1818–90), a Canadian-American carriage maker from Hamilton, Ontario, dug the first well at Oil Springs, near Sarnia, two years before the more famous 1859 strike at Titusville, Pennsylvania. Canadian production peaked in 1900 at 913,498 barrels, most of it coming from Ontario, but because of depleting resources production fell to only 243,614 barrels in 1907. To encourage production the Canadian government in 1904 passed legislation that established a premium of 1.5 cents/Imperial gallon (gal) on domestically produced petroleum. It remained in effect until 1925. Yet Canada's liquid fuel production remained almost nonexistent. For comparison, the United States in 1912 produced 220 million barrels of petroleum, global production that year reached 351 million barrels. Canadian natural gas production increased yearly for several decades, however, and did not fall until the depression of the 1930s. In 1907 Canadian natural gas production reached 15.2 billion ft^3, all of it coming from three provinces: Ontario 12.5 billion ft^3, Alberta 2.5 billion ft^3, and New Brunswick 0.2 billion ft^3.[8]

Haanel recognized these shortcomings, and in summary reports of 1909–12 he discussed the possibility of developing Canada's oil shale and oil sands as alternative petroleum sources. The Geological Survey had studied oil shales

6 Eric Nystrom and S. A. Anrep, "Investigation of Peat Bogs and Peat Fuel Industry of Canada During the Season 1908–9," Bulletin No. 1, *Report 30*, Mines Branch (Ottawa: Government Printing Bureau, 1909); see also note 5. Construction completed on Fuel Testing Station in 1910.

7 See citations in notes 5 and 6 and Ignatieff, *A Canadian Research Heritage* (note 3), p. 22.

8 Ignatieff, *A Canadian Research Heritage* (note 3), p. 38; E. H. Wait, "Petroleum and Natural Gas in Eastern Canada," *Investigations of Mineral Resources and the Mining Industry*, Mines Branch (Ottawa: Government Printing Bureau, 1930): 35–38; S. C. Ells, "Recollections of the Development of the Athabasca Oil Sands," *Information Circular 139*, Mines Branch (Ottawa: Government Printing Bureau, 1962); E. Haanel, "Summary Report of the Mines Branch of the Department of Mines for the Calendar Year Ending December 31, 1912," *Report 224*, Mines Branch (Ottawa: Government Printing Bureau, 1913), p. 11; F. G. Clapp et al., "Natural Gas and Petroleum Resources of Canada," Vols. 1 and 2, *Report 291*, Mines Branch (Ottawa: Government Printing Bureau, 1930).

in New Brunswick since their discovery by the geologist-physician Abraham Gesner (1797–1864) in 1849 and in Nova Scotia since their discovery by local residents in 1859. But with the subsequent oil discoveries in Ontario and the United States, interest in oil shale ceased until March 1908 when the Albertite, Oilite, and Cannel Coal Company of New York sent 45 tons to the Pumpherston Oil Company's shale works at Midcalder, Scotland.[9] It obtained encouraging oil yields, some as much as 40 gallons per ton (gal/ton) and well above the 15 gal/ton acceptable minimum.[10] These results led Haanel to suggest using oil shale as a source of fuel oil for the navy, an undertaking the British navy also considered, but much to his disappointment no industrial development had started by 1911. Very little testing of oil shales occurred until after World War I, and the Mines Branch concluded that high mining costs and low oil yield made recovery uneconomic compared with well-produced oil.[11]

Knowledge of Alberta's oil sands in the Peace and Athabasca River vicinities dated from 1778, to the New England fur trader and explorer Peter Pond (1739/40–1807), the first non-native to travel there. But serious investigation of the oil sands deposits began only after Sydney C. Ells (1880–1969), a Canadian mining engineer, joined the Mines Branch in 1912 and the Geological Survey in 1913.[12]

---

9 R. W. Ells, "Joint Report on the Bituminous or Oil Shales of Nova Scotia and New Brunswick," *Report 55*, Mines Branch (Ottawa: Government Printing Bureau, 1909); E. Haanel, "Summary Report of the Mines Branch of the Department of Mines for the Calendar Year Ending December 31, 1912," *Report 63*, Mines Branch (Ottawa: Government Printing Bureau, 1910), p. 131; A. A. Swinnerton, "Fifty Years of Fuel Testing and Research," *Technical Memorandum 136*, Mines Branch (Ottawa: Government Printing Bureau, 1957), pp. 1–15; A. A. Swinnerton, "Preliminary Report on the Investigation of Oil Shales," Summary Report of Investigations Made by the Mines Branch During the Calendar Year Ending December 31, 1921, *Summary Report 586*, Mines Branch (Ottawa: Government Printing Bureau, 1923), pp. 239–338.

10 R. W. Ells, "Joint Report" and B. F. Haanel, "The Fuel Resources of Canada" (note 4), p. 305.

11 Ignatieff, *A Canadian Research Heritage* (note 3), p. 39; A. A. Swinnerton, "Fifty Years of Fuel Testing and Research," pp. 1–15; A. A. Swinnerton, "Preliminary Report on the Investigation of Oil Shales," p. 239, both note 9.

12 Gordon E. Taylor, "Address of Welcome," *Proceedings Athabasca Oil Sands Conference*, ed. K. A. Clark et al. (Edmonton, Alberta, September 1951), p. 166. The four oil sands regions and their reserves in billion barrels were: Athabasca (9,000 square miles) 600–900; Cold Lake (4,062 square miles) 164; Wabasca (1,937 square miles) 54; and Peace River (1,875 square miles) 51. Later estimates (1979) in billion barrels are: Athabasca 868.6, Cold Lake 270.5, Wabasca 118.8, Peace River 91.8, R. F. Meyer, C. T. Steele, and J. C. Olson eds., *The Future of Heavy Crude and Oil sands* (New York: McGraw-Hill, 1981), p. 35. Canada's oil sands are the third largest proven oil reserves in the world after Venezuela and Saudi Arabia. Iran ranks fourth. Wabasca is sometimes included with the Athabasca oil region.

FIGURE 6.1 The Athabasca tar sands in Alberta, Canada 1900–30
SOURCE: LIBRARY AND ARCHIVES CANADA, MIKAN ID NUMBER 3592868. PUBLIC DOMAIN

The oil sands, consisting of unconsolidated sand, clay, silt, and a viscous asphaltic oil or bitumen, varied in thickness up to 200 ft (61 m). Overburden (ground cover), sometimes several hundred feet deep, covered the oil sands although those bordering the Athabasca River and some of its tributaries remained exposed or had very light overburden. Open-pit methods made these deposits easily recoverable. The bitumen existed as a thin film surrounding individual sand grains and usually constituted 12–18% by weight of the oil sands. Carbonization (destructive distillation), the method generally used to separate bitumen's heavy oil components, gave about one-half that amount (6–9%) as light oils.[13] The oil sands at that time had only one application, highway paving material, although Ells believed in their value as a future liquid fuel source. His 1916 summary report pointed out that northern Alberta's McMurray district, located within the largest and northernmost of

Later studies (2014) give the oil sands area as 54,000–87,000 square miles and the proven reserves economically recoverable as 162 million barrels (of Canada's total 168 million barrels) or 166/168 million barrels (of Canada's 171/173 million barrels). See capp.ca/oil /what are the oil sands.

13 B. F. Haanel, "The Fuel Resources of Canada" (note 4), p. 305. Some of the oil sands contained richer concentrations of liquid bitumen (with an 8–25% water content), containing up to 80% bitumen on a dry basis. The oil sands generally contain sand (83%), bitumen (10–12%), water (4%), clay (3%).

FIGURE 6.2 Sidney Ells (left), Karl Clark (right)
SOURCE LEFT: CANADA DEPT. OF MINES AND TECHNICAL SURVEYS/LIBRARY AND ARCHIVES CANADA, PA-014454. SOURCE RIGHT: GLENBOW MUSEUM ARCHIVES, ND-3-4596A

Alberta's three oil sands regions (the Athabasca) 250 miles north of Edmonton, contained about 750 square miles of oil sands. But their heavy overburden and lack of uniform quality eliminated 80% of the area's outcrops from further consideration.[14] The McMurray district oil sands required an efficient process for separating bitumen from sand, and in 1916–17 Ells worked for several months with the Mellon Institute of Industrial Research in Pittsburgh and at the University of Alberta in Edmonton to develop a hot water froth flotation method similar to present separation techniques. Although Ells continued his field work until retirement in 1945, the Mines Branch never pursued the Ells-Mellon method because of its poor economics.[15]

Karl Clark (1888–1966), father of the oil sands industry, at the University of Alberta and from early 1921 with the newly established (1919) Scientific and

14 S. C. Ells, "Use of Alberta Bituminous Sands for Surfacing of Highways," *Report 684*, Mines Branch (Ottawa: Government Printing Bureau, 1927), 37 pp. Fort McMurray, founded in 1870, is in the McMurray district. It merged with the nearby town, Waterways, and its name changed to McMurray in 1947 and back to Fort McMurray in 1962.

15 Ignatieff, *A Canadian Research Heritage* (note 3), p. 37. S. C. Ells, "Estimated Cost of Producing Solid and Liquid Hydrocarbons from Bituminous Sand," *Report 727*, Mines Branch (Ottawa: Government Printing Bureau, 1931), pp. 140–45.

Industrial Research Council of Alberta, invented a second hot water separation method in 1920. For four weeks during the summer of 1925 Clark tested his method in a small 10-ton/d continuous operation pilot plant in Edmonton. Later, in August 1929, after dismantling the idle plant, Clark and the Research Council working jointly with the Mines Branch, reconstructed and successfully operated the pilot plant in the oil sands region on the Clearwater River, just south of Fort McMurray, and the small town that grew around Fort McMurray, across from Waterways, Alberta.[16]

Clark's method required adding oil sand, and a sodium silicate solution to facilitate separation, to a mixer and heating. The oil sand-silicate mixture along with a stream of hot water (85° C) passed into a trough and then to a heated separation cell in which most of the sand settled and the froth containing bitumen and fine sand floated to the surface. After collecting the froth, mixing it a second time, and feeding the mixture to a steam-jacket settling tank, the fine sand separated. A steam-heated dehydrator dried the bitumen and discharged it to a final settling tank. Analysis of the bitumen showed that it contained about 5% inorganic mineral matter and 1% water. As a result of his two-year, $30,000 (Canadian) study, Clark estimated an oil sands separation cost of 63.8¢/bbl of bitumen for an oil sands separation plant treating 1,200 tons/d and operating 200 days per year (d/yr).[17]

Several industrial firms experimented with direct distillation of the oil sands to obtain petroleum. The Woodhall-Duckham Company of London, England, and UOP of Chicago used the Dubbs cracking method, the Kansas City (Missouri) Testing Laboratory investigated the Cross cracking method.[18] But the oil sands' potential as a liquid fuel source remained untapped.

In 1924 Canada's petroleum supplies increased somewhat with the November strike at Turner Valley, Alberta, a community of 1,200, about 40 miles

16 K. A. Clark and D. S. Pasternack, *Eleventh Annual Report of the Research Council of Alberta, Report 26* (Edmonton: King's Printer, 1930), pp. 41–62; K. Clark, "Oil Recovery Session," *Proceedings Athabasca Oil Sands Conference* (note 12), pp. 183–84. Mary Clark Sheppard, *Oil Sands Scientist: The Letters of Karl A. Clark* (Edmonton: University of Alberta Press, 1989).

17 Ells, "Estimated Cost of Producing" (note 15); T. E. Warren and K. W. Bowles, "The Bituminous Sands of Alberta, Canada, as a Source of Liquid Fuels," *Transactions of the Fuel Economy Conference: The Hague* 1947, Vol. 2 (London: Lund Humphries, Limited, 1948), pp. 540–49; T. E. Warren and K. W. Bowles, "The Bituminous Sands of Alberta, as a Source of Liquid Fuel," *The Engineering Journal*, 30 (December 1957): 597–600. In this article all costs that relate to Canadian expenditures are in Canadian dollars.

18 M. L. Boyd and D. S. Montgomery, "A Study of the Athabasca Bitumen from the Abasand Quarry, Alberta, Canada," Pt. 1, *Research Report R78*, Mines Branch (Ottawa: Government Printing Bureau, 1961), pp. 7–8.

FIGURE 6.3 Alberta's oil sands

SOURCE: N. A. STROM AND R. B. DUNBAR, "BITUMEN RESOURCES OF CANADA," IN THE FUTURE OF HEAVY CRUDE AND OIL SANDS, R. F. MEYER, C. T. STEELE, AND J. C. OLSON, EDS. (NEW YORK: MCGRAW HILL, 1981), P. 49

southwest of Calgary. The Calgary Petroleum Products Company (acquired by Imperial Oil, Calgary, in 1921 and renamed Royalite Oil Company) had discovered oil there ten years earlier, on 14 May 1914, but many of the wells produced little oil, and with the outbreak of World War I on 1 August 1914 the boom ended quickly. Royalite made a much bigger and longer lasting strike in 1924. For nearly five years, until the middle of 1929, it yielded 500–600 bbl/d before production fell significantly and reconditioning efforts failed. Another strike occurred in 1929, a short-lived 700 bbl/d discovery, before the major strike of 16 June 1936. Turner Valley's intensive development really began on that date, and within three and one-half years Alberta's production reached 7.6 million bbl/yr.[19] But while petroleum production increased in Alberta, and the decline continued in Ontario, total Canadian consumption increased significantly. In the four-year period 1920–24 gasoline consumption increased from 2.8 million barrels to 6.7 million barrels. In 1924 Canadians operated 500,000 marine and land internal combustion vehicles. Yet in this same period Canadian crude petroleum production fell from 196,251 barrels to 160,773 barrels while imports rose from 8.9 million barrels to 16 million barrels.[20]

The geologist Charles Camsell (1876–1958), Deputy Minister of Mines and chairman of the Dominion Fuel Board, expressed serious concern that Canada, especially Ontario and Quebec, had experienced fuel shortages during World War I and the early twenties. When he and B. F. Haanel, now Chief Engineer, Division of Fuels and Fuel Testing, attended the London World Power Conference in the summer of 1924, they visited the depleted oil fields at Pechelbronn in Alsace and Wietze in Hanover to inspect an underground seepage method used for final oil recovery. Development of these fields had occurred during World War I, and they continued producing in 1919.[21]

Camsell made other visits abroad to investigate how European nations were dealing with their lack of significant domestic petroleum resources. In autumn 1927 he accompanied Charles Stewart (1868–1946), Minister of Mines, on an extensive tour of British and German fuel processing centers. The tour included visits to twelve low-temperature carbonization plants, a small Bergius coal hydrogenation unit at Britain's Fuel Research Station in Greenwich, Franz Fischer's synthesis gas (coal gasification) center at the Kaiser Wilhelm Institute for Coal Research at Mülheim, Ruhr, and IG Farben's (BASF) newly operating industrial-size coal hydrogenation plant in Leuna, near Merseburg in eastern Germany. The coal conversion operations particularly interested Stewart and Camsell because they offered a way of transform-

19 B. F. Haanel, "The Fuel Resources of Canada and Their Utilization for the Production of Power and Other Purposes" (note 4); B. F. Haanel, "The Bearing of High and Low Temperature Carbonisation and Synthetic Fuel Processes on Canada's Fuel Problems," *Transactions of the Fuel Conference, World Power Conference*, 1928, 3 Vols. (London: Percy Lund Humphries & Co., Ltd., 1929), Vol. 2, pp. 1–26.

20 Ed Gould, *The History of Canada's Oil and Gas Industry* (Saanichton, B. C.: Hancock House Publishers, 1976), pp. 73–82; S. G. Coultis, "Production Methods in the Turner Valley Field," *Canadian Mining and Metallurgical Bulletin*, 22 (1929): 895–901.

21 Ignatieff, *A Canadian Research Heritage* (note 3), pp. 85–86; Walter Ruehl, "Oil Mining in Germany," *Proceedings Athabasca Oil Sands Conference*, in Clark et al. (note 12), pp. 162–65.

ing Canada's abundant but low-quality coals into much more valued liquid fuel.[22]

The following year Camsell and Haanel traveled to London again to attend the 1928 World Power Conference, where Haanel presented a lengthy paper on Canada's fuel problems and coal conversion processes. Haanel pointed out that like some European nations Canada possessed large quantities of coal and little petroleum. But more than 90% of these coals were lower grade sub-bituminous and lignite, located in Canada's eastern and western regions 1,000–2,000 miles from its most populous and industrialized centers of Ontario and Quebec. The United States' coalfields lay much closer, 400–500 miles, and provided Canada with the major portion of its solid fuels. In fact 50% of the 35 million tons of coal consumed annually came from abroad.

Haanel noted that for liquid fuels Canada almost totally depended (98%) on imports from the United States and other countries and that her requirements were increasing steadily from year to year. In 1927 Canada produced 0.48 million barrels of crude petroleum while importing 23 million barrels. She now ranked second per capita to the United States in the number of motor vehicles and internal combustion engines in use. By 1929 the cost of imported crude petroleum had risen to $50 million (Canadian) annually. Canada could solve its problem of distantly located coal resources and little petroleum, Haanel wrote, if it developed economic processes for converting solid fuels into liquids or gases and then transporting them economically through pipelines. Because carbonization or other heat treatments of coal, shale, and oil sands produced only small quantities of liquid fuels, about 30–40 gal/ton maximum, Haanel did not think they could relieve Canada's dependence on foreign supplies. Only the Fischer-Tropsch synthesis or the more advanced Bergius hydrogenation process produced sufficient liquid fuel from coal. But given the ample availability and low cost of imported petroleum, neither process had any chance of succeeding commercially at that time. If the petroleum situation should change, as Haanel believed, Canada should be ready to develop a synthetic fuel industry.[23] These remarks concluded Haanel's 1928 paper and in effect launched the Mines Branch's search for the most promising process for converting solid fuels to liquid fuels. High-pressure hydrogenation of coal and bitumen seemed to provide a solution.

---

22 Charles Stewart, "Low Temperature Carbonization and Coal Liquefaction in Europe," *Canadian Mining Journal*, 48 (1927): 983–85.

23 B. F. Haanel, "The Fuel Resources of Canada" (note 4), p. 312; B. F. Haanel, "The Bearing of High and Low Temperature Carbonisation and Synthetic Fuel Processes on Canada's Fuel Problems" (note 19), pp. 1–26, on pp. 8–9.

3 Synthetic Fuel Research at the Mines Branch

Research on high-pressure hydrogenation began in 1929 when Thomas E. Warren (1904–74) joined the Fuels and Fuel Testing Division of the Mines Branch. It continued until 1939 when more immediate wartime projects on chemical explosives and on coal, oil, and natural gas resources brought the synthetic fuel program to a temporary halt.[24] These were the Depression years, and superintendent John McLeish (1874–1961), who had succeeded Eugene Haanel in 1921, faced severe budgetary restraints beginning in 1930. Funding had increased throughout the 1920s rising from $429,660 for fiscal year ending March 31, 1921 to $826,610 for 1929–30, but fell to $405,062 for 1934–35, or about 50% of the total expenditure for the peak year 1929–30. In 1937 the government eased its restrictions on funds, and the Mines Branch's expenditures rose to $473,090 in 1938 and $455,000 in 1939.[25]

Up to 1929 the Fuels and Fuel Testing Division had consisted of five groups. When it moved into its new and larger Fuel Research Laboratories that same year, Chief Engineer B. F. Haanel added a sixth group called Special Investigations. It would look into current German research on coal-derived liquid fuels and in particular test the suitability of hydrogenating Canadian solid fuels such as coal, peat, oil shale, and oil sands.[26] Hydrogenation produced the higher octane gasoline required for the rapidly expanding automobile industry, it cost less than other processes, and at that time was more advanced than the F-T synthesis. Warren also had expertise in high-pressure research. He had obtained the ScD at MIT in 1929 after spending three years in the Laboratory

24 Ignatieff, *A Canadian Research Heritage* (note 3), p. 103.

25 Ignatieff, *A Canadian Research Heritage* (note 3), pp. 29, 59, 61, 100, 102. With passage of the Mines and Resources Act of 1936 the Mines Branch became the Bureau of Mines. In 1949 its name reverted to the Mines Branch. The 1936 act also renamed the Fuels and Fuel Testing Division, the Fuels Division. During the war years the Bureau's regular expenditures averaged $434,785 but were supplemented with large special wartime expenditures totaling $5,644,016. Total staff also increased during the 1920s from 98 in 1920 to 188 in 1930 and dropped only to 180 in 1937. McLeish made the reductions in travel, field work, and equipment, not in staff and allowed the Bureau to maintain its full scope of activities, including synthetic fuels. There were salary cuts of 10% in 1932 but restorations of 5% in 1935 and in 1937. At this time the chief of the Bureau of Mines' salary range was $4,500–5,400; divisional chiefs $4,440–4,980; and a stenographer with 44 years' experience had an annual salary of $1,680.

26 B. F. Haanel, "The Fuel Research Laboratories at Ottawa and the Scope of Their Investigations," *The Canadian Mining and Metallurgical Bulletin*, 22 (October 1929): 1197–1214; B. F. Haanel, "New Fuel Research Laboratories," *Memorandum Series 34*, Mines Branch (Ottawa: Government Printing Bureau, 1928).

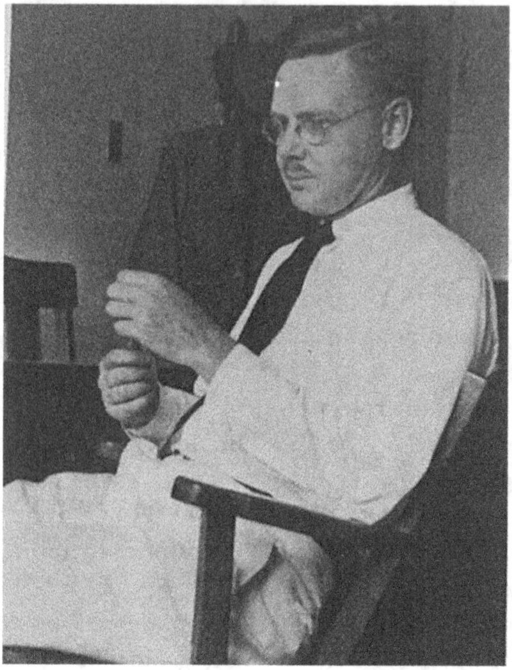

FIGURE 6.4 Thomas E. Warren in 1938
SOURCE: ALEXANDER IGNATIEFF, *A Canadian Research Heritage* (OTTAWA: CANADIAN GOVERNMENT PUBLICATION CENTRE, SUPPLY AND SERVICE, 1981), P. 80

of Applied Chemistry that Robert T. Haslam (1888–1961) directed. Haslam later became a key figure in Standard Oil's (NJ) high-pressure petroleum hydrogenation program.

As head and sole member of Special Investigations, Warren carried out his initial studies in 1930–31 at the new Fuel Research Laboratories using bitumen samples obtained from northern Alberta's oil sands. While not the first, his studies were among the earliest on converting bitumen to liquid fuel. In 1922 the Fuel Testing Station sent a small bitumen sample to Bergius's Rheinau-Mannheim laboratory, and the results reported in 1926 gave 30% gasoline out of an 80% total conversion. Eight years later in 1930 Edward H. Boomer and A. W. Saddington at the University of Alberta carried out a second study, hydrogenating bitumen in batch experiments with small, 1-L and 2-L capacity autoclaves.[27] Prior to 1930 almost all research on bitumen had investigated

27 Edward G. Boomer and A. N. Saddington, "On the Hydrogenation of Bitumen from the Bituminous Sands of Alberta," *Canadian Journal of Research*, 2 (1930): 376–83; Boomer

its use as a road paving material and its refining into petroleum products by pressure-cracking processes.[28]

Warren's tests on 200-g samples had clear objectives: to establish the effects of temperature, pressure, time, and catalysts on the hydrogenation reaction. The reactor, a 1-L, electrically heated, air-cooled, chromium-nickel steel autoclave, had inlet and outlet tubes for adding and removing liquids and gases, a pyrometer for measuring temperature, and an externally rotated stirring paddle. A 0.48 cm (0.19 inch) sheet steel wall surrounded the apparatus providing protection from explosion. He placed all controls, valves, gauges, rheostats, ammeter, and pyrometer (thermocouple), outside the wall. A fan and air duct inside the wall prevented any accumulation of gas from leaks in the apparatus.

From the results of over 60 batch tests Warren established that maximum conversion of bitumen to crude oil occurred at 410° C and 100 atm pressure, in 30 minutes neglecting heating and cooling periods, with a catalyst of finely powdered iron (III) oxide or calcium oxide (CaO), 5% by weight of bitumen. These conditions gave a crude oil yield of 90.5% that contained 18.2% unrefined gasoline (bp to 200° C). No solid carbon formed, and either oxide completely removed any sulfur, a well-known catalyst poison. Warren's batch experiments did not provide meaningful figures comparing possible industrial costs of cracking versus hydrogenating the bitumen, or of hydrogenating bitumen instead of coal. Alberta's bitumen would give cheaper gasoline, he said, only if the bitumen with its higher hydrogen content proved more amenable to hydrogenation than coal and differed little in price from coal.[29]

The extensive set of batch experiments provided encouraging data for small-size hydrogenation of bitumen, and in 1932 Warren prepared to test a larger, continuous-operating reactor. Kenneth Bowles (1906–91) had participated in Mines Branch research from 1931 while still a chemical engineering

---

and Saddington, "On the Hydrogenation of Bitumen from the Bituminous Sands of Alberta II," *Canadian Journal of Research*, 4 (1931): 517–39.

28 S. C. Ells, "Bituminous Sands of Northern Alberta," *Report 632*, Mines Branch (Ottawa: Government Printing Bureau, 1926), pp. 4–11; S. C. Ells, "Core Drilling Bituminous Sands of Northern Alberta," *Report 694*, Mines Branch, (Ottawa: Government Printing Bureau, 1928), pp. 28–35; K. A. Clark and D. S. Pasternack, *Eleventh Annual Report of the Research Council of Alberta* (note 16).

29 T. E. Warren, "Experiments on the Hydrogenation of Alberta Bitumen and On the Effect of Pressure on the Pyrolysis of Methane," *Investigations of Fuels and Fuel Testing 1930 and 1931, Report 725*, Mines Branch (Ottawa: Government Printing Bureau, 1933), pp. 112–35. Mines Branch researchers reported some of their results as volume% and others as weight%. For their investigations volume% is always higher. I have used weight% throughout, except when only volume% was available. 1 American gal = 3.79 L, 1 Imperial gal = 4.55 L.

FIGURE 6.5 Continuous liquid phase pilot plant 4 liter capacity. Left: Hydrogen compression system. Right: (1) Magnetic pump, (2) Condenser, (3) High-pressure receiver, (4) Reaction chamber
SOURCE: ALEXANDER IGNATIEFF, *A Canadian Research Heritage* (OTTAWA: CANADIAN GOVERNMENT PUBLICATION CENTRE, SUPPLY AND SERVICE, 1981), P. 88

student at Queen's University in Kingston, Ontario, and upon graduation in 1933 he began a nearly twenty-year association with Warren on the study of high-pressure hydrogenation. The new reactor was a 4-L, heat-treated, chrome-nickel (18–8) steel cylinder closed at one end and whose other open end they sealed with a cover held by eight studs. As in previous tests, Warren and Bowles reacted bitumen from Alberta at 428–452° C, and 183–187 atm for 3–6 hr. In each experiment they used about 3 kg of reactant and a catalyst of molybdic oxide supported on coke, adding 260–280 g, or about 10% by weight of reactant, to the continuous reactor.

An iron-constantan thermocouple placed in the reactor measured temperature. A pressure gauge gave readings in the line connecting the hydrogen storage cylinders and the reactor. Five other thermocouples and two other gauges measured temperatures and pressures at different points in the apparatus. To ensure operational safety, Warren and Bowles constructed all high-pressure and high-temperature apparatus of chrome-nickel steel and tested each with water at twice the expected working pressure. Once again, they placed all high-pressure equipment behind a 0.64 cm (0.25 in) sheet steel wall.

Considering the total amount of bitumen tested, Warren and Bowles obtained a 73.3% volume yield (solid-to-oil conversion). But as a percentage of net reactant (total amount of bitumen added minus the partly converted oil left in the reactor) they obtained a much more impressive 102% volume yield. Of the oil synthesized, 90% boiled below 300° C; half of it, the gasoline fraction, boiled below 210° C. Tests in an internal combustion engine using bitumen-derived gasoline gave only a 53-octane number and indicated the need for refining to produce a marketable gasoline. Because hydrogenation at the Fuels and Fuel Testing Division had not progressed beyond the single stage, or liquid phase, Warren and Bowles expected the low octane rating.[30] To improve the oil's quality and its distillation range required a second stage, or vapor phase hydrogenation, in which they recirculated the liquids, treated them with a catalyst, and reacted the liquids again with hydrogen gas at high pressure. Pier at IG Farben, only a few years earlier in 1927 had invented, developed, and tested the two-stage hydrogenation in IG Farben's industrial-size plant in Leuna.[31]

The Fuels and Fuel Testing Division's single-stage operation also prevented Warren and Bowles from making any detailed economic analysis of their experiments. They had available some cost figures from Germany and Britain, and in a 1932 paper Warren summarized the cost figures and discussed the feasibility of further developments in Canada.[32] According to estimates that Heinrich Brückmann, director of the Bergius plant in Rheinau-Mannheim provided in January 1928, hydrogenating one metric ton (2,200 lb) of coal cost RM 71. His estimate corresponded to a synthetic gasoline cost in 1928 of about 26¢/US gal (31¢/Imperial gal) at a German refinery or 40–50¢/US gal at an American refinery, both based on a 50% yield of gasoline from coal.[33] Gasoline obtained from crude petroleum at American refineries cost 7–9¢/US gal.[34] In Britain, Imperial Chemical Industries' (ICI) Billingham pilot plant in 1931 reported significantly lower, preliminary cost figures for gasoline synthesized

30 T. E. Warren and K. W. Bowles, "Description of an Apparatus for Continuous Hydrogenation and Experiments on Coal Tar, Bitumen, and a Suspension of Powdered Coal in Coal Tar," *Report 737*, Mines Branch (Ottawa: Government Printing Bureau, 1934), pp. 86–106.

31 Stranges, "Friedrich Bergius" (note 1), pp. 643–67.

32 T. E. Warren, "Status of Hydrogenation of Petroleum, Bitumen, Coal Tar and Coal," *Chemical News*, 144 (25 March 1932): 193–99. Exchange rate 1932: Can $1.00 = US $0.881.

33 Heinrich Brückmann, "Coal Liquefaction and Its Importance in World Economy," *The Mining Journal*, 160 (28 January 1928): 69–70; 160 (4 February 1928): 94; 160 (11 February1928): 116. Exchange rate for 1928: 1 RM = 28¢ US. For units conversions use below.

1 metric ton = 7.5 barrels = 2,200 lb
1 bbl = 42 American gal = 35 Imperial gal
1 cubic foot = 0.028 cubic meters

34 Arno C. Fieldner, "Recent Developments in the Production of Motor Fuels from Coal," *Fuel*, 7 (1928): 492–501.

from bituminous coal: 14–15¢/Imperial gal that included the 4¢/Imperial gal cost of coal.[35] Warren pointed out that the British taxed imported gasoline at a rate of 16¢/Imperial gal and that if Britain exempted synthetic gasoline from an import tax, ICI could operate an industrial-size plant profitably. In fact the outlook for gasoline synthesized from coal looked particularly promising because in the period 1927–31 synthetic gasoline's cost had dropped from 31¢/Imperial gal to 14¢/Imperial gal.

For the production of coal-derived petroleum in Canada, Warren estimated that the expense of building and operating a small hydrogenation plant with a daily capacity of 35 Imperial gal ranged from $700 to $2,500/bbl or 2.45¢/Imperial gal. He assumed $2,000/bbl/d and 15% depreciation and maintenance cost. Hydrogen costs including its production from reacting steam with natural gas, compression, and transportation, added 3.02¢/Imperial gal. Combined cost of the hydrogenation plant and hydrogen therefore totaled 5.47¢/Imperial gal. Including ICI's figure for coal costs, synthetic gasoline cost 10¢/Imperial gal, cheaper than all other estimated synthetic gasoline costs and comparable to gasoline costs at an American petroleum refinery.[36]

In Canada, the bitumen found in Alberta's oil sands was the most widely available raw material for hydrogenation, not coal. An Alberta Research Council study in 1930 had estimated a cost of $1.80 to mine and transport to a separation plant sufficient oil sand to give one ton of bitumen and $4.20/ton to separate the bitumen. Bitumen's $6.00/ton cost made it a much more expensive raw material than coal, but the Fuels and Fuel Testing Division believed bitumen's higher hydrogen content would result in a lower hydrogenating cost. Without further laboratory and large-size tests, however, the Fuel and Fuels Testing Division could not estimate the cost of synthesizing gasoline from bitumen nor make any meaningful cost comparisons between the two materials, coal and oil sands-bitumen. Hydrogenation's economic success required a large-size operation. In fact, the Fuels and Fuel Testing Division preferred a combination of cracking and topping processes to a small hydrogenation unit. Raw materials included lignite and peat from central Canada, bituminous coal from Alberta, and imported petroleum and indigenous bituminous coal or coal tar from the Maritime Provinces, whereas oil shale's high production cost made it an unlikely source.

In the United States, Standard Oil already had begun hydrogenating crude petroleum at its industrial-size plants in Bayway, New Jersey, and Baton Rouge,

35 Imperial Chemical Industries, *J. Soc. Chem. Ind.*, 50 (1931): 669; "Coal Hydrogenation," *Fuel*, 10 (1931): 481–84.

36 Warren, "Status of Hydrogenation" (note 32).

Louisiana. Warren expected this to happen in Canada, with hydrogenation's first use in central Canada's refineries that presently were cracking imported petroleum. Full development would come after appreciable depletion of natural petroleum supplies globally, moving gradually as in Germany and the United States from hydrogenation of petroleum to coal, tar, and other raw materials.[37]

While considerable disagreement on hydrogenation's economics existed, near unanimity prevailed regarding the availability of petroleum despite the late-1930–31 discovery of the huge East Texas oil fields near Kilgore, the largest ever discovered until the Middle East discoveries in Saudi Arabia beginning in March 1938. By 1937 global consumption of petroleum had increased to twice that of 1924 and three times that of 1920. Global production that year (1937) totaled 2 billion barrels with proved reserves of about 20 billion barrels. A recent study also had shown that in the United States the discovery rate declined during the Depression years 1930–37 to 50% the discovery rate during 1924–30. Although the United States' share of global production increased slightly but steadily and now stood at 63%, American consumption in the last few years exceeded the amount discovered in new fields.[38]

B. F. Haanel noted these trends, and in the preface to Warren and Bowles' long 1938 report, he wrote: "It is not too soon, therefore, to take under consideration the consequences of a possible deficiency of petroleum. As the demand approaches the potential supply it is probable that the price of petroleum products will rise until substitutes can compete economically.… The principal raw materials available for the production of substitute liquid fuels are coal and natural bitumens occurring in shales and sands."[39] Mines Branch research on hydrogenating bitumen, peat, and different coals in fact had begun by 1930–31, and in the years immediately preceding World War II, Warren, Bowles, and Ross E. Gilmore (1888–1978) published their results in a series of papers. Their publications described in detail the last hydrogenation exper-

37 Warren, "Status of Hydrogenation" (note 32), p. 198; Clark and Pasternak, *Eleventh Annual Report of Research Council of Alberta* (note 16).

38 "World Oil Needs Pass 2 Billion Barrels," *Oil and Gas Journal*, 36, no. 33 (30 December 1937): 52–57. T. E. Warren and K. W. Bowles, "Tests on the Liquefaction of Canadian Coals by Hydrogenation," *Report 789*, Mines Branch (Ottawa: Government Printing Bureau, 1938) 106 pp., on p. v; "Drilled Oil Reserves Are Now at All-Time Peak," *National Petroleum News*, 30, no. 40 (5 October 1938): 16; "Oil Output Rises 13.40% in 1937," *World Petroleum*, 9, no. 2 (February 1938): 67–69; K. Dijk and K. Mulders, "The World Oil Situation To-Day," *Petroleum Times*, 39, no. 995 (5 February 1938): 161–63.

39 Warren and Bowles, "Tests on the Liquefaction of Canadian Coals" (note 38), p. v.

iments the Fuels Division carried out before World War II temporarily halted their research.[40]

4 Wartime Developments on Synthetic Fuel

At the time World War II began in Europe, Warren's group had done an impressive amount of high-pressure research but only in the laboratory. The program had yet to reach the pilot plant stage and in no way could contribute to the wartime production of liquid fuel. By contrast, in Germany IG Farben and other industrial firms already had operating several industrial-size coal and coal tar high-pressure hydrogenation plants that grew to twelve before the war's end and produced about 128 million barrels of petroleum. In Britain, no extensive coal hydrogenation industry developed, with ICI's 1,000,000 bbl/yr plant at Billingham the only plant constructed.[41] Elsewhere in Europe the Italian government through its state supported Azienda Nazionale Idrogenazione Combustibili (ANIC) operated two crude petroleum hydrogenation plants in Bari and in Livorno (Leghorn). No plants comparable to the European coal hydrogenation plants operated in Canada or in the United States other than the Bureau of Mines' small 45.5 kg/d laboratory-size plant at its Pittsburgh location.[42]

By 1942 the deteriorating liquid fuel situation in Canada rekindled interest in Alberta's oil sands as an alternative oil source. Turner Valley production fell from 10 million barrels in 1942, to 9.5 million barrels in 1943, to 8 million barrels in 1944, and continued to decline through the postwar years. Until the petroleum and natural gas discoveries at Leduc, Alberta, on 17 February 1947,

---

40 T. E. Warren and R. E. Gilmore, "Hydrogenation Tests on Canadian Coal," *Industrial and Engineering Chemistry*, 29 (1937): 353–58; T. E. Warren, K. W. Bowles, and R. E. Gilmore, "Hydrogenation Tests on Canadian Coal," *Industrial and Engineering Chemistry*, 31 (1939): 1021–25; T. E. Warren, K. W. Bowles, and R. E. Gilmore, "Multiple Tests on Catalysts for Coal Hydrogenation," *Industrial and Engineering Chemistry*. Analytical Edition, 11 (1939): 415–19. The ash found in coal could vary from 3 to 10% and contained about 10 metallic oxides such as iron (III) Fe_2O_3, aluminum Al_2O_3, and silicon SiO_2.

41 B. F. Haanel, "The Fuel Resources of Canada" (note 4), p. 312 and "The Bearing" (note 21), pp. 8–9.

42 Stranges, "Friedrich Bergius" (note 1), pp. 643–67 and "Synthetic Petroleum" (note 1), pp. 617–25. In the late 1930s the two plants in southern Italy, at Bari and Livorno (Leghorn), used hydrogenation to convert Albanian crude petroleum to gasoline, see G. Coppa-Zuccari, "L'Hydrogenation des Combustibles en Italie," *Chimie et Industrie*, 41 (1939): 1061–64.

the Canadian petroleum industry had remained at a stand-still, with no large discoveries occurring after the 1936 Turner Valley strike.[43] The Canadian government never carried out a detailed assay of the oil sands bitumen content, and because it required this knowledge before building a bitumen separation plant, the government organized a drilling and sampling program in the oil sands area.

From 1943 to 1947 the government employed the Boyles Brothers Drilling Company of Vancouver to survey and drill and to send oil sands core samples to the Fuels Division Laboratories where Warren and his group assayed the oil sands samples for bitumen content. The survey established that Alberta's Mildred-Ruth Lakes area, on the Athabasca River's west bank twenty miles north of Fort McMurray-McMurray (244 miles north of Edmonton), had the richest and thickest samples, containing 12–18% by weight bitumen under a light overburden. One three-square mile region contained 224 million barrels of bitumen per square mile.[44]

The government asked UOP of Chicago to evaluate refining the bitumen to motor and aviation gasoline. UOP carried out laboratory and pilot plant tests and in 1943 developed an elaborate cracking, isomerization, and alkylation process that gave a motor gasoline of 73–74-octane number and aviation gasoline with an 87.5-octane number unleaded and greater than 100 leaded. But because of low volume yields, 22% gasoline (17% fuel oil, 21% by weight coke) and low quality because of high sulfur concentrations, conventional refining of the bitumen appeared uneconomic.[45]

During this same period the Fuels Division resumed its hydrogenation research, using for the first time a two-stage process to prepare aviation gasoline from bitumen. In the liquid phase Warren and Bowles hydrogenated bitumen in 10-L reactor at 433° C, 200 atm, calcium oxide catalyst, bitumen feed rate of 2 L/hr, and hydrogen flow rate of 142 L/hr. Examination of the gasoline fraction showed that the sulfur concentration had fallen significantly from 5.3% in the original bitumen to 1%. Treating the gasoline over a fixed-bed catalyst of tungsten-nickel sulfide (NiS) in a small vapor phase reactor at 475° C

43 Ed Gould, *The History of Canada's Oil and Gas Industry* (note 20).

44 K. W. Bowles, "Summary of Work of High-Pressure Chemistry Section 1929–1958," *Technical Memorandum 138*, Mines Branch (Ottawa: Government Printing Bureau, 1958), pp. 1–32.

45 "Processing Athabasca Tar Sand Bitumen for Consolidated Mining and Smelting, Waterways, Alberta, Canada." *Confidential Report by Research and Development Laboratories of Universal Oil Products Co.* (Chicago, Illinois, 1943). The leaded gasoline contained 4.8 cm^3 lead tetraethyl/gal.

and 137 atm reduced further the sulfur concentration to 0.02%. These experiments clearly established by 1947 that hydrogenating bitumen in two stages produced a gasoline with half the 0.05% sulfur concentration permitted in aviation gasoline.[46] By then the wartime liquid fuel shortages were a distant memory. With the spending of the program's remaining funds and the discovery of oil and natural gas at Leduc the government discontinued oil sands exploration.[47]

Research on hydrogenating bitumen to give gasoline continued into the period 1951–53, with one group of laboratory experiments designed to eliminate the troublesome liquid phase conversion. Warren and Bowles succeeded in doing this by first distilling the bitumen to give coke and then hydrogenating the coker distillate in a low-pressure vapor phase reactor at 420° C, 68 atm, and a fixed bed catalyst of cobalt molybdate ($CoMoO_4$) on an alumina support. They obtained more favorable high-volume yields of oil, 101–102% with 0.2% sulfur content, than those assumed in previous estimates. Further tests on catalyst deterioration at 68 atm showed that below 427° C the catalyst retained almost its full activity for an indefinite period, but its activity declined steadily above that temperature because of coke deposits on it. The rate of deposit increased with temperature, and as the catalyst's activity decreased so did the oil's quality. At maximum catalyst activity, Warren and Bowles obtained volume yields of 14–23% gasoline with bp to 200° C, 35–38% gas oil with bp range 200–325° C, and 29–39% heavy oil with bp above 325° C. Gasoline quality ranked slightly below that of regular grade automotive fuel, the gas oil of borderline acceptability for diesel fuel, and the heavy oil suitable for fuel oil.[48]

5 Postwar Developments on Oil Sands Separation

When the European theater of the war ended on 8 May 1945, the Allies anxiously awaited investigation of Germany's successful synthetic fuel industry, having learned little about it since 1939. To acquire this information the Allied governments sent several investigative teams: the United States' Technical Oil Mission (TOM) which with British investigators established the Combined

46 Bowles, "Summary of Work" (note 44), p. 14. They used 50.3 cm³ of catalyst.

47 Ignatieff, *A Canadian Research Heritage* (note 3), p. 123; "Drilling and Sampling of Bituminous Sand of Northern Alberta," Vol. 1, "Results of Investigations (1942–1947);" Vol. 2, "Detailed Drilling and Sampling Records;" Vol 3, "Cross-Sections and Plans of Area Drilled;" *Report 826*, Mines Branch (Ottawa: Government Printing Bureau, 1949).

48 Bowles, "Summary of Work" (note 44), pp. 17–19, 23–24.

Intelligence Objectives Sub-Committee (CIOS), the British Intelligence Objectives Sub-Committee (BIOS), and Field Intelligence Agency Technical (FIAT), to the plant sites to interrogate German scientists and collect data on plant operations. Warren went to Germany as a member of BIOS and spent 4–21 September 1945 inspecting four hydrogenation plants and six F-T plants in western Germany, in the British zone of occupation, nine of them in the Ruhr.

IG Farben had constructed the first coal hydrogenation plant in 1927 in Leuna, which by 1931 had an annual capacity of 300,000 metric tons (2.25 million barrels) of synthetic petroleum. Other companies, both government and privately owned, among them Brabag, Gelsenberg Benzin AG, and Hydrierwerke Pölitz AG, joined IG Farben in producing synthetic petroleum. By 1944 the industry had grown to twelve large plants. At their peak in 1944, the German hydrogenation plants produced over 3 million metric tons of synthetic petroleum of which 2 million metric tons were high quality 100-octane leaded aviation and motor gasoline. They provided 50% of Germany's liquid fuel consumption in World War II. Germany's second synthetic petroleum process, the F-T or synthesis gas process, had reached industrial development in 1935. The much smaller F-T plants contributed only 12–15% of the total liquid fuel supply. Because hydrogenation produced a higher quality gasoline, the Germans constructed the last of their nine F-T plants in 1939 while construction of hydrogenation plants continued into the war years.[49]

Warren's work with BIOS took him from Ottawa for two and one-half months. Before his departure the National Research Council of Canada and the United States Bureau of Mines had discussed expanding the Canadian program with the construction of a large coal hydrogenation pilot plant in Ottawa. But Warren's departure delayed the expansion, and upon his return a government committee of seven officials in 1946 introduced a new plan that called for erecting a much smaller plant in the existing Fuels Division Laboratories.

The United States Bureau of Mines was launching its own synthetic fuel program at this time (1946), and the discussions led to the Bureau's decision to concentrate on treating coal, while the Mines Branch continued its work on hydrogenating bitumen from the oil sands. Also, because of hydrogenation's high cost, Warren's group confined its research to those areas not included in the Bureau's much more comprehensive program, in particular to designing a plant operating at 1,360 atm, or about twice the highest pressures used in

49 Stranges, "Friedrich Bergius" (note 1), pp. 643–67 and "Synthetic Petroleum" (note 1), pp. 617–25; T. E. Warren, *Inspection of Hydrogenation and Fischer-Tropsch Plants in Western Germany during September 1945*, BIOS Final Report No. 82, Item No. 30 (London: British Intelligence Objectives Subcommittee, 1945), pp. 1–28.

the German plants.[50] Thus in the decade following World War II, the Fuels Division's high-pressure research acquired a twofold purpose: hydrogenation of bitumen and pilot plant investigations at very high pressures.

Hydrogenating at pressures as high as 1,360 atm seemed like a move designed to increase costs. Yet, as Warren and Bowles pointed out, the introduction of higher pressures resulting in greater efficiency already had occurred in plant design and operation during the last twenty years. Warren and Bowles also noted that the high-pressure unit's operating cost represented less than one-third of the plant's total operating cost. Of the plant's four unit operations, hydrogen production, hydrogenation, gasoline refining, and power production, only the hydrogenation unit required high-pressure equipment, and except for the chromium-plated reactors much of the hydrogenation unit consisted of ordinary structural and low-pressure equipment. In their opinion the cost of increasing the pressure influenced the plant's cost very little and the greater efficiency at higher pressures more than balanced the higher cost.[51] Despite the enthusiasm for the 1360 atm plant, almost ten years passed (1955–56) before the Fuels Division managed to construct and use part of the plant to hydrogenate bitumen, and another two years before the plant's completion in April 1958, six years after Warren left the Mines Branch.[52]

The Department of Mines, particularly Ells of the Geological Survey, had investigated the oil sands intermittently since 1913, but only in the years immediately following World War II did the time seem right for their development. The war had shown how easily nations could consume large quantities of liquid fuel, and both the Bureau of Mines and the Mines Branch saw synthetic liquids as insurance against possible shortages especially in the event of future wars. Canada suffered serious fuel shortages throughout the war and had to import large quantities of anthracite from the United States. The government recognized that during such crises reliance on imports posed a threat to the country's survival. The Leduc discovery of oil and natural gas in February 1947 improved tremendously Canada's liquid fuel resources, yet the Fuels Division still believed that circumstances called for a review of the oil sands situation. Warren and Bowles presented the oil sands case in a paper read at the Fuel Economy Conference in The Hague in September 1947.

50 Ignatieff, *A Canadian Research Heritage* (note 3), pp. 122, 252. 1 atm = 14.7 pounds per square inch (psi).

51 T. E. Warren and K. W. Bowles, "Oil from Coal," *Transactions of the Canadian Institute of Mining and Metallurgy*, 55 (1952): 383–86.

52 Bowles, "Summary of Work" (note 44), p. 25.

A major reason for reviewing the potential of 1,500 square miles of oil sands had arisen, they said, because both imports and the price of petroleum in Canada continued to increase in the postwar years. In the Prairie provinces alone (Alberta, Manitoba, Saskatchewan) a market existed for the gasoline and other liquid fuels obtained from bitumen. Imports from Texas and Oklahoma were expensive because of the long rail trip, but their high transportation cost guaranteed considerable protection for domestic liquid fuel obtained from bitumen and other raw materials. In 1946 Prairie refineries had treated 13 million barrels of petroleum or about 47% more than in 1940. But in 1946 Prairie production had fallen to 619 million barrels. Imports accounted for the difference between production and consumption and had risen steadily from 192,552 barrels in 1941 to 6.2 million barrels in 1946. Synthetic liquid fuel produced from bitumen or some other indigenous material, according to Warren and Bowles, might eliminate the difference, especially if no discoveries of new Canadian petroleum fields occurred.

As a second reason, Warren and Bowles pointed out the favorable results obtained in the 1943–47 core-drilling in selected areas on both banks of the Athabasca River, especially on the Mildred-Ruth Lakes' west bank. Samples taken up to the end of 1946 indicated that the area drilled contained 350 million barrels of bitumen and no other resources either sufficiently plentiful or inexpensive to convert into liquid fuel.

Warren and Bowles gave as their third reason for examining the oil sands the experience both government and industry had acquired in bitumen separation and refining. Investigations had begun nearly 20 years earlier in 1929 with construction of the jointly operated Mines Branch-Alberta Research Council hot-water separation plant on the Clearwater River. In 1930 the International Bitumen Company (Canada) built a second hot-water separation plant at Bitumount on the Athabasca River's east bank, 54 miles downriver from Fort McMurray-McMurray. It ceased operation in 1940. Oil Sands Limited (Canada) acquired the small plant in 1942 and with a $500,000 investment from the Alberta government constructed an experimental plant and refinery in 1944. The Alberta government took over the plant in 1948, and the following year it completed a 500 ton/d plant. Abasand Oils Limited in 1933–36 built a third plant on the Horse River near Fort McMurray-McMurray. Max Ball (1885–1954), an American geologist, engineer, and former one-time chairman of the United States Geological Survey's Oil Board, founded and served as Abasand's president. Abasand operated the semi-industrial plant from 1936 until early 1943 when the Canadian government acquired it and invested $1.5 million in constructing an industrial-size pilot separation plant and refinery with a 600 bbl/d capacity. A fire on 25 June 1945 completely destroyed the plant.

The valuable information obtained from captured German synthetic fuel documents was the final reason Warren and Bowles gave for reviewing the oil sands situation. The documents showed that synthetic fuel plants at Pölitz and Lützkendorf, both of which started operation in 1940, had hydrogenated bitumen-like raw materials to give high yields of gasoline. But the conversion was expensive and required preliminary liquid-phase hydrogenation. Warren and Bowles pointed out that Alberta's bitumen contained a considerable amount of asphaltene and inorganic material which if removed by destructive distillation would permit hydrogenation over an active catalyst such as tungsten sulfide. This eliminated the troublesome liquid phase while improving the reaction's yield and economics. Bitumen, they concluded, had several advantages over other materials as a source of gasoline and liquid fuel. Its reserves surpassed Canada's natural gas and heavy crude oil reserves, it had fewer and less important uses than natural gas, and it cost less than coal to convert to liquid fuel. Alternatively, Warren and Bowles noted serious disadvantages to developing a bitumen-based liquid fuel industry, particularly the oil sands' distant location from petroleum and gasoline markets, no large-size method of separating bitumen from sand existed, and refining the bitumen to gasoline required special techniques.[53]

6 The Blair Report, Athabasca Conference, and the State of Canadian Fuel Research

During the immediate postwar years of concern over petroleum supplies, the Alberta government organized a major conference, the Athabasca Oil Sands Conference of 1951, to discuss development of Alberta's oil sands as a future source of liquid fuel. Over 100 delegates from Canada, the United States, and Europe came to Edmonton in September 1951 for the four-day conference that included field trips to oil sands deposits near Fort McMurray-McMurray, the separation pilot plant at Bitumount, and the Leduc oil fields. Delegates representing government, industry, and university heard papers on the geology of oil sands formation, mining methods, oil sands separation, transportation, and refining. The Mines Branch sent several staff members to the conference,

53 Warren and Bowles, "The Bituminous Sands" (note 17), pp. 540–49. To convert from metric tons to barrels, Warren and Bowles assumed that 1 metric ton = 7.5 barrels and 1 cubic foot = 0.028 cubic meters.

including Warren who presented the Fuels Division's latest results from its two-stage hydrogenation of bitumen.[54]

The Athabasca Conference followed the recently released Blair *Report* of December 1950 that investigated the economics and possible development of oil sands separation processes.[55] Sidney M. Blair (1897–1981), an Athabasca Conference technical director and consultant petroleum engineer associated with the Alberta government's Oil Sands Project, wrote the report in which he discussed operation of a proposed oil sands plant separating 20,000 bbl/d of bitumen, although delegates at the conference and later usually mentioned a plant separating 100,000 bbl/d. The report estimated total bitumen reserves recoverable from the oil sands by all known methods as high as 300 billion barrels and indicated that bitumen clearly represented a major potential source of Canadian liquid fuel. Yet, no attempt at commercial oil sands development occurred until the 45,000 bbl/d Great Canadian Oil Sands (GCOS) plant 25 miles (40 km) north of Fort McMurray-McMurray came on stream in 1967.[56]

Industry's lack of interest in oil sands development in the 1950s resulted from two factors: the high initial cost of the mining equipment and of erecting an integrated oil sands recovery and refining plant and the tremendous increase of Canadian production of petroleum and natural gas during this period. According to the Blair Report, a bitumen-based liquid fuel complex required an open-pit mine, a hot-water separation unit, a fluidized distillation unit, a desulfurization unit using mild hydrogenation, and product marketing. Mining equipment for removing overburden, excavating, crushing, and conveying cost $1.93 million; the three processing units excluding interest and profit on the capital added $43.6 million. Using these figures, Blair estimated that a barrel of desulfurized coker distillate cost $3.10, calculated to include mining $0.55, processing $1.53, and transportation from the oil sands fields to the Great Lakes terminal in Chicago $1.02. Alternatively, the distillate when catalytically cracked to give gasoline of 91-octane number, fuel gas, light and heavy cyclic oils, had a market value of $3.50. A plant producing 20,000 bbl/d

54 T. E. Warren, F. L. Booth, R. E. Carson, and K. W. Bowles, "Hydrosulphurization of Coker Distillate from Athabasca Bitumen," in Clark et al. (note 12), pp. 289–305.

55 Sidney M. Blair, *Report on the Alberta Bituminous Sands, Government of the Province of Alberta* (Edmonton, Alberta: A. Shnitka, King's Printer, 1951), 85 pp., on pp. 28, 29, 55; *Report* submitted on 12 December 1950; D. S. Pasternack, G. W. Hodgson, and K. A. Clark, "Oil Recovery from the Alberta Oil Sands by the Hot Water Washing Method," in Clark et al. (note 12), pp. 200–06.

56 Ignatieff, *A Canadian Research Heritage* (note 3), p. 253; Arthur M. Johnson, *The Challenge of Change: The Sun Oil Company, 1945–1977* (Columbus, Ohio: The Ohio State University Press, 1983).

FIGURE 6.6
Blair *Report* 1951, cover
SOURCE: GOVERNMENT MINISTRIES, GOVERNMENT OF ALBERTA, PUBLIC DOMAIN

or 6 million barrels annually earned a profit of $0.40/bbl and represented a 5–6% yield on the total investment. The Blair *Report* had concluded "that the economics… of producing marketable oils from this source correspond with the economics of the oil industry."[57]

Despite its generally optimistic tone Blair had not based his report on production costs of a large-size operating plant. At a time of abundant Canadian petroleum and natural gas supplies, industrialists hesitated to invest the millions of dollars required to build a bitumen recovery and refining plant. In 1949 Canada had produced only 21 million barrels of crude petroleum while importing 74 million. By 1960, however, production had increased to 190 million barrels. Imports also increased, but far less dramatically, rising to 126 million barrels in 1960. Production figures for natural gas showed the same trend and appear in Table 6.1 with the petroleum figures.[58]

During the 1950s–60s, while little industrial effort to develop the oil sands occurred, the Fuels Division continued its policy of working on the oil sands and other low-grade crudes "with long-term objectives in view." The Fuels Division adopted this policy because it did not consider the oil sands a conventional crude petroleum, and it intended to fulfill one of its principal objectives by carrying out research on such marginal yet strategically important

57 Blair, *Report on the Alberta* (note 55), pp. 61–80. Quote is on p. 79.
58 Ignatieff, *A Canadian Research Heritage* (note 3), p. 253.

TABLE 6.1 Canadian production, imports, and exports of crude oil and natural gas

Year	Crude oil barrels (millions)			Natural gas cubic feet (billions)		
	Production	Imports	Exports	Production	Imports	Exports
1945	8	57		48		
1949	21	74		60		
1960	190	126	42	523	6	91
1970	461	208	241	2,277	12	768

SOURCE: ALEXANDER IGNATIEFF, *A Canadian Research Heritage* (OTTAWA: CANADIAN GOVERNMENT PUBLICATION CENTRE, SUPPLY AND SERVICE 1981), P. 253

resources. In fact, of the three organizations most likely to conduct research on fuel technology, government, industry, and university, the Dominion and Provincial governments had the most active programs. According to Alexander Ignatieff, former Fuels Division chiefs, and Alan E. Cameron (1890–1977), then president of the Nova Scotia Technical College, Canadian universities offered no fuel technology or chemical engineering courses that emphasized hydrocarbon fuels. Even at the graduate level the universities offered few courses, and students with such research interests left for study in the United States. There, the Massachusetts Institute of Technology, Carnegie Institute, Pennsylvania State University, and the University of Illinois had well established fuel research programs. The few existing Canadian programs included two joint efforts by the Research Council of Saskatchewan and the University of Saskatchewan on low-temperature lignite carbonization; the Nova Scotia Research Foundation's sponsored studies on coal geology and chemistry; the largest provincial program, the Alberta Research Council's basic coal research that started in 1919; and the oldest fuel research program, the Mines Branch's Fuels Division that had evaluated coal and other fuels beginning in 1907. Only Imperial Oil Ltd., which had an active research program on petroleum and natural gas, and government organizations, particularly Alberta's Research Council and the Fuels Division, conducted liquid and gaseous fuel research.[59]

59 R. E. Gilmore and Alan E. Cameron, "Fuel Research Facilities in Canada," *Transactions of the Canadian Institute of Mining and Metallurgy*, 54 (951): 75–84. A. Ignatieff, "Fuel Technology in Canada," *J. Inst. Fuel*, 33 (1960): 223–58; on pp. 228, 232.

FIGURE 6.7 Canada Mines Branch, Booth Street, Ottawa, early 1950s
SOURCE: ALEXANDER IGNATIEFF, *A Canadian Research Heritage* (OTTAWA: CANADIAN GOVERNMENT PUBLICATION CENTRE, SUPPLY AND SERVICE, 1981), P. 135

The Fuels Division's total staff size grew from 71 to 107 during this period. But the number of researchers working on bitumen and other fuels increased only from 19 to 22. Warren had left in 1952 to take a position first with the Cyanamid Company and then as director of the Saskatchewan Research Council. Douglas S. Montgomery (1916–2019), who joined the Mines Branch in 1948, became head of the Fuels Division's Hydrocarbons Group. This group consisted of an Analytical, Fundamental, and Applied Research section and had responsibility for the overall bitumen research.

The Applied Research Section, also called the Hydrogenation Section and in 1958 the High-Pressure Chemistry Section, continued to work on the hydrogenation of bitumen, heavy gas oils, and low-grade petroleum residues after completing the 1,360 atm plant in 1955. Bowles, Frederick L. Booth (1916–2008), and W. H. Merrill considered of particular importance vapor phase experiments at 68, 204, 340, and 680 atm to test a fixed bed catalyst of cobalt molybdate on an alumina support. Their experiments showed a three times greater throughput at 680 atm than at 68 atm and that to increase quality and yield the catalysts required larger pore diameters. By 1962 the High-Pressure Chemistry Section had succeeded in preparing a much better catalyst. In one

FIGURE 6.8 Pressure vessels originally designed for 1360 atm for vapor phase hydrogenation, Booth Street, Ottawa
SOURCE: ALEXANDER IGNATIEFF, *A Canadian Research Heritage* (OTTAWA: CANADIAN GOVERNMENT PUBLICATION CENTRE, SUPPLY AND SERVICE, 1981), P. 259

series of pilot plant experiments with bitumen coker distillate and pressures that varied from 68 to 240 atm, the catalyst at 204 atm deactivated about ten times slower than at 68 atm.[60]

Two years later the High-Pressure Chemistry Section showed that vapor phase hydrogenation of bitumen coker distillate gave good quality jet and diesel fuel. The gasoline tested poorly, but reforming by catalytic cracking upgraded it to satisfactory. Bowles, Merrill, and W. A. O. Herrmann who had worked since 1936 on coal hydrogenation in Germany, therefore tested a new, combined liquid-vapor phase continuous unit known in Germany as

60 A. R. Aitken, W. H. Merrill and M. P. Pleet, *Canadian Journal of Chemical Engineering*, 42 (1964): 234–38.

the Combi system.[61] Its design allowed the vaporized liquid products to pass directly to the vapor phase reactor without any cooling or drop in pressure, which supposedly improved the two-stage hydrogenation with distillation between the stages. The plant, constructed from 1960 to 1965, had one liquid phase and one vapor phase reactor, 3.96 m long, a 1 bbl/d throughput, and operated at pressures up to 680 atm. Industrial hydrocracking processes used a fixed-bed catalyst and usually treated clean feedstocks such as heavy gas oils or de-asphalted residues to avoid poisoning the catalyst with small amounts of sediment or asphalt. Bowles, Merrill, and Herrmann's experiments specifically tested hydrocracking or hydrogenating very low-quality distillates such as the oil sands.

The demand for lighter distillates, particularly in the gasoline and jet fuel range, increased steadily in the 1960s. Jet fuel accounted for only 5% of all crude oil refined, but with greater military and commercial consumption expected, demand would rise faster than for other major petroleum products. Using a Weyburn crude oil obtained from Shell Oil's refinery in Oakville, Ontario, Bowles, Merrill, and Herrmann conducted several Combi experiments at 445° C, 40 atm, and with three different catalysts. They initially tested only the liquid phase and obtained the same approximate volume yields with each catalyst: gasoline, bp to 200° C, 28.5%; jet fuel, bp 200–280° C, 13.3%; diesel fuel, bp 280–325° C, 24.7%; residuum, bp +325° C, 33.4%. By 1968 the High-Pressure Chemistry Section had tested bitumen at pressures from 68 atm to 238 atm.[62]

7 Canada's Changing Energy Situation

During the 1950s–60s the energy consumption pattern in North America and much of the industrialized world changed from coal dependence to a petroleum and natural gas dependency. The oil discoveries at Leduc (1947), Redwater (1948), and other fields in Alberta that transformed Canada from major

61 Ignatieff, *A Canadian Research Heritage* (note 3), p. 262, Wilhelm Urban, "Erdöl Verbeitung in der Scholvener 300 at Kombi-Hydrierkammer," *Erdöl und Kohle*, 8 (1955): 780–82. Hydrocracking is a process in which heavy hydrocarbon molecules at high temperature and pressure and in the presence of hydrogen gas and catalyst are cracked (broken apart) to give lighter fragments. Hydrogen then reacts with the fragments to give more valuable liquid distillates.

62 D. H. Quinsey, W. H. Merrill, W. A. O. Herrmann, and M. P. Pleet, "Combined Liquid and Vapor-Phase Hydrocracking of Heavy Oil," *Canadian Journal of Chemical Engineering*, 47 (1969): 418–21; Ignatieff, *A Canadian Research Heritage* (note 3), p. 262.

importer to major petroleum producer occurred at this time. In 1951 petroleum production globally reached 11.7 million bbl/d or 4.25 billion bbl/yr. The United States produced about half of that amount, 6.1 million bbl/d, an increase of 13.7% from 1950. Annual production in Canada leaped 65% to 48.1 billion barrels in 1951 from 29.1 billion barrels in 1950.[63] In the western provinces, mainly Alberta which provided about 80% of Canada's petroleum, daily production rose from 20,000 to 125,000 barrels following the Leduc discovery. The greater production had increased the domestic share of Canada's total petroleum requirement of 400,000 bbl/d from 8% to 33% and reduced the amount spent on imports that came mainly from the United States. Alberta's petroleum saved the government $9 million in 1948, $100 million in 1950, and $150 million in 1951.[64] But while production rose, so did consumption in Canada and globally. Estimated Canadian consumption of petroleum and petroleum products in 1951 reached 154 million barrels or about 11 barrels per capita compared with 4.7 barrels per capita in 1939. This placed Canada second to the United States, the largest consuming nation with an annual consumption of 2.61 billion barrels or 17 barrels per capita.[65] Canada remained second in 1958, but per capita consumption had risen to 16.8 barrels.[66]

Although Canada had ample endowments of solid, liquid, and gaseous fossil fuels, their distant location from the population and industrial centers, always had presented an economic problem. Table 6.2 makes this abundancy clear but does not show the 48.1 trillion tons of coal located mainly in the western provinces and the estimated 100–300 billion barrels of oil in the Athabasca oil sands. Western production capacity totaled one million bbl/d, enough to meet Canada's liquid fuel requirement of 270 million bbl/yr, but actual production reached only half that amount.[67]

Despite these abundant reserves, during the 1970s Canada's supply of high-quality crude oil declined at a rate that exceeded its production from low-grade heavy oil, oil sands, and coal.[68] Potential and proved reserves remained high, however. In 1980 Canada had 100 billion barrels of proved and potential

63 G. S. Hume, "Canada's Petroleum and Natural Gas Resources," *Transactions of the Canadian Institute of Mining and Metallurgy*, 55 (1952): 299–301.

64 Oliver B. Hopkins, "Canada's Petroleum Industry," *Transactions of the Canadian Institute of Mining and Metallurgy*, 55 (1952): 98–107.

65 Hume, "Canada's Petroleum" (note 63), p. 299; Ignatieff, *A Canadian Research Heritage* (note 3), p. 232.

66 Ignatieff, "Fuel Technology in Canada" (note 59), p. 223.

67 Ignatieff, "Fuel Technology," p. 227.

68 D. S. Montgomery, "Problems and Challenges of Securing Fossil Fuel Energy Sources for the Future," *Chemistry in Canada*, 32 (May 1980): 23.

TABLE 6.2 Canadian fuel production and reserves

	Production 1958	Proved reserves at end of 1958
Crude oil and natural gas liquid (thousands of imperial barrels)		
Northwest Territories	466	52,409
British Columbia	1,963	36,534
Alberta	116,948	2,995,190
Saskatchewan	44,910	531,409
Manitoba	5,871	27,500
Ontario	777	6,978
New Brunswick	15	77
Total	170,950	3,650,097
Natural gas (million standard cubic feet)		
Northwest Territories	337	29,427
British Columbia	62,511	1,687,626
Alberta	257,446	20,222,824
Saskatchewan	44,332	1,165,003
Manitoba	515	2,345
Ontario	15,350	186,783
Quebec		83
New Brunswick	124	961
Total	380,615	23,295,052

SOURCE: CANADIAN PETROLEUM ASSOCIATION

high-quality crude oil, enough for 125 years, and another 300 years supply from proved recoverable oil sands and heavy oil. Canada's supply-to-demand ratio for liquid fuel exceeded the supply-to-demand ratio of every developed nation in the Western world. Yet one-third of Canada's crude oil supply, costing more than $7 billion annually, came from politically unstable foreign sources.[69]

Already by the 1950s–60s the Mines Branch and other Canadian government agencies recognized that any long-term Canadian energy plan had to consider development of the oil sands and heavy oil reserves to achieve energy

69 J. P. Gallagher, "Security of Energy Supply for Canada," *Chemistry in Canada*, 32 (May 1980): 28. Heavy oil is the source of creosote, diesel oil, fuel oil, and heavy crude oil.

self-sufficiency and political independence.[70] George S. Hume (1893–1965), Director General of Scientific Services, Department of Mines and Technical Surveys in Ottawa, had proposed their development in a report published in 1952, shortly after release of the Blair *Report* (1951) and the Athabasca Oil sands Conference (1951).

"The publication of an exhaustive report by S. M. Blair on the commercial feasibility of development, has again focused attention on one of Canada's richest undeveloped resources. The time seems to be close when plans will be laid for large-scale operations, and the evidence being gradually accumulated by scientific investigations points to economic possibilities in the very near future.... The limitations of the market for oil from Western Canada in the immediate future need not be a prime concern since it is obvious to all familiar with this problem that some years must elapse before any plant would be in a position to produce significant supplies from the bituminous sands.... Some agreement with the United States would be of great mutual benefit in obtaining development of the bituminous sands by insuring an outlet when a plant is brought into production."[71]

The United States was a very likely market for oil obtained from the oil sands because by 1958 it had become a net petroleum importer. In fact, the continually deteriorating balance between petroleum consumption and domestic production in the United States led to three government-industry projects to develop a high-pressure coal liquefaction (hydrogenation) industry: Gulf Oil's Solvent Refined Coal Process (SRC) in Fort Lewis, Washington, in 1962, Hydrocarbon Research's H-Coal Process in Catlettsburg, Kentucky, in 1963, and Exxon's Donor Solvent Process (EDS) in Baytown, Texas, in 1966. The projects received further incentive from the Arab oil embargo of 20 October 1973–19 March 1974 which quadrupled petroleum prices from about $2.00–2.50/bbl to $10.00–12.00/bbl and the Iranian hostage crisis of 4 November 1979–20 January 1981 which more than tripled prices to $40.00/bbl. But the severe global recession of 1981–82 and the price of oil that began to fall in 1983 and plummeted to less than $10.00/bbl by 1986, ended the United States' budding synthetic fuel industry.[72]

During the 1960s Canada's significant increase in petroleum consumption intensified further government's and industry's efforts to develop the oil sands.

---

70 Gallagher (note 69).

71 Hume (note 63), p. 301.

72 Stranges, "Synthetic Petroleum" (note 1), pp. 617–25.

Estimates of the three largest known deposits, (including Abasca) the largest in North America, indicated the presence of about a trillion (1×10^{12}) barrels of bitumen, 10% of which had less than 50 meters overburden and lay close enough to the surface for open-pit mining. The remainder, about 9 billion (9×10^9) barrels, lay deeper and recoverable by in situ production methods.[73]

Great Canadian Oil Sands Ltd. (GCOS), a subsidiary of Sun Oil Company (established in 1919 in Montreal), and since its 1979 merger with Sun Oil called Suncor with headquarters in Calgary, in 1963–67 constructed the first industrial-size open-pit mining and upgrading plant. Built at a cost of $235 million and located on the banks of the Athabasca River about 25 miles north of Fort McMurray-McMurray, plant production by 1990–91 reached 51,000–60,000 bbl/d.[74] In 1978 Syncrude Canada Ltd., a conglomerate of federal government and business interests such as Imperial Oil and Petroleum North America, with headquarters in Fort McMurray, completed a second open-pit mining and upgrading plant on a site immediately adjacent to the Suncor lease. The $2.2–2.5 billion Syncrude plant had a design capacity more than twice the older Suncor plant, or 125,000 bbl/d of crude oil. In 1984, Syncrude invested another $1.2 billion to improve the plant's efficiency and increase its capacity to 130,000 bbl/d.[75] The two plants by the mid-1980s

73 Maurice A. Carrigy, "New Production Techniques for Alberta Oil Sands," *Science*, 234 (19 December 1986): 1515–18. The reserves of oil (bitumen) obtainable from open pit methods, about 26×10^9 barrels, are equal to the United States' light oil reserves. There are altogether at least 300×10^9 recoverable barrels of bitumen of which $150–200 \times 10^9$ barrels can be processed into synthetic oil. The Persian Gulf area contains in place 350×10^9 barrels of oil of which 168×10^9 barrels are recoverable.

74 In 1979 Sun Oil Ltd. and GCOS formed Suncor. Sun Oil owns 98% of Suncor, 2% is publicly owned. Because of the economics of surface mining the Suncor plant operates continuously, 24 hours a day, 365 days a year. After mining, the process consists of treating the oil sands in extraction plant with steam, hot water, and caustic soda, followed by a two-cycle centrifuging process that separates the mineral matter from raw bitumen with a 92% efficiency, and then delayed coking to give a synthetic crude oil. About five tons of oil sands produce one barrel of synthetic crude. Grant D. Mossop, "Geology of the Athabasca Oil Sands," *Science*, 207 (11 January 1980): 145–52; Sam Fletcher, "Alberta's Oil Sand Area High in Reserves Potential," *Houston Post* (15 March 1981): 240; "Canada's Synfuels: A Cautious Phoenix," *Chemical Week*, 134 (22 February 1984): 40, 42.

75 The participants and their percentage interests are: Esso Resources Canada, 25%; Canada-Cities Service, 17.6; Gulf Canada Resources, 13.4; Petro-Canada Exploration, 12; Alberta Energy Company, 10; Alberta Government, 8; Hudson's Bay Oil and Gas, 5; Petrofina Canada Exploration, 5; Pan-Canadian Petroleum, 4. The governments of Canada, Alberta, and Ontario bought in when Atlantic Richfield withdrew in 1975; Mossop, "Geology of the Athabasca" (note 74), pp. 149, 152.

produced more than 170,000 bbl/d of oil, about 15% of Canada's total production.[76]

Several oil companies had more ambitious projects underway in the late 1970s, but all fell victim to the severe 1981–82 recession with its high interest rates and high inflation and the beginning of the oil price collapse. Esso Resources Canada, established in 1978 in Calgary, in 1982 abandoned a $12 billion heavy oil extraction plant near Cold Lake, Alberta.[77]

The Alsands consortium, consisting of Shell Canada in Calgary, Shell Explorer in Houston, Petro-Canada (PC), Canada's national oil company, established in Calgary in 1975, and four other organizations, on 1 May 1982 canceled plans for a $13.1 billion mining and extraction plant scheduled for completion in the mid-1980s at a site about 12 miles north of Fort McMurray.[78]

Yet, only two years after the apparent demise of these mega-projects, circumstances had changed sufficiently to warrant renewed interest in oil sands development. The release of a twenty-three-year oil supply forecast by the Alberta Energy Resources Conservation Board (1971–2013) in Edmonton, in January 1984 contributed to the renewed interest. According to the report, Alberta could expect a 35% drop in its conventional oil reserves in the next ten years.[79] The oil sands, therefore, became the likely replacement for 75,000 bbl/d of conventional oil supplies.

Energy forecasts also predicted strong markets for crude oil by 1990 with prices high enough to sustain commercial operations such as oil sands projects.[80] In view of these favorable forecasts, several oil sands expansions occurred. Shell Canada and some of its former Alsands partners expanded to 12,500 bbl/d their 1,500 bbl/d in situ steam recovery plant constructed in 1979 at the Peace River oil sands, although by May 2017 Shell had reduced or sold most of its oil sands interests. In 1983, Petro-Canada entered into partnership with British Petroleum (BP) Canada in Calgary to launch a $200 million oil

76 Carrigy (note 73).

77 "Canada's Synfuels" (note 74), p. 40.

78 Mossop, "Geology of the Athabasca" (note 74), p. 151; "Canada's Synfuels" (note 74), p. 40. The Canadian government established Petro-Canada as a Crown Corporation (government owned) in 1975. It remained a Crown Corporation until the government began selling shares in Petro-Canada in July 1991. It operated as a share-holder owned, private, independent company July 1991–2009. Suncor Energy with headquarters in Calgary purchased Petro-Canada June-August 2009. Petro-Canada has a capitalization of $43.3 billion Canadian.

79 "Canada's Synfuels" (note 74), p. 40.

80 Carrigy (note 73); "Canada's Synfuels" (note 74), 40; *Chemical Week*, 133 (10 August 1983): 61.

sands project at Wolf Lake, southeast of the Athabasca deposit. The two companies increased the initial 7,000–8,300 bbl/d output achieved in September 1985 and 1986 to 13,000 bbl/d by 1990. A third oil company, Canadian Natural Resources Limited in Calgary, in 1985 operated a 5,000 bbl/d oil sands plant at Wolf Lake-Primrose-Primrose East areas that in 2013 increased production to 109,000–120,000 bbl/d.[81]

During this series of booms and busts in the petroleum industry, the Mines Branch, renamed the Canada Centre for Mineral and Energy Technology (CANMET) in 1975, continued its long-term development policy on upgrading Canada's bitumen and low-grade oil reserves. Since 1970 it had worked on a unique bitumen hydrogenating process called Canmet. Whereas all earlier commercial bitumen and heavy oil upgrading plants, such as Suncor and Syncrude, separated bitumen from the oil sands with mining technology and upgraded the bitumen by coking, CANMET upgraded the separated bitumen by hydrocracking. The hydrocracking process broke the C-C bonds in large (heavy) hydrocarbon molecules such as bitumen and added hydrogen gas (H_2) to the cracked fractions to increase the H/C ratio, thereby producing smaller, lighter, lower-boiling hydrocarbon molecules, namely gasoline, kerosene, and diesel oil. Metallic sulfide catalysts (MoS_2, WS_2, NiS, CoS) and other metallic combinations (Ni-Mo deposited on Al_2O_3) increased the yield.

CANMET's single stage, coke free, thermally stable, catalytic hydrocracking process operated at a lower pressure (P = 120–140 atm, T = 380–480° C) than most industrial hydrocracking processes because it continuously fed an inexpensive 1% by weight coal/iron additive to the incoming bitumen or crude oil that lowered the pressure and gave a high conversion rate of bitumen or crude oil to the lower-boiling hydrocarbon fractions, boiling below 524° C. By the early 1980s, CANMET researchers at the government's Energy, Mines and Resources (EMR) Department's laboratories in Ottawa had conducted over 20,000 hours of pilot plant studies in two, 1-bbl/d units hydrogenating Athabasca bitumen, Lloydminster heavy crude, Cold Lake bitumen, Boscan (Venezuela) residue, and Hondo crude. All these substances had high pitch

81 *Shell Canada Peace River in Situ Expansion*, Carmon Creek Project, Public Information Document, 2009, pp. 1–17. Petro-Canada also considered the construction of a 50,000–70,000 bbl/d plant in the Athabasca Oil Sands near Fort McMurray. Its expected partner was NOVA, the parent company of Husky Oil. NOVA operated all natural gas pipelines systems in Alberta. "Canada's Synfuels" (note 74), p. 40. For British Petroleum see "Talisman Energy," *Encyclopedia.com*. BP London sold 57% of its interest in BP Canada in early 1992 to Talisman Energy in Calgary (it became Talisman Energy) which then sold the Wolf Lake assets to Amoco Canada Petroleum Company Limited in Calgary in April 1992.

concentrations, and in each case CANMET converted about 90% of the pitch, a dark brown-black mixture of hydrocarbons and resinous compounds obtained from the distillation of coal tar, into lower boiling fractions.[82]

CANMET expanded its Canmet process in the 1980s. It assigned Petro-Canada the exclusive right to commercialize the hydrocracking process in 1979. Partec Lavalin (Toronto and Calgary), a division of Lavalin Inc. in Montreal, and one of Canada's largest engineering firms, joined Petro-Canada's commercialization program in 1981; and in July 1985 Petro-Canada, Partec Lavalin, and CANMET completed at Petro-Canada's Montreal refinery a $90 million, 5,000 bbl/d demonstration plant that by June 1989 provided convincing engineering and economics data for possible commercialization. Additional tests supported commercialization of the Canmet process, and upon their completion in 2003, CANMET projected a construction cost of $1.409 billion for a 100,000 bbl/d plant, with a Canmet hydrocracking slurry phase reactor operating at 93% conversion. The plant would produce a high-quality, low-sulfur synthetic crude for $12.57/bbl.[83]

No Canadian hydrocracking plant construction occurred at that time, but four years later in 2007, UOP, renamed Honeywell UOP in Des Plaines, Illinois, in 2005, acquired the exclusive rights to license the Canmet hydrocracking process. Two years later, in 2009, UOP combined the Canmet hydrocracking process with its unionfining and unicracking processes and developed a new advanced commercial-size conversion process using molybdenum-based dispersed catalysts such as MoS_2 and other sulfide catalysts Fe_2S_3, NiS, WS_2, and CoS_2. UOP called its process SRC Uniflex Process or UOP Slurry Hydrocoking Heavy Oil Conversion Process and has applied it to a wide range of heavy oil feedstocks, converting residues from conventional crude oils and bitumen

82 G. Lunin, A. E. Silva, and J. M. Denis, "The Canmet Hydrocracking Process: Upgrading of Cold Lake Heavy Oil," *Chemistry in Canada*, 33 (March 1981): 17–21; G. Lunin, J. Lamarre and K. Witham, "Canmet Hydrocracking-A Canadian Story," *Third International Conference on Heavy Crude and Oil sands* (1985): 1–21. None of the articles on the process specifies its pressure.

83 Lunin, Lamarre, and Witham (note 82), costs are in 1985 US dollars; Lunin, Silva, and Denis (note 80), 17; "Advance Process Control Helps in Success of Canmet Demo.," *Process Industries Canada*, 71 (April 1987): 17; J. P. Paquette, A. E. Silva, and G. Muir, "Canmet Hydrocracking Process; Demonstration Plant Operating Experience," presented at the 37th Canadian Engineering Conference, Montreal, Quebec, May 1987. Partec Lavalin established in Montreal in 1936, sold in 1991 to the Canadian (Montreal) engineering company SNC (est. 1911 and named after its owners, Arthur Surveyer, Emil Nenniger and Georges Chênevert) and renamed SNC-Lavalin.

to transportation fuels, especially distillates, and high quality synthetic crude oil.[84]

8 Conclusion

In their search to discover why some scientific ideas, technologies, or inventions succeed while others fail, historians of science and technology have asked whether the innovation's intrinsic quality or social-economic factors determine its success or failure. For high-pressure hydrogenation both factors require careful and thorough examination. In Germany, the process succeeded technologically six years before the Nazi government rose to power in 1933 and proceeded to subsidize industrial development. Industrialization may not have occurred otherwise. ICI's technologically successful process received similar economic incentives from the British government. The United States Bureau of Mines' program succeeded technologically, but because of unfavorable economics and the abundance of cheap petroleum, hydrogenation never achieved industrialization. In Japan the opposite happened. During World War II the Japanese government provided every possible economic and social incentive, but technologically high-pressure hydrogenation failed because of insufficient pilot plant research and development. In Canada, the Mines Branch's plan of long-term development of its low-grade bitumen and crude oil reserves met with considerable technological success, but uncertain economics resulted in a cautious and limited approach to industrialization and to licensing its hydrocracking process.

CANMET's process of converting low-grade hydrocarbon fuels to high-quality synthetic liquid fuels satisfied the first criterion that determines the success or failure of scientific or technological innovation.[85] External factors

84 UOP SRC (Supplemental Rectification with Compression), "Advanced Slurry Hydrocracking Process: Technology for Heavy Oil Conversion," Maureen Bricker, James McGehee, Robert Haizmann, Paul Zimmerman, Alak Bhattacharyya, Lorenz Bauer, and Beckay Mezza, UOP LLC, A Honeywell Company, 1 UOP Research Center, Des Plaines, IL 60017–5016. Daniel Gillis, 5th NCUT Upgrading and Refining Conference 2009, 14–16 September 2009, Edmonton, Alberta, Canada. Al-Attas, T. A.; Ali, S. A.; Zahir, M. H.; Xiong, Q.; Al-Bogami, S. A.; Malaibari, Z. O.; Razzak, S. A.; Hossai, M. M. "Recent Advances in Heavy Oil Upgrading Using Dispersed Catalysts," *Energy Fuels*, 2019. [Google Scholar] [CrossRef].

85 In May 1988, the Solv-Ex Company of Albuquerque, New Mexico, claimed that the residual sands or tailings are far richer in metals and metallic compounds than suspected. Solv-Ex which has a patented process for separating the bitumen from the oil sands, ear-

slowed its industrial development. CANMET's hydrocracking process, nevertheless, ranks as Canada's most significant contribution to the science and technology of petroleum refining.

---

lier in 1986 had worked with Shell of Canada on an oil sands project. Its preliminary analysis of the separated sand showed significant amounts of gold, silver, and titanium dioxide, a valuable pigment used in paints, whose combined value, Solv-Ex says, would pay for the extraction costs. Bitumen recovery in effect could be "free and clear of all mining costs." The world market price of oil would not matter, Solv-Ex claimed, because after Saudi Arabia it will be the lowest cost producer. See Robert Metz, "New Potential for Oil sands," *Houston Chronicle*, (14 September 1988), p. 1 B; "Solv-Ex Confirms Recovery of Titanium Coproduct," Solv-Ex Corporation Press Release (11 October 1988), Form 8-K (31 August 1988), and Form 10-K (30 June 1988).

CHAPTER 7

The United States Bureau of Mines' Synthetic Fuel Program 1920s–50s: German Connections and American Advances

1 Introductory

Following the end of World War II the United States government under the code name Project Paperclip (1945–59) secretly brought about 1,600 German scientists to the United States to assist in various military projects. Hints and discussions of this operation have appeared in the well-known accounts of the writers-scientists Leslie Simon (1900–83) US army Major General and scientist, physicist Samuel Goudsmit (1902–78), Clarence Lasby historian at the University of Texas, and Israeli historian-politician Michel Bar-Zohar (b. 1938). For the most part their presentations focused on the arrival and scientific contributions of Germany's rocket scientists and to a lesser extent on the German nuclear scientists. They say nothing about the wartime contributions of Germany's synthetic fuel scientists.[1] The US Bureau of Mines had long followed developments in Germany's synthetic fuel industry, and to advance its own fledgling postwar synthetic fuel program the Bureau joined with interested US industrial representatives to organize the Technical Oil Mission (TOM) to Germany.[2] TOM, who with British synthetic fuel scientists functioned as sin-

1 Leslie Simon, *German Research in World War II* (New York: John Wiley & Sons, Inc., 1947); Samuel Goudsmit, *Alsos* (New York: H. H. Schuman, 1947); Michel Bar-Zohar, *The Hunt for German Scientists 1944–60* (New York: Hawthorn Books, Inc, 1967); Clarence G. Lasby, *Project Paperclip* (New York: Atheneum, 1971). Although all of these books relate to Project Paperclip, each has a different objective. Simon gives a straightforward account of German World War II research, Goudsmit and Bar-Zohar discuss the Allies' locating and capturing of German scientists, especially the nuclear and rocket scientists, whereas Lasby provides the first detailed account of Paperclip. A later book on Paperclip, Linda Hunt, *Secret Agenda: The United States Government, Nazi Scientists, and Project Paperclip 1945–1990* (New York: St. Martin's Press, 1991), criticizes the US government's ignoring or covering up some of the German scientists' Nazi connections or party membership. Only her book mentions the postwar arrival in the United States of one of the German synthetic fuel scientists.

2 Cornelius S. Snodgrass, director of the Foreign Refining division of the Petroleum Administration for War, also directed the TOM which was part of CIOS. Albert E. Miller, "The Story of the Technical Oil Mission, Preprint to the Twenty-fifth Annual Meeting of the Ameri-

 | DOI:10.1163/9789004690912_008

gle intelligence-gathering agency called CIOS, arrived in Germany in February 1945 and for several months conducted a systematic and thorough investigation of the German synthetic fuel plants, collecting thousands of documents and interrogating many of the synthetic fuel scientists who operated the synthetic fuel plants. Upon completing the interrogations in the months after the war ended TOM and CIOS released the numerous reports they had compiled.[3]

The reason why the synthetic fuel scientists failed to appear in the accounts of Project Paperclip, therefore, has little to do with unavailability of information but resulted largely because of an almost total lack of interest in their work. Germany's aerospace, rocket, and nuclear scientists, not its synthetic fuel scientists, had captured the government's, military's and public's attention. Simon and Goudsmit may have overlooked the German synthetic fuel scientists because they published their books in 1947, the same year the synthetic fuel scientists began arriving in the United States to work for the Bureau. But a more accurate assessment of why the published accounts of Project Paperclip ignored the synthetic fuel scientists is that most of the publications appeared well before the United States' 1973–74 oil crisis and well before any widespread or serious concern arose over the country's rapidly increasing petroleum consumption and growing dependence on imports to meet demand. Yet, on closer inspection, signs warning of possible future liquid fuel shortages already existed.

During the war years, or more precisely from 1941 to 1946, the United States had consumed tremendous amounts of petroleum for both military and military-related purposes, doubling its agricultural use of gasoline in this five-year period. In the decades after the war, consumption became reckless.

---

can Petroleum Institute," Chicago, Illinois, 14 November 1945; W. C. Schroeder, "Technical Oil Mission Studies German Petroleum Research Activities," *Oil and Gas Journal*, 44, no. 29 (24 November 1945): 112, 115–16; Anthony N. Stranges, "Synthetic Petroleum from Coal Hydrogenation," *Journal of Chemical Education*, 60 (1983): 617–25.

3 Schroeder (note 2). The TOM sent the captured German synthetic fuel plant documents and interviews with German scientists to the Combined Intelligence Office Subcommittee (CIOS) in London for translating and abstracting. CIOS prepared 141 microfilm reels and after moving its operation to Washington, DC, produced another 164 reels. CIOS and two other intelligence gathering teams, British Intelligence Objectives Subcommittee (BIOS) and Field Intelligence Agency Technical (FIAT), printed and released more than 1,400 reports on the German synthetic petroleum plants. Many of these reports and interviews are on the TOM reels. Copies of the microfilm reels and the intelligence committees' reports are in the Synthetic Fuel Collection, Texas A&M University Archives. See also Henry H. Storch, Norma Golumbic, and Robert B. Anderson, *The Fischer-Tropsch and Related Syntheses* (New York: John Wiley & Sons, Inc., 1951), p. vii.

The United States was not only the world's biggest producer but the biggest user, accounting for 45% of global consumption in 1947. The Soviet Union, the United States' closest competitor, consumed only 11% of the total. From 1946 to 1947 the United States burned 25% more oil for electric power generation, installed 29% (829,470) more domestic oil-burning furnaces, registered a 10% gain in motor vehicle licenses, and experienced a 33% rise in aviation gasoline consumption.

Production no longer matched consumption, but the development of the large Middle East oilfields of Saudi Arabia, Kuwait, (and Iran after 1953) led to increasing imports and a quick and inexpensive solution to the nation's supply and demand problem.[4] Oil imports were 20 million barrels more than exports in 1948, the first time this had happened since 1920–22. By 1965 the import-export imbalance had risen to 800 million barrels.[5] Other than the Bureau of Mines, hardly any United States petroleum producer considered synthesizing petroleum from the country's huge coal deposits to help alleviate the imbalance. During these years of cheap, abundant, and recklessly consumed petroleum, Germany's large and successful wartime synthetic petroleum industry remained almost a long-forgotten chapter in the annals of science and technology.

Arno C. Fieldner, the Bureau's chief chemist in Washington, DC, established the direction of the Bureau's fuel research during the first thirty years of the Bureau's synthetic fuel program. Fieldner was a world-renowned authority on coal combustion, whose technological style of coal research emphasized the potential applications of the research. He keenly observed international developments in coal research and made their study an essential and important part of the Bureau's synthetic fuel program. International developments continued to interest the Bureau during the 1930s–40s, and after the war, the seven German synthetic fuel scientists who came to the United States under Project Paperclip assisted in the construction and operation of the Bureau's coal hydrogenation (liquefaction) and Fischer-Tropsch (F-T) demonstration plants in Louisiana, Missouri. The two plants were the centerpiece of the Bureau's $87.6 million postwar synthetic fuel program. During their period of operation from 1949 to 1953 they successfully converted American coals to synthetic fuel, and although the plants operated much more efficiently than

4 J. A. Krug, "Natural Resources Programs," *United States Department of Interior*, Annual Report of the Secretary for fiscal year ended 30 June 1948.

5 Hans H. Landsberg and Sam H. Schur, *Energy in the United States* (New York: Random House, 1968), p. 188.

the older German plants, the economics of the conversion became a matter of considerable debate. The Bureau accomplished its objective, however. It had demonstrated the technological feasibility of converting domestic coals to petroleum-like liquids, but the newly elected congress, showing little interest in continuing the program, ended its financial support and permanently shut down the Missouri operation in 1953.

As part of its postwar synthetic fuel research program the Bureau also resumed its pre-World War I oil shale program, and from 1949 to 1955 it operated an oil shale demonstration plant at the Naval Oil Shale Reserve in Rifle, Colorado. The Bureau aimed to improve oil shale retorting (distillation) and refining methods used to separate and upgrade the oil present in shale. Shale retorting differs from coal-to-oil conversion processes because it is a distillation not a synthesis and gives a derived not a synthesized oil. Congress provided funds for the oil shale research in the Bureau's synthetic fuel budget, and I have included its discussion to provide a complete description of the Bureau's postwar liquid fuel demonstration program.

Since the termination of the Bureau's synthetic fuel program in 1953, the synthetic fuel industry, like most alternative energies in the United States and globally, has endured a series of fits and starts that continued to plague it since the early 1980s.The history of synthetic fuel shows that shortly after its invention in 1913 the enthusiasm the process generated in petroleum-poor Germany during the early World War I years largely disappeared once Germany acquired Romania's oil fields. Similarly, in the United States the East Texas petroleum discoveries that began on 5 October 1930 and continued throughout the decade ended the Bureau's small pilot program. The Bureau's promising start following World War II also ceased within a decade with the discovery of new petroleum deposits and the failure of the anticipated energy crisis to materialize.

The historical record has demonstrated that after more than a century of research and development synthetic fuel did not emerge as an important alternative energy source. Despite the technological success of synthesizing oil from coal, the neglect resulted directly from government's and industry's uninterest in making a firm and a long-term commitment to the economic survival of synthetic fuel research and development. The synthetic fuel industry experienced intermittent periods of intense activity internationally in times of crises, only to face quick dismissal as unnecessary or uneconomical upon disappearance of the crises. In the United States the Bureau's hope that industry would transform its accomplishments at the demonstration stage into a full-fledged industrial development never materialized.

2 Early Bureau of Mines, European, and Japanese Research and Development of Derived and Synthetic Liquid Fuel

The Bureau of Mines had shown interest in obtaining oil from shale and coal almost from its inception. A congressional act of 6 May 1910 passed during President William Howard Taft's administration established the Bureau as a branch of the Department of the Interior (established 1849) with headquarters in Washington, DC. It aimed to conduct and develop the mining technology program that the United States Geological Survey, a branch of the Department of the Interior established in 1879, had begun, specifically fuel testing and investigating mining methods and mine explosions. Joseph A. Holmes (1859–1913) of the Geological Survey, a Cornell University-educated mining engineer and champion of mineral resource conservation and mine safety, served as the Bureau's first director. Under Holmes and succeeding directors, Vannoy H. Manning (1861–1932), Frederick Cottrell (1877–1948), and H. Foster Bain (1871–1948), the Bureau by 1921 grew to thirteen mining experiment stations including Bartlesville, Oklahoma; Berkeley, California; Birmingham, Alabama; Butte, Montana, and established twelve offices, the most important of which was at its Pittsburgh, Pennsylvania, Central Experiment Station and nearby Bruceton mine.[6]

During this period of growth, the Bureau in 1913 began its research on deriving oil from oil shale. Its laboratory program to establish the quantity and quality of the derived oil followed the appearance of several gloomy pre-World War I reports predicting that the United States lacked sufficient petroleum reserves to meet its future needs. One of the reports that David T. Day (1859–1925) of the Geological Survey published in 1908, estimated that by the end of 1923 the United States would exhaust its mid-continent fields. A second report, the National Conservation Commission Report of 1909, predicted negligible production in ten years from fields in New York and Pennsylvania. A third report, that of the geologist Ralph Arnold (1875–1961) who in 1914 surveyed the country's future oil wealth, indicated the exhaustion of fields in Kansas, Kentucky, Tennessee, Louisiana, and Texas by 1926.[7]

The Geological Survey, whose examining and mapping of oil shale sites had begun with Day and ongoing for fourteen years, contributed to the Bureau's

6 Fred Wilber Powell, *The Bureau of Mines* (New York: D. Appleton and Company, 1922), pp. 1, 4, 76–80.

7 *Federal Oil Conservation Board*, Public hearing, 27 May 1926 (Washington, DC: Government Printing Office, 1926), p. 6; Victor C. Alderson, *The Oil Shale Industry* (New York; Frederick Stokes Company, 1920), pp. 39–40.

laboratory program by establishing the extent and quality of the sites. The state governments and universities in Colorado and Utah including the Colorado School of Mines in Golden and the University of Utah in Salt Lake City also cooperated with the Bureau and the Geological Survey. The Bureau continued its research into the 1920s, and in March 1925 it received $90,000 in federal funds to erect and operate a semicommercial (pilot) retorting plant at Rulison, west of the United States Navy's oil shale reserve at Rifle in western Colorado. The Bureau aimed to apply fundamental laboratory principles to low temperature (450–550° C) oil shale retorting and refining of the oil on a semicommercial scale and thereby determine the reliability of the Navy's oil shale reserves in times of national emergency. It expected the information obtained from operating a small experimental plant to benefit a future oil shale industry by providing a basis for the best retort design and refining technique. Plant operation began in September 1926 but continued only to 30 June 1927 when Congress failed to appropriate additional funds.[8] Oil shale research did not resume until the closing years of World War II.

At the time of the Rulison plant's opening, the four-member Federal Oil Conservation Board, which President Calvin Coolidge (1872–1933) had established on 19 December 1924, wrote in its report that "They (oil shale) form an almost unlimited reserve and may, therefore, be taken as the final protection in the matter of our essential supply."[9] What were the estimated United States oil shale reserves to which the Oil Conservation Board referred? In 1926 the board estimated 394,344 billion tons of oil shale that yielded on average 11.5 gallons oil/ton, or a theoretical recovery of 108,425 million barrels oil that in 1928 the Bureau revised to 92 trillion barrels oil.

The predicted cost of extracting oil from shale, $1.00/bbl, compared very favorably with petroleum's $2.50–3.00/bbl cost, but the favorable economics failed to compensate for the shale oil's lower quality. While the Bureau believed that further research on oil shale retorting and oil refining would improve the product, it concluded that at the present time shale oil remained

8 Alderson (note 7), p. 45; *Report II of the Federal Oil Conservation Board*, 16 January 1928 (Washington, DC: Government Printing Office, 1928), p. 16. All the experimental projects ceased at that time, with little interest shown in oil shale treatment until the passage of the Synthetic Liquid Fuels Act near the end of World War II (5 April 1944).

9 *Report of the Federal Oil Conservation Board*, 6 September 1926 (Washington, DC: Government Printing Office, 1926), p. 12. See also Joseph Marchese, *Government and Energy: The Demonstration Program for Synthetic Liquid Fuels* (PhD diss., Princeton University, 1983), pp. 8–19. The board consisted of four members: Hubert Work (Chairman), Secretary of the Interior; Dwight F. Davis, Secretary of War; Curtis D. Wilbur, Secretary of the Navy; Herbert Hoover, Secretary of Commerce.

an unsuitable substitute for natural petroleum, except possibly in emergencies. When the operation ended in 1927 the Oil Conservation Board reported that the Bureau had completed its work and that the results obtained would benefit any oil shale industry developed in the future.[10]

In addition to initiating the Bureau's program, the gloomy pre-World War I reports led to the filing of numerous mining claims on virtually all the public oil shale lands of the Green River formation in Colorado, Utah, and Wyoming, the new Albany formation in Indiana, Kentucky, and Ohio, and in Nevada. About 100 oil shale companies using approximately 30 different retorts came into existence during the period 1916–25 but just as quickly passed away, leaving only 7 small experimental and semicommercial plants operating in Nevada, Colorado, and Utah. Most of the plants discontinued operation because of either equipment failure or a poorly designed process. For the few that achieved a satisfactory stage of development, high actual production costs compared to predicted costs forced their shutdown by the end of the 1920s. The most successful of the privately financed experimental oil shale projects were the 100 ton/d Catlin Shale Products Company plant in Elko, Nevada, and the 50 ton/d Continental Oil Shale Mining and Refining Company plant in Rio Blanco County, Colorado. Both began operating shortly after the war ended but had ceased operation by the end of the decade. Ten years earlier oil shale enthusiasts, such as Victor Alderson (1862–1946), president of the Colorado School of Mines, had remarked that if oil is king, oil shale is heir apparent. By 1920 oil was still king and oil shale no longer heir apparent.[11]

The Bureau's program for the production of liquid fuel from coal, lignite, and coal tar began in 1924 when Fieldner traveled to Europe to investigate recent developments in coal research. Fieldner had joined the Technologic Branch of the Geological Survey in 1907 after graduating from Ohio State University and had been a founding member of the Bureau's fuels chemistry laboratories. During World War I he headed the Bureau's gas mask investigations. When the War Department's Chemical Warfare Service took over this

---

10 *Report II of the Federal Oil Conservation Board* (note 8), p. 16; *Federal Oil Conservation Board Complete Record of Public Hearings*, 10 and 11 February 1926 (Washington, DC: Government Printing Office, 1926), p. 118.

11 L. W. Schramm "Shale Oil, Mineral Facts and Problems," *Bureau of Mines Bulletin 667* (Washington, DC: Government Printing Office, 1975), pp. 963–88; Alderson (note 7), pp. 2, 40–45, gives a list of some of the companies and the types of retorts they used, for example, horizontal, vertical, intermittent, continuous. See also A. J. Kraemer and F. E. Buchan, "Development of Oil-Shale Resources Undertaken by US Bureau of Mines," *Chemical and Engineering News*, 23, no. 17 (10 September 1945): 1523–7 and William Beaver, "The US Failure to Develop Synthetic Fuel in the 1920s," *The Historian*, 53 (1991): 241–45.

activity in mid-1918, Fieldner served as director of its Gas Mask Research Division from 1 July 1918 to 11 January 1919. He developed methods for testing gas masks and gas absorbents while directing the research of the division's more than 1,000 scientists.

After his wartime service ended in 1919, Fieldner became the Bureau's supervising chemist and superintendent of its Pittsburgh Experiment Station before his appointment as Chief Chemist, Chief Engineer of the Experiment Stations Division, with headquarters in Washington. By this time Fieldner had an international reputation as the inventor and developer of several techniques for testing and analyzing coal, coke, and gas samples and for his knowledge of the physical and chemical properties of coal and other fuels.[12]

Fieldner's European inspections included several British and German low-temperature coal carbonization (LTC) plants and Bergius's 2 metric ton/d industrial-size coal hydrogenation pilot plant in Rheinau-Mannheim. Because of the increasing importance of liquid fuels, especially gasoline and fuel oil, the British and Germans were investigating possible ways of establishing a home source of petroleum in the event of another war and loss of foreign supplies. The newer LTC processes that they developed in the years 1906–16 heated coal in the absence of air to 450–700° C, rather than the 900–1200° C of the older high-temperature carbonization (HTC), to give tar and smokeless coke or semicoke fuels. The lower temperature reduced tar decomposition, producing 20–30 gal tar/metric ton coal versus 10–12 gal/metric ton coal for HTC, and when distilled the tar gave maximum gasoline yields of 3–6 gal/metric ton coal instead of 2.5–3.0 gal/metric ton coal. Bergius had invented the coal hydrogenation process in 1913. He reacted coal with hydrogen gas at high pressure and high temperature rather than carbonizing or decomposing the coal and significantly increased the liquid yield per metric ton coal from a few gallons of liquid fuel to a few barrels. His Rheinau-Mannheim plant was the world's only operating coal hydrogenation plant in the post-World War I years.

At the time of Fieldner's European inspections Britain and Germany as well as a number of countries with very few petroleum sources within their boundaries had ongoing coal conversion programs. In Britain the government had begun its program to alleviate liquid fuel shortages in 1915, a few months after the outbreak of World War I on 1 August 1914, by establishing the Department of Scientific and Industrial Research. Two years later the government appointed a Fuel Research Board with Sir George Beilby as Director of Fuel Research and Chairman of the Board, and in 1919 it opened a Fuel Research

12 "Dr. Arno C. Fieldner, Melchett Medalist 1942," *J. Inst. Fuel*, 16 (1942): 4–20, on p. 4.

Station in East Greenwich (London) to develop a commercially feasible LTC process. In addition to liquid fuel the government expected the easily ignited smokeless coke or semicoke to replace the smoky soft coal burned in the widely used but highly inefficient British open grate stoves. Eliminating this major source of air pollution was an important part of the Fuel Research Station's program. After the war, several industrial LTC plants were operating in Britain. They included Fusion Corporation Ltd.'s small, 3–5 long ton/d plants with rotary retorts, at Middlewich, Cheshire, and Burghlee Colliery, Loanhead, near Edinburgh; Low Temperature Carbonization Ltd.'s 36-long ton/d plant with a vertical, externally heated Parker retort (increased to 200 long ton/d in 1928) at Barugh, near Barnsley in Yorkshire; and Glasgow Gas Corporation's 100 long ton/d plant with an internally heated McLaurin retort at Dalmarnock, near Glasgow. By 1937 Low Temperature Carbonization Ltd. was the largest operation in England with four plants that carbonized 500,000 long tons of coal annually, producing 250,000 barrels of crude tar oil and light oil.

Germany's fuel engineers had shown little interest in using LTC to produce a solid smokeless domestic fuel. Instead they investigated LTC for the production of badly needed gasoline, fuel oil substitutes, and a dense metallurgical coke. Germany consumed 1 million metric tons of petroleum (gasoline, kerosene, diesel oil, and lubricants) in 1922, and although it produced only 45,000 metric tons Germany had sufficient coalfield capacity to meet its petroleum demand. Reducing petroleum imports required carbonizing 34 million metric tons of coal or more than half of the 63 million metric tons that Germany's central electric power stations, large industrial power plants, and briquette works consumed annually.[13]

Early in World War I, Franz Fischer, director of the recently established (1914) Kaiser Wilhelm Institute for Coal Research at Mülheim-Ruhr, began experimenting on LTC. Fischer and his co-workers at the institute followed

13 A. C. Fieldner, "Low-Temperature Carbonization of Coal," *Bureau of Mines Technical Paper 396* (Washington, DC: Government Printing Office, 1926), pp. 1–46, on pp. 21, 25, 36; Fieldner, "The Production of Liquid Fuels from Coal and Lignite," *Bureau of Mines Statement* (Washington, DC: Government Printing Office, 4 August 1943), also in *Universal Oil Products Company Library Bulletin*, 18, no. 34 (25 August 1943). Tons are US (short) tons unless specified as long tons or metric tons; gallons are US unless specified as Imperial gallons. For units conversions use below.

1 metric ton = 2,200lb (1,000 kg)
1 long ton = 2,240 lb (1,018 kg)
1 short ton = 2,000 lb (909.1 kg)

a chemical style of coal research that focused on understanding the composition of coal through chemical analyses. When Fieldner arrived in Germany, Fischer already had designed, a small laboratory-size rotary retort and a larger 50 metric ton/d industrial-size rotary retort in operation at the August Thyssen Works near Essen. He arranged for Fieldner's inspection of the Thyssen retort, a larger 100 metric ton/d rotary retort at the Mathias Stinnes Colliery at Karnap near Essen, and a 54 metric ton/d Fellner-Ziegler rotary retort in Gelsenkirchen. The three retorts had operated experimentally for months, Gelsenkirchen's for 30 months, but none currently operated regularly because of unsolved technical problems. German scientists and engineers, such as Bergius and Fischer, throughout the 1920s continued to investigate the expanded commercial utilization of low-temperature tar and of semicoke from LTC as a pulverized fuel in carbonization plants.[14]

Fieldner did not visit any LTC plants in France, Italy, or other European countries that lacked domestic sources of petroleum, but he carefully followed the active conversion programs that most of them had underway. The French, who from 1920 to 1926 produced 50 to 70% of their domestic petroleum consumption, had only small oil shale and coal carbonization industries. They concentrated more on using the water gas reaction to synthesize methanol (methyl alcohol) and other organic liquid fuels.[15]

$C + H_2O \rightarrow CO + H_2$

Coke + Steam → Carbon Monoxide + Hydrogen (water gas reaction)

$CO + 2\,H_2 \rightarrow CH_3OH$

Carbon Monoxide + Hydrogen → Methanol (or other organic products)

14 Fieldner, "Low-Temperature Carbonization" (note 13), pp. 2–3, 14–15, 17–18. The Germans also produced a low-volatile semicoke that when mixed with high-volatile gas coals gave a dense metallurgical coke. The Thyssen, Fellner-Ziegler, Bamag-Meguin retorts were industrial-size modifications of Fischer's small externally heated rotary retort. The Fusion retorts were similar. Glasgow Gas used an internally heated continuous MacClaurin retort. See "Low-Temperature Carbonization at Glasgow," *Colliery Engineer*, 2 (1925): 546–53 and *Gas Journal*, 171 (1925): 757–58. For a summary of European developments in coal to oil conversion to May 1925 see Richard Lessing, *The Conversion of Coal into Oils* (London: Ernst Benn Limited, 1925). This is a translation of Franz Fischer's *Die Umwandlung der Kohle in Öle* (Berlin: Verlag von Gebrüder Borntraeger, 1924).

15 Henrie de Peyerimhoff, "The Coal Situation in France," *Proceedings of the Third International Conference on Bituminous Coal*, 1930 (Pittsburgh: Carnegie Institute of Technology, 1931), Vol. I, pp. 48–52. Most of the French research on LTC, as in Germany, was for the production of gasoline, fuel oil substitutes, and dense metallurgical coke. See Fieldner, "Low-Temperature Carbonization" (note 13), p. 3.

A long tradition of French research on the water gas reaction dated from the late 1800s–early 1900s studies of Marcelin Berthelot at the Collége de France; Paul Sabatier and co-workers John Baptiste Senderens at the Catholic University of Toulouse, Alphonse Mailhe, and Marcel Marat at the University of Toulouse; and Etienne Audibert (1888–1954) at the recently established (June 1924) Société Nationale de Recherches sur le Traitement de Combustibles in Senlis, twenty miles north of Paris.[16] General Georges Patart (1869–1944), Inspector General of the French Bureau of Explosives in Paris, continued and expanded the studies on synthesizing methanol and in a 1925 lecture before the Société d' Encouragement pour l'Industrie National summarized his work up to that time.[17]

Patart had a notable reputation in French industry. Earlier he had directed the 1 April 1920 agreement between France's War Minister and its Minister of Industrial Reconstruction and BASF on the use of the synthetic ammonia patents that the French government had acquired under the June 1919 Treaty of Versailles. He also directed the synthetic ammonia plant that the Compagnie Nationale de l'Azote planned to construct at the National Powder Works in Toulouse.[18] By 1921, following BASF's successful wartime application of van't Hoff's and le Chatelier's equilibrium principle to the industrial synthesis of ammonia, Patart began a similar study to establish the most favorable equilibrium conditions for methanol's industrial synthesis. Both syntheses used almost identical apparatus, but Patart believed the methanol synthesis easier because it did not require the reaction of pure gases.

Assisted by the War Ministry's Bureau of Explosives and the Ministry of Commerce's Petroleum and Gasoline Bureau, Patart in 1925 constructed a small methanol unit at the Oxylite Company's plant in Asnieres near Paris. Experiments in which he reacted a 2:1 hydrogen: carbon monoxide volume (m^3) mixture at 300–600° C, 150–200 atm, and an 8–10% metallic oxide-metal catalyst such as zinc oxide-copper, produced 426 g of methanol.[19] He

16 Paul Sabatier, *La Catalyse en Chimie Organique* (Paris: Librairie Polytechnique, 1913), trans. E. Emmet Reid, *Catalysis in Organic Chemistry* (New York: D. Van Nostrand, 1922) contains numerous references to the publications of Berthelot, Sabatier, and the others; E. Audibert, "Synthetic Motor Spirit from Mixtures of Carbon Monoxide and Hydrogen," *Fuel*, 5 (1926): 170–77.

17 Charles Lormand, "Industrial Production of Synthetic Methanol," *Industrial and Engineering Chemistry*, 17 (1925): 430–32.

18 J. R. Partington and L. H. Parker, *The Nitrogen Industry* (London: Constable and Company Limited, 1922), pp. 180–81.

19 Georges Patart, French Patent 540,343 of 19 August 1921, published 12 July 1922; Lormand (note 17), pp. 430–31.

estimated an annual production cost of 5–7¢/L (18–26¢/gal) for a large plant having a synthesis gas capacity of 7.5 metric ton/d (3,280 kL/d), which compared favorably with the 1925 United States bulk (tank) price of 65¢/gal for 95% pure methanol (67¢ if 97% pure), the equivalent of 3s 6d-4s/Imperial gal in Britain.[20] Patart's process cost one-third that of distilling methanol from wood residues and cheaper, therefore, than producing fuels by either liquefying or carbonizing coal.

Patart and Paul Dumanois (1885–1964), France's Chief Engineer of Air Service, also discussed French synthetic methanol research in papers presented before the first International Conference on Bituminous Coal held in November 1926 at the Carnegie Institute of Technology in Pittsburgh. Patart described a semicommercial plant with a methanol production of 150–200 kg/d (ca. 160 kL/d) that the Paulene Brothers Establishment operated at Vitry-sur-Seine. Dumanois focused on recent research by the French air force and the Automobile Club of France on synthesizing methanol for aviation and motor fuel.[21]

France's carbonization plants produced only 436,000 metric tons of coal tar in 1925. Distilling the tar to separate benzol, a liquid mixture yielding benzene and toluene, provided about 5% of France's motor fuel requirements,[22] although at the time French industrialists had under construction in France's northern coalfield area and Lorraine new carbonizing plants giving 50 kg of tar/metric ton of coal.[23] Benzol distilled from coal tar also was an important source of liquid fuel in Italy, Belgium, and Bohemia (Czechoslovakia). In fact, during the World War I years (ca. 1915–19) nearly all European countries had laws that required the coal tar industries to remove benzol from the tar. Switzerland in 1915 and France in 1920 passed laws that prohibited exporting

20 Lormand (note 17), p. 432; 1 kL = 264 gal; corresponding cost figures are 105 francs/ kL, 130 francs/kg; T. P. Hilditch, *Catalytic Processes in Applied Chemistry* (New York: D. Van Nostrand Company, 1931), p. 88.

21 Georges Patart, "The Industrial Transformation of Bituminous Coal into Organic Technical Products," *Proceedings of the International Conference on Bituminous Coal*, 1926 (Pittsburgh: Carnegie Institute of Technology, 1927), pp. 132–60, on p. 137; Paul Dumanois, "Methyl Alcohol as Motor Fuel," *Proceedings of the International Conference on Bituminous Coal*, 1926, pp. 161–64. Patart's article is also in *Fuel*, 6 (1927): 160–70.

22 Jean Bing, "The Gas and Tar Industries in France," *Proceedings of the Second International Conference on Bituminous Coal*, 1928, 2 Vols. (Pittsburgh: Carnegie Institute of Technology, 1929), Vol. 2, pp. 766–88, on p. 774.

23 Jean Bing, "Bituminous Coal as a Source of Chemical Products," *Proceedings of the International Conference on Bituminous Coal*, 1926 (Pittsburgh: Carnegie Institute of Technology 1927), pp. 404–18, on pp. 409–12.

coal tar, and in 1923 France and Italy once again made benzol removal compulsory.[24]

In Asia research on the production of liquid fuel from LTC got underway in 1921. In that year the Japanese government established the Imperial Fuel Research Institute in Kawaguchi (Honshû), a northern suburb of Tokyo and appointed Yoshikiyo Ộshima its first director. LTC research expanded under Ôshima, and by 1924 he and Yoshisada Ban had operating a small LTC unit that carbonized 26–30 kg of coal/hr. Japan, like England, Germany, France, Italy, and other European countries, had no domestic petroleum sources, and the Imperial Fuel Research Institute, as part of its low-temperature conversion of coal to a semicoke solid fuel, produced tar for liquid fuel tests in diesel engines.[25]

3 Liquid Fuel Developments in the United States

To replace the petroleum imports lost because of the British sea blockade during World War I the Germans had started construction on coal carbonization and coal hydrogenation pilot plants. By 1925–26 they concentrated on synthesizing liquid fuel from coal. Numerous patents and publications on the catalytic synthesis of alcohols and hydrocarbons from water gas and on BASF's development of the Bergius high-pressure coal hydrogenation process appeared in those years.[26] Bergius, Fischer, and Patart presented noteworthy papers at the first International Conference on Bituminous Coal that Fieldner organized in Pittsburgh in 1926.[27] The conference, coming only months after the Federal Oil Conservation Board's two controversial 1926 reports on the American petroleum industry's health, stimulated American interest in synthesizing liquid fuel from coal and appeared to underscore the need for the Bureau to take a leadership role in synthetic petroleum production should the United States experience a shortage of natural petroleum.

---

24 Bing (note 23), pp. 409–12.

25 Yoshisada Ban "The Low-Temperature Carbonization Plant at the Imperial Fuel Research Institute," *Proceedings of the Second International Conference on Bituminous Coal*, 1928, 2 Vols. (Pittsburgh: Carnegie Institute of Technology, 1929), Vol. 1, pp. 303–11.

26 See Chapter 11, footnote 88.

27 Friedrich Bergius, "The Transformation of Coal into Oil by Means of Hydrogenation," *Proceedings of the International Conference on Bituminous Coal*, 1926 (Pittsburgh: Carnegie Institute of Technology, 1927), pp. 102–31; Franz Fischer, "The Synthesis of Petroleum," *Proceedings of the International Conference on Bituminous Coal*, 1926, pp. 234–36; Patart (note 21), pp. 132–60.

The Board's most significant finding was its statement on the country's petroleum production, consumption, and reserves. In 1925 the United States produced 70% of the world's total petroleum supply, but it consumed about 70% of the total.[28] A nine fold increase in automobile production, which jumped from 2.5 million in 1915 to 21 million in 1925, caused the major increase in petroleum consumption.[29] Petroleum production in 1925 had reached an all-time high of 750 million barrels, up from 280 million barrels in 1915, but at the current consumption rate the Board believed that demand would outstrip supply for the first half of 1926. Production in fact had declined from 2.35 million barrels on 30 May 1925 to 1.9 million barrels on 11 February 1926, or at an annual rate of 27.6%. The Board placed reserves at about 5.3 trillion barrels in 1925 and indicated that if consumed at the 1925 rate, the reserves would last from 3 to 7.5 years. The Board listed nine possible provisions to insure future petroleum supplies, one of which called for the production of liquid fuels from oil shale and coal carbonization (distillation), and called on the Bureau to expand research on petroleum production and refining.[30]

The Bureau believed the synthesis of methanol from water gas would have immediate industrial possibility. Until 1923, nearly all the methanol produced in the United States and globally came from the distillation of wood. BASF industrialized the water gas-methanol synthesis that year. It produced methanol at a cost of 18–27¢/gal, marketing it for 45¢/gal. and within a short time, the United States began importing increasing amounts of inexpensive synthetic methanol. Imports rose to 500,000 gallons in 1925,[31] about 10% of domestic production, while methanol's market price fell from $1.10/gal in 1923 to 70¢/gal in 1925.[32] The pattern continued in 1926, and to protect American methanol producers against cheaper German imports, Congress, on 27 November 1926 passed a 12–18¢/gal tariff increase, based on the selling price of domestic output from wood distillation.[33]

---

28 *Report of the Federal Oil Conservation Board* (note 9), p. 6.

29 *Report II of the Federal Oil Conservation Board* (note 8), p. 86.

30 *Report of the Federal Oil Conservation Board* (note 9), pp. 8, 9, 13. The board mentioned the need to develop more efficient automobile engines to conserve fuel, but said nothing explicitly about conservation.

31 Alfred T. Larson, "Some Observations of Ammonia and Methanol Catalysts," *Transactions of the Electrochemical Society*, 71 (1937): 345–46; Hilditch (note 20), pp. 89–90.

32 Jasper E. Crane, "Development of the Synthetic Ammonia Industry in the United States," *Industrial and Engineering Chemistry*, 22 (1930): 795–99.

33 Carleton Ellis, *Hydrogenation of Organic Substances*, 3rd ed. revised, (New York: D. Van Nostrand Company, 1930), p. 652; see provisions of Section 315 of the Tariff Act that went into effect 27 December 1926; "Tariff Act," *Chemical Age*, 16 (15 January 1927): 77; Otto Wil-

An immediate consequence of the emerging synthetic methanol industry led Fieldner and other Bureau scientists in 1927 to begin a small laboratory study of the thermodynamics and kinetics of synthesizing methanol, intending to establish the reaction conditions for maximum production.[34] By this time American industry also moved into synthetic methanol production. Lazote Inc., a DuPont Ammonia Corporation affiliate in Belle, West Virginia, developed a high-pressure water gas synthesis of methanol in 1925 and two years later began the industrial production of synthetic methanol in the United States. By the end of the 1920s two more companies, Commercial Solvents Corporation in 1929 at its plants in Terre Haute, Indiana, and Peoria, Illinois, and Carbide and Carbon Chemicals Corporation in 1927 in Niagara Falls, New York, started synthesizing methanol but from carbon dioxide and hydrogen. The two gases were by-products of corn fermentation, a process used to manufacture butanol and acetone.[35] European countries such as Belgium, Italy, France, and in Japan, also had underway extensive research and development programs on synthetic methanol.[36]

European and American fuel researchers, as a result of the invention and development of the newer synthetic processes in the mid-1920s, soon recognized that in the event of future liquid fuel shortages neither oil shale distillation nor LTC, despite the earlier enthusiasm for each, could solve their liquid fuel requirements. The extensive European research on LTC had impressed Fieldner, but LTC did not produce enough liquid fuel per ton of coal to sat-

---

son, "Chemicals in the New Tariff Law," *Industrial and Engineering Chemistry*, 22 (1930): 800–04.

34 R. L. Brown and A. E. Galloway, "Methanol from Hydrogen and Carbon Monoxide," *Industrial and Engineering Chemistry*, 20 (1928): 960–88; Brown and Galloway, *Industrial and Engineering Chemistry*, 21 (1929): 310–13 and 322 (1930): 175–6; Crane (note 32); D. F. Smith, "Equilibrium Conditions in the Formation of Hydrocarbons and Alcohols from Water Gas," *Industrial and Engineering Chemistry*, 19 (1927): 801–03; A. C. Fieldner and R. L. Brown, "Comparative Study of Values Obtained in Synthesis of Methanol," *Industrial and Engineering Chemistry*, 20 (1928): 1110–112; D. F. Smith and L. L. Hirst, "Reactions that Occur on a Methanol Catalyst," *Industrial and Engineering Chemistry*, 22 (1930): 1037–40; D. F. Smith and B. F. Branting, "Equilibrium between Methanol, Carbon Monoxide, and Hydrogen," *Journal of the American Chemical Society*, 51 (1929): 129–39.

35 J. C. Woodruff, "Synthetic Methanol from Butyl Fermentation Gases," *Industrial and Engineering Chemistry*, 19 (1927): 1147–50; "The Chemical Kaleidoscope for 1929," *Industrial and Engineering Chemistry*, 22 (1930): 4–10, on p. 7; "Synthetic Methanol Produced on Commercial Scale," *Chemical and Metallurgical Engineering*, 34 (1927): 265; Crane (note 32); Ellis (note 33).

36 "Italian Production of Synthetic Methanol," *Chemical Age*, 22 (11 January 1930): 24; "French Production of Synthetic Methanol Increasing," *Chemical Age*, 22 (8 March 1930): 212; "Synthetic Methanol in Belgium," *Chemistry and Industry*, 48 (1929): 1095.

isfy the United States' demand even when taking into account the country's tremendous coal deposits. According to a 1917 report by the geologist Marius R. Campbell (1858–1940) of the United States Geological Survey, the United States possessed about 3.5 trillion tons of coal lying within 3,000 feet of the surface and that up to 1918 the depletion amounted only to 0.5% of the original total. Fieldner believed coal was "the logical source of liquid fuel substitutes for petroleum gasoline" because most of it, "constituting over 60% of the world's reserve, will still be in the ground" even after "all the petroleum available by drilling and pumping methods becomes exhausted in the United States."[37] Yet, in 1924 the United States obtained an insignificant quantity of gasoline from HTC in by-product coke-ovens, amounting to 95 million gallons or 1.25% of the 7.6 billion gallons required that year. Including the quantity of ethanol (ethyl alcohol) derived from coal gas increased the gasoline yield to 3.3%. Even carbonizing the United States' entire 1923 bituminous coal output of 545.5 million tons in by-product coke ovens would have produced 1,335 million gallons of gasoline, or 18% of the 1924 requirement.

LTC did not give significantly higher yields. In tests carried out at Britain's Fuel Research Station in East Greenwich, Fischer's institute in Mülheim, and the Bureau of Mines, 1 ton of bituminous coal produced 6 gallons of gasoline. At an expected annual consumption rate of 160 million tons of bituminous coal this amounted to 960 million gallons of gasoline, 13% of the United States' annual requirement or in Fieldner's estimate, 20–30%.[38] The United States had only two privately owned LTC plants in operation between 1920 and 1930, and they produced smokeless fuel (coke, charcoal) from bituminous coal or lignite, not liquid fuel. The Pittsburgh Coal Carbonization Co., using a modified Wisner carbonization process, treated 325 tons of bituminous coal per day; Lehigh Briquetting Co. in Dickinson, North Carolina, used the Lurgi process to carbonize 500 tons of lignite per day. Each plant had an average gasoline yield of about 10 gal/ton of reactant. By 1930 few fuel researchers in the United States and Europe considered oil shale distillation and coal carbonization particularly valuable sources of liquid fuel and had turned instead to the newer and higher yielding coal hydrogenation and F-T synthesis as possible coal-based alternatives.[39]

---

37 A. C. Fieldner and R. L. Brown, "Complete Utilization of Coal and Motor Fuel," *Oil, Paint and Drug Reporter*, 108 (5 October 1925): xii-xiii on p. xii; see also M. R. Campbell, "The Coal Fields of the United States," Professional Paper 100-A, *United States Geological Survey* (1917), p. 25.

38 Fieldner and Brown (note 37), p. xii.

39 Fieldner (note 13). For documents supporting LTC over liquefaction see W. A. Bristow, "Petrol from Coal," *Gas Journal*, 203 (1933): 211–12; Bristow, "Production of Oil from Coal in

4 Development of the F-T Synthesis

In 1913 BASF patented a process for the catalytic hydrogenation (reduction) of carbon monoxide to give hydrocarbons other than methane.[40] Because of World War I and priority given to industrializing the ammonia and methanol syntheses, BASF never continued its initial work. In the meantime Fischer and Tropsch at the Kaiser Wilhelm Institute in Mülheim upon learning of BASF's patent, decided to test its claims and began investigating the catalytic reduction of carbon monoxide at various temperatures and pressures. The experiments continued into the 1920s,[41] and by 1928 Fischer and Tropsch had modified reaction conditions once again.[42] Fischer continued his investigations into the 1930s, constructing a small experimental plant in Mülheim in 1932.[43]

In 1933 Fischer's decade-long research resulted in the construction of the first large pilot plant for synthesizing hydrocarbons from carbon monoxide and hydrogen. Ruhrchemie AG, a company Ruhr coal industrialists founded in 1927, envisioned the F-T synthesis as an outlet for its surplus coke, and upon acquiring the patent rights to the synthesis in 1934, constructed a plant in Oberhausen-Holten, Ruhr. The plant operated at the same conditions as

---

Great Britain," *Petroleum Times*, 30, no. 750 (1933): 191–92. In computing total production costs of the two processes, Bristow claimed LTC was cheaper by $2.5 million per year, although LTC production was only 7 million Imperial gallons gasoline versus 10 million from liquefaction.

40 BASF, German Patent 293, 787 of 8 March 1913; British Patent 20,488 of 10 September 1913; French Patent 468,427 of 13 February 1914; US Patent 1,201,850 of 17 October 1916, assigned to A. Mittasch and C. Schneider.

41 Fischer and Tropsch, "Über die Herstellung" (note 26). The reaction produced no hydrocarbons. Their first publications on this research were in 1923.

42 Franz Fischer and Hans Tropsch, German Patent 484,337 of 22 July 1925; Fischer, "The Synthesis of Petroleum," *Proceedings of the International Conference on Bituminous Coal*, 1926 (Pittsburgh: Carnegie Institute of Technology, 1927), pp. 234–46. See their publications in note 26.

43 Franz Fischer, Helmut Pichler, and Rolf Reder, "Überblick über die Möglichkeiten der Beschaffung geeigneter Köhlenoxyd-Wasserstoff-Gemische für die Benzinsynthese auf Grund des heutigen Standes von Wissenschaft und Technik," *Brennstoff Chemie*, 13 (1932): 421–28; Franz Fischer, Otto Roelen, and Walter Feisst, "Über die nunmehr erreichten technischen Stand der Benzinsynthese," *Brennstoff Chemie*, 13 (1932): 461–68; Franz Fischer and Herbert Koch, "Über den Chemismus der Benzinsynthese und über die motorischen und senstigen Eigenschaften der dabei auftretenden Produkte (Gasol, Benzin, Dieselöl, Hartparaffin)," *Brennstoff Chemie*, 13 (1932): 428–34; Herbert Koch and Otto Horn, "Vergleichende Untersuchung über das motorische Verhalten eines synthetischen Benzins nach Franz Fischer (Kogasin I) und eines Erdöl-Benzins," *Brennstoff Chemie*, 13 (1932): 164–67.

Fischer's experimental plant and had an annual capacity of 1,000 metric tons (7,240 barrels) of motor fuel (gasoline, diesel) and lubricating oil.[44] By 1936 the F-T plants constructed or under construction had a total annual capacity of 145,000 metric tons (1.05 million barrels) of gasoline, diesel oil, and other petroleum hydrocarbons.[45] Total annual plant capacity increased to 300,000 metric tons (2.17 million barrels) by 1937–38, rose to 525,000 metric tons (3.7 million barrels) by the end of 1938 and to 740,000 metric tons (5.4 million barrels) at the outbreak of World War II in September 1939. Actual peak production reached 570,000 metric tons (4.15 million barrels) in 1944. The older F-T plants operated at 1 atm and 180–200° C, whereas newer medium pressure plants operated at 5–15 atm and 180–200° C.

Japan and France in the 1930s also had industrial-size F-T plants in operation or under construction. The Japanese military sponsored Japan's synthetic fuel program which ranked as the most ambitious outside of Germany. Three plants, located at Miike, near Ômuta (Kyûshû), Takikawa (Hokkaidô), and Amagasaki (Honshû), that the Mitsui Mining Company and the government's Imperial Fuel Development Company jointly owned, began producing synthetic liquid fuel for the military in 1941–42 and continued through the war years. Each plant had a design capacity of 30,000–50,000 metric tons/yr, but because of technical problems their production totaled only about 50,000 metric tons of liquid fuel when the war ended in 1945. Mitsui-Imperial Fuel

44 "Synthetic Petrol by the Fischer Process," *Gas World*, 105 (24 October 1936): 362–63; H. H. Storch, "Synthesis of Hydrocarbons from Water Gas," in *Chemistry of Coal Utilization*, ed. H. H. Lowry, 3 Vols. (New York: John Wiley & Sons 1945 and 1963) Vol. 2, p. 1800. The motor fuel's estimated average cost was 22 pfennigs/kg. According to Arno C. Fieldner and American engineers the wartime cost of either the F-T or the hydrogenation process was 20–30¢/gal. See Fieldner, "Frontiers of Fuel Technology," *Chemical and Engineering News*, 26, no. 23 (7 June 1948): 1700–01.

45 *Report on the Petroleum and Synthetic Oil Industry of Germany* (London: Ministry of Fuel and Power, 1947), pp. 82–90; "Germany's Home Production of Motor Fuels," *Gas World*, 104 (9 May 1936): 421; Franz Fischer, "The Conversion of Coal into Liquid Fuels," *Chemical Age*, 35 (24 October 1936): 353–55; "The Fischer Process," *Chemical Age*, 35 (31 October 1936): 367. For units conversions use below.

1 short ton (US) = 6.7 bbl
1 metric ton = 7.3 bbl
1 long ton = 7.4 bbl
1 bbl = 42 US gal, 34.97 Imperial gal, or 159 L
1 US gal = 3.78 L
1 Imperial gal = 1.2 US gal
For crude oil 1 bbl = 300 lb or 136 kg

started construction on two other plants in Manchuria, but completed neither of them. All the Japanese F-T plants operated under Ruhrchemie licensing and were similar in design to the German F-T plants.[46]

The Société Courriéres-Kuhlmann constructed France's single F-T plant at Harnes in the Pas-de-Calais district about 6 miles south of Lille. The plant used French coal to produce synthesis gas, an iron (II) oxide (FeO) catalyst, and had 24 operational reactors. Its workforce numbered 120, and by June 1937 reached one-half its annual capacity of 20,000 metric tons of liquid hydrocarbons, including gasoline and diesel oil. Two US naval officers attached to CIOS visited the plant on 26–27 September 1944 and reported that the F-T plant had operated continuously since its construction except for a brief shutdown in 1942 because of bomb damage. They recommended that the plant continue to operate and that the US navy have sole access to the good quality diesel oil that the plant produced.[47]

F-T development in Britain lagged far behind the continent's, with only four small experimental plants in operation. One of them was at the University of Birmingham where in a series of experiments in 1920–27 Oliver C. Elvins and Alfred W. Nash obtained a 2–3% maximum conversion to liquid hydrocarbons. The other British F-T plants were the 4 Imperial gal/d and the 150 Imperial gal/d plants at the East Greenwich Fuel Research Station[48] and the privately owned Synthetic Oils Ltd. 136–180 Imperial gal/d plant at Bedlay Colliery, near Glasgow.[49]

Like the British program, the Bureau's work at this time remained at the laboratory scale. David F. Smith, Joseph D. Davis, and D. A. Reynolds presented their preliminary results at the American Chemical Society's Detroit meeting of 6–10 September 1927. Their investigations continued until 1930, during

46 Anthony N. Stranges, "Synthetic Fuel Production in Prewar and World War II Japan: A Case Study in Technological Failure," *Annals of Science*, 50 (1993): 229–65; Storch (note 44), p. 1801; "Catalysts for the Fischer-Tropsch Process," *Chemical Age*, 39 (19 November 1938): 391; "Japan Seeks N.E.I. Oil for Co-Prosperity Sphere," *Oil and Gas Journal*, 39, no. 33 (26 December 1940): 82–83, 108–10, 112–13.

47 H. A. Schade, E. Foran, and R. C. Aldrich, *The Fischer-Tropsch Process*, CIOS VI-22, X-18, X-22, XV-5, Item No. 20 (26, 27 September 1944), 28 pp.; R. C. Aldrich, *Synthetic Lubricating Oil Production in France*, CIOS XVIII-5, Item No. 30 (24 March 1945), 6 pp.; Étienne Hirsch, "La Fabrication D'Essence Synthétique de la Société Courriéres-Kuhlmann," *Memories de la Société des Ingénieurs Civils de France*, 90 (1937): 756–67; Storch (note 44), p. 1801.

48 W. W. Myddleton, "The Production of Hydrocarbon Oils from Industrial Gases: The Process of Synthetic Oils Ltd.," *J. Inst. Fuel*, 11 (1938): 477–84; *Report of the Fuel Research Board for the Year Ended March 1937* (London: His Majesty's Stationery Office, 1938), pp. 141–47.

49 Myddleton (note 48).

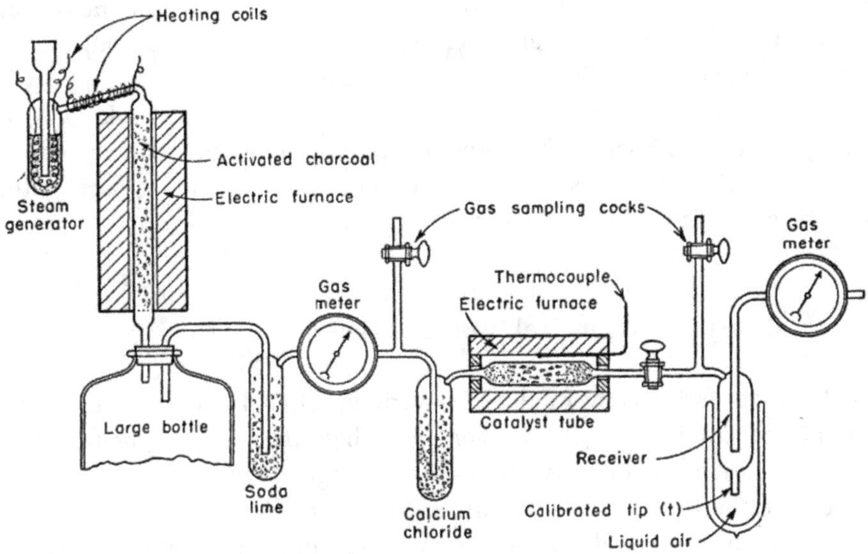

FIGURE 7.1 First Bureau of Mines F-T apparatus 1928
SOURCE: D. F. SMITH, J. D. DAVIS, AND D. A. REYNOLDS, *Industrial and Engineering Chemistry*, 20 (1928): 462–64. REPRODUCED IN HENRY H. STORCH, NORMA GOLUMBIC, AND ROBERT B. ANDERSON, *The Fischer-Tropsch and Related Syntheses* (NEW YORK: JOHN WILEY & SONS, INC., 1951), P. 119

which time they tested iron-copper and cobalt-copper-manganese oxide catalysts at 1 atmosphere and temperatures ranging from 200–300° C. But with the discovery of very large petroleum reserves in Oklahoma and East Texas in 1930 and a significant reduction in the Bureau's appropriations throughout the 1930s depression, the Bureau discontinued further F-T work until 1942.[50]

By that time both the national and international scene had changed dramatically. The price of gasoline had increased steadily since 1938 (continuing into the postwar years), and with the United States' entry into World War II on 7 December 1941 the government considered the development of synthetic liquid fuel processes a matter of national security.[51] It awarded the Bureau an $85,000 appropriation for synthetic fuel research in 1942, which according to the Bureau's director, Royd R. Sayers (1885–1965), enabled the Bureau to train a number of men "in the intricacies of the process who would assist

50 Storch, Golumbic, and Anderson (note 3), p. 131.

51 H. H. Storch, "Review of Development of Processes for Synthesis of Liquid Fuels by Hydrogenation of Carbon Monoxide," *Chemical Engineering Process* 44, no. 6 (1940): 469–80.

or direct its industrial application whenever needed." At this time the Bureau resumed F-T research in a small experimental plant it constructed adjacent to the Central Experiment Station. The plant had a daily production capacity of 100 pounds (ca. 14 gallons) of gasoline and tested improved cobalt catalysts nearly identical to Fischer's. The production of gasoline from Alaskan coals to supply American air bases in Alaska and in other remote places became the key objective of the Bureau's research.[52]

5 Early Research on Coal Hydrogenation

Coal hydrogenation converts different varieties of coal into synthetic petroleum by reacting coal with hydrogen gas at high pressure and high temperature. Friedrich Bergius was the first to hydrogenate coal successfully and to foresee the industrial development of a synthetic fuel industry. Largely because of the mass production of automobiles, petroleum consumption globally had increased steadily in the period 1900–10.[53] For Germany, a country with tremendous coal deposits but little natural petroleum, Bergius saw in coal hydrogenation a way to eliminate Germany's almost total dependence on imported petroleum to meet its ever-increasing liquid fuel requirements.

High-pressure research was a relatively new field of investigation, and it appeared very promising to many chemists in the first decade of the twentieth century.[54] To acquire the necessary research experience, Bergius spent eighteen months, in 1908–09, working with two of its leaders, Walther Nernst in Berlin and Fritz Haber in Karlsruhe. A short time later he moved to a small plant in Hanover and in 1910 began hydrogenation experiments on an artificial coal he prepared from cellulose and then on natural coals.[55]

52 "Bureau of Mines to Study Liquid Fuel from Coal," *Coal Age*, 47, no. 8 (August 1942): 48.

53 Arno C. Fieldner, "Future Trends in Automotive Fuels," *Industrial and Engineering Chemistry*, 18 (1926): 1009–14; R. Wigginton, "Notes on Developments in Fuel Technology," *Fuel*, 4 (1925): 95.

54 F. Sherwood Taylor, *A History of Industrial Chemistry* (New York: Arno Press, 1972), pp. 428–34.

55 Friedrich Bergius, "Production of Hydrogen from Water and Coal from Cellulose at High Temperatures and Pressures," *J. Soc. Chem. Ind.*, 32 (1913): 462–67; Bergius, *Die Anwendung koker Druck bei chemische Vorgangen und eine Nachbildung des Enstehungsprozesses der Steinkohle* (Halle: Wilhelm Knapp, 1913); Bergius, German Patents 301,231 of 9 August 1913 and 304,348 of 17 September 1919 (filed 6 May 1913); British Patents 18,232 of 1 August 1914) and 5,021 of 31 March 1915.

By the summer of 1913, Bergius had successfully hydrogenated natural coals and in his patent that year showed for the first time that the reaction produced synthetic petroleum. Of all the coals tested, Bergius found that the younger coals (lignite, brown, and bituminous) gave the best yield, about 85% conversion to synthetic petroleum. They had a carbon content of 85%, and the reaction raised their hydrogen content from 6% to nearly 15%. His discovery proved crucial to understanding coal hydrogenation. Bergius could hydrogenate coals with less than 85% carbon but not those whose carbon content exceeded 85%.[56]

World War I revealed Germany's critical need for petroleum. To alleviate the shortage Bergius tried to establish an industrial-size coal hydrogenation process for the production of synthetic petroleum, but his efforts failed. The Konsortium für Kohlenchemie in 1916 gave him RM 30 million ($5.4 million) to construct a large plant at Rheinau near Mannheim, but Bergius never solved the major operating problems until long after the war had ended. Development also languished, particularly during the later years of the war, because Germany had obtained access to Romanian oilfields. The conversion of coal to synthetic petroleum became of minor importance.[57]

BASF, which on 9 December 1925 merged with seven other chemical companies to establish the German industrial giant IG Farben, improved the synthesis considerably. Within two years after reaching its 1925 financial agreement with Bergius to continue the coal hydrogenation research, Matthias Pier, BASF's director of research, succeeded in separating Bergius's low-yielding, uncatalyzed, one-stage operation into two: hydrogenation of coal to a heavy oil in the first or liquid phase, followed by decomposition of the heavy oil to gasoline-size molecules in the second or vapor phase. Pier also directed the successful search for the catalysts that accelerated each phase of the synthesis, discovering the insensitivity of metallic sulfides such as tungsten sulfide to sulfur, one of the worst catalyst poisons and almost always present in coal.[58]

From the first coal hydrogenation plant BASF constructed at Leuna in April 1927, which by 1931 had the capacity to produce annually 300,000 metric

56 Friedrich Bergius, "An Historical Account of Hydrogenation," *Proceedings of the World Petroleum Congress*, 2 Vols. (London: Office of the Congress, 1934), Vol. 2, pp. 282–89. See also patents listed in note 55.

57 Anthony N. Stranges, "Friedrich Bergius and the Rise of the German Synthetic Fuel Industry," *Isis*, 75 (1984): 643–67; "Early Difficulties of Hydrogenation," *The Iron and Coal Trades Review*, 135 (3 December 1937): 937.

58 "High-Pressure Hydrogenation at Ludwigshafen-Heidelberg," FIAT *Final Report No. 1317*; Joint Intelligence Objectives Agency (Dayton, Ohio, 1951), I, pp. 24–26, 32.

tons of synthetic petroleum, the German coal hydrogenation industry in 1944 had grown to twelve large plants.[59] At their peak in 1944, the German hydrogenation plants produced over three million metric tons (21.6 million bbl) of synthetic petroleum of which two million metric tons (14.6 million bbl) after adding lead were high quality aviation and motor gasoline of 100-octane number. These plants provided much of the German military's fuel in World War II.[60] The cost of hydrogenating coal was high, RM 190/metric ton, or the equivalent of 24¢/gal of gasoline, more than double the price of imported gasoline. For Germany, with only a limited supply of domestic natural petroleum available during 1930s–40s, no alternative remained other than the construction of synthetic petroleum plants. In this way Germany utilized its naturally abundant supplies of bituminous and brown coal.[61]

British research on coal hydrogenation began in late 1923, shortly after J. Ivan Graham and Hugh G. Shatwell at the University of Birmingham's Mining Research Laboratory visited Bergius's Rheinau-Mannheim plant. For five years Graham, Shatwell, and other Birmingham researchers worked with small 200 g coal samples in a batch-type process and investigated the influence of different catalysts on the reaction. Investigations on economical and efficient ways of utilizing British coals and their solid, liquid, and gaseous products had begun a few years earlier in 1919 at the government's East Greenwich Fuel Research Station, and in 1923 research on coal hydrogenation also began there. While the Birmingham group worked with a small batch-type autoclave, by 1926 the Fuel Research Station had advanced to the construction of a continuous unit that hydrogenated a bituminous coal-oil paste input of 1 kg/hr.

The Fuel Research Station's program continued for sixteen years, until 1939. During this period, in April 1931, Imperial Chemical Industries (ICI) acquired the patent rights to coal hydrogenation and in 1935 constructed a £ 5.5 million ($27.5 million) industrial-size plant at Billingham. Billingham was the world's first industrial-size bituminous coal hydrogenation plant. It had an annual production capacity of 150,000 long tons (1.28 million barrels) of gasoline. During World War II Billingham produced about 750,000 long tons of 87-octane number gasoline mainly from creosote oil, but in the postwar years the high cost of coal and creosote made operation uneconomical and resulted in the Billingham plant's shutdown in 1958.[62]

---

59 *Report on the Petroleum* (note 45), pp. 46–48.

60 *Report on the Petroleum* (note 45), pp. 1–2, 50.

61 See conversion figures in note 45.

62 Anthony N. Stranges, "From Birmingham to Billingham: High-Pressure Coal Hydrogenation in Great Britain," *Technology and Culture*, 26 (1985): 726–57.

France and Japan began coal and coal tar hydrogenation programs in the 1930s. They constructed smaller plants that never produced large quantities of synthetic petroleum, and in the Japanese case were far less successful than the German and British plants. The Société des Carburants Synthétique des Mines de Béthune in 1935 completed construction of a two-stage semicommercial plant in the Pas-de-Calais region of northern France. In 1936, its first year in operation, the plant converted about 7,600 metric tons of coal to 3.24 million liters of gasoline with a 76.5-octane number. Also in 1935, the Compagnie Francais des Essences Synthétiques constructed a larger, 48 metric ton/d two-stage plant at Liévin in northern France. France's coal hydrogenation program suffered from high plant construction costs estimated at FF 200 million in 1936 and from increases in the price of lignite in the late 1930s. Both plants discontinued operation after World War II; the Compagnie Francais dismantled the Liévin plant in 1964.[63]

The Japanese constructed two out of ten proposed World War II hydrogenation plants. One of the plants, a South Manchuria Railway-Imperial Navy plant at Fushun, Manchuria, had a design annual capacity of 10,000 metric tons of coal. The plant went on stream in 1939, but until 1942 when it began hydrocracking gas oil and kerosene the plant produced only 1,460 kL of synthetic petroleum. The Chôsen Jinzô Sekiyu KK plant at Agochi, Korea, completed in 1939, had a design annual capacity of 20,000 metric tons of coal. It failed to produce any synthetic petroleum before its conversion to a synthetic methanol plant in 1942.[64]

6 Coal Hydrogenation and F-T Synthesis at the Bureau of Mines, the Synthetic Liquid Fuels Act of 1944

The Bureau had done limited research on coal hydrogenation, the third coal conversion process that Fieldner investigated when in Europe, but the expense

63 W. R. K. Wu and H. H. Storch, "Hydrogenation of Coal and Tar," *Bureau of Mines Bulletin 633* (Washington, DC: Government Printing Office, 1968), pp. 1–195 (p. 4); "Calais Petrol-from-Coal Works to be Abandoned," *Chemical Age*, 91 (15 February 1964): 265; "Coal Hydrogenation-France," *Chemical Trade Journal*, 102 (1938): 109. During the 1930s laboratory and small-scale research on the hydrogenation of heavy oils and petroleum residues went on in several other countries such as Romania and Holland. See C. Candea and A. Marscall, "Hydrogenation du Petrole Roumain," *Proceedings of the Second World Petroleum Congress*, 2 Vols. (Paris: Journal des Economistes, 1937), Vol. 2, pp. 227–36 and J. C. Vlugter, "Hydrogenation de Residue, Notamment de Residus du Cracking, en Carburants et en Essence d'Aviation," *Proceedings of the Second World Petroleum Congress*, pp. 277–80.

64 Stranges (note 46).

FIGURE 7.2
James G. King, chief chemist of British Fuel Research Station (left) and Arno C. Fieldner in London 1924
SOURCE: SHERIE MERSON AND TIM PALUCKA, *A Century of Energy*, NATIONAL ENERGY TECHNOLOGY LABORATORY (FORMERLY US BUREAU OF MINES), US DEPARTMENT OF ENERGY, 2010, P. 84

of producing hydrogen gas and constructing high-pressure apparatus discouraged its active development. Fieldner pointed out in 1928 that synthetic petroleum obtained from coal hydrogenation at a cost of 40–50¢/gal showed its "utter impossibility of approaching competition with petroleum gasoline costing 7 to 9 cents at the refinery."[65] "A forward-looking national fuel policy would seek to delay the day of making liquid fuels from coal as long as possible by reserving the higher value fuels of natural gas and petroleum for those uses which cannot be met so efficiently by the direct combustion of coal."[66]

Fieldner's argument no doubt contributed to the Bureau's delay in investigating high-pressure coal hydrogenation, but German and British advances in the mid-1930s convinced the Bureau at least to evaluate their work. In eastern Germany BASF's Leuna plant by 1931 was producing annually 300,000 metric tons of synthetic petroleum while ICI's recently opened Billingham plant (1935) had the capacity to produce annually 150,000 long tons of gasoline. The Bureau, therefore, decided to test the suitability of American coals for

65 Arno C. Fieldner, "Recent Developments in the Production of Motor Fuels from Coal," *Fuel*, 7 (1928): 492–501 (p. 497); Stranges (note 2).

66 Arno C. Fieldner, "300 Years of American Fuels," *Industrial and Engineering Chemistry*, 27 (1935): 983–88, on p. 988.

hydrogenation, expecting the data to have value at some future time when the United States might have to supplement its non-renewable petroleum reserves.[67]

Coal hydrogenation began in 1936 at the Pittsburgh Central Experiment Station under the direction of Fieldner, Henry H. Storch (1894–1961), and Lester L. Hirst (1903–63). They conducted both batch and continuous type experiments. Typical batch experiments carried out in a horizontal, rotating, stainless steel autoclave (36 cm × 7 cm), hydrogenated 50–100 g of lignite and bituminous coal. They studied the fundamentals of coal hydrogenation, such as the influence of temperature, pressure, time, solvent, coal composition, and catalysts on the reaction. Catalysis ranked of particular significance in every hydrogenation program, and the Bureau's catalytic tests included tin (II) sulfide (SnS), tin-iodine compounds, molybdic oxide, and an aluminum oxide-molybdic acid mixture. The two-stage continuous unit hydrogenated 45–90 kg (100–200 pounds) of pulverized coal in 24 hours and produced about 8 gallons of gasoline. Its electrically heated, stainless steel converters ranged in size from 1.8 m to 2.5 m in length, 3.0 cm to 7.5 cm inner diameter and operated at 400–450° C and 210–240 atm. The continuous unit supplemented the research by providing the flowsheet data required for the design of any proposed industrial-size operation.

The program at the Pittsburgh Central Experiment Station established several important conclusions. It showed first of all the amenability of American coals to hydrogenation. Bituminous and sub-bituminous coals, which contained less than 85% carbon in the dry ash-free state, gave the best results. Yields of liquid products ranged from 10 to 55%, or 105–138 gal/ton, and analyses of 13 coals hydrogenated revealed that 20% boiled in the gasoline range of 200–300° C, the remaining 80% boiled between 305–345° C. The gasoline had good quality with a 73.5-octane number and tested satisfactorily in automobiles, tanks, and airplanes.[68]

---

67 *Report on the Petroleum* (note 45), pp. 46–48 and Stranges (note 62).

68 L. L. Hirst, C. D. Haw, G. C. Sprunk, I. I. Pinkel, R. L. Boyer, J. R. Scheffer, R. H. Kallenberger, H. A. Hamilton, and H. H. Storch, "Liquid-Phase Hydrogenation of Pittsburgh Seam Coal," *Industrial and Engineering Chemistry*, 31 (1939): 869–77; H. H. Storch, L. L. Hirst, C. H. Fisher, and G. C. Sprunk, "Hydrogenation and Liquefaction of Coal, Part 1," *Bureau of Mines Technical Paper 622* (Washington, DC: Government Printing Office, 1941), pp. 1–110; H. H. Storch, L. L. Hirst, C. H. Fisher, H. K. Work, and F. W. Wagner, "Hydrogenation of High-Temperature Tar from By-Product Coke Ovens," *Industrial and Engineering Chemistry*, 33 (1941): 264–74; H. H. Storch, L. L. Hirst, P. L. Golden, I. I. Pinkel, R. L. Boyer, J. R. Schaeffer, and R. H. Kallenberger, "US Bureau of Mines Experimental Plant," *Industrial and Engineering Chemistry*, 29 (1937): 1377–80; H. H. Storch and A. C. Fieldner, "Coal

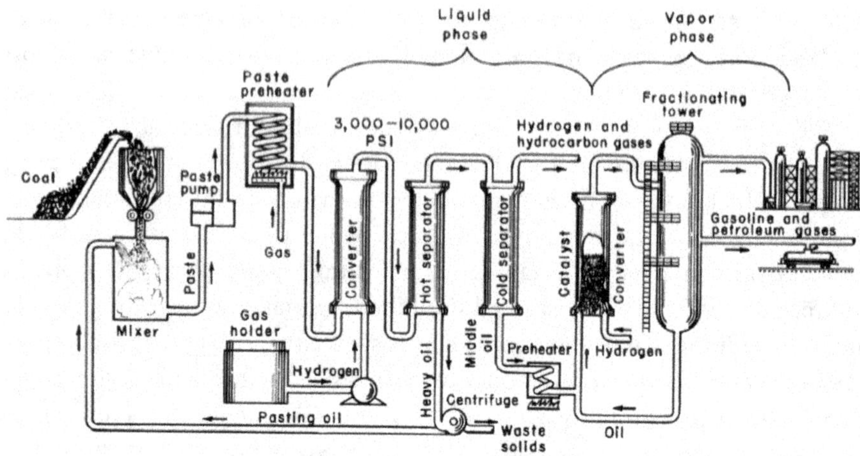

FIGURE 7.3 Bureau of Mines diagram of Bergius-IG Farben coal hydrogenation process 1930s
SOURCE: SHERIE MERSON AND TIM PALUCKA, *A Century of Energy*, NATIONAL ENERGY TECHNOLOGY LABORATORY (FORMERLY US BUREAU OF MINES), US DEPARTMENT OF ENERGY, 2010, P. 90

The Bureau continued and gradually expanded its research program in the early 1940s. From 1942 to 1944 it constructed and successfully operated a small 10 gal/d coal hydrogenation pilot plant and a research and development laboratory on a 12-acre site near Bruceton, the location of its experimental coal mine, 13 miles south of Pittsburgh. F-T investigations also resumed at Bruceton with the operation of a pilot plant producing daily 0.25–5 gallons of liquid products. The F-T pilot plant, located in an engineering laboratory for process improvement, operated for eight months during which time the Bureau experimented with iron and cobalt catalysts in six testing units. The Bureau hoped that its synthetic fuel research would lead to the development of better cat-

---

Hydrogenation, US Bureau of Mines Experimental Plant," *Mechanical Engineering*, 61 (1939): 605–11; L. L. Hirst, A. Eisner, J. H. Field, H. M. Cooper, R. F. Abernathy, and H. H. Storch, "Hydrogenation and Liquefaction of Coal, Part III," *Bureau of Mines Technical Paper 646* (Washington, DC: Government Printing Office 1942), pp. 1–27; H. H. Storch, C. H. Fisher, C. D. Hawk, A. Eisner, "Hydrogenation and Liquefaction of Coal, Part IV," *Bureau of Mines Technical Paper 654* (Washington, DC: Government Printing Office, 1943), pp. 1–50; Arno C. Fieldner, Henry H. Storch, and Lester L. Hirst, "Bureau of Mines Research on the Hydrogenation and Liquefaction of Coal," *Bureau of Mines Technical Paper 666* (Washington, DC: Government Printing Office, 1944), pp. 1–69; Arno C. Fieldner, "Synthetic Liquid Fuels Program," *National Petroleum News*, 37, no. 36 (5 September 1945): R709, R711–2, R714–5.

FIGURE 7.4
US Bureau of Mines coal hydrogenation converter 1930s
SOURCE: SHERIE MERSON AND TIM PALUCKA, *A Century of Energy*, NATIONAL ENERGY TECHNOLOGY LABORATORY (FORMERLY US BUREAU OF MINES), US DEPARTMENT OF ENERGY, 2010, P. 92

alysts, simplification of plant engineering designs, and efficient methods of removing the heat of reaction, particularly from the F-T synthesis.[69]

By 1944 the war also had demonstrated convincingly the United States' rapid consumption of huge quantities of petroleum. The 78th Congress acknowledged the possibility of a petroleum shortage, and on 5 April 1944 it acted. Mainly through the efforts of Michael Straus (1897–1970), Director of the War Resources Council; Harold Ickes (1874–1952), Secretary of the Interior; and their two Democratic spokesmen from coal-bearing states, Senator Joseph O' Mahoney (1884–1962) of Wyoming and Representative Jennings Randolph (1902–98) of West Virginia, chairman of the House Mining Committee, Congress appropriated $30 million over five years ($5 million for 1945) for the Bureau to begin semicommercial development of coal hydrogenation, F-T synthesis, and oil shale mining and distillation. Public Law 290, the Synthetic Liquid Fuels Act of 1944, directed the Department of the Interior, Bureau of Mines, to carry out the production of synthetic liquid fuels from coal and other

69 J. A. Markovits, K. C. Braun, J. T. Donovan, and J. H. Sandaker, "Special Equipment in the Coal Hydrogenation Demonstration Plant," *Bureau of Mines Report of Investigations 4584* (Washington, DC: Government Printing Office, 1950), pp. 1–40.

FIGURE 7.5 Bureau of Mines truck equipped to run on synthetic gasoline produced from coal hydrogenation, 1941
SOURCE: SHERIE MERSON AND TIM PALUCKA, *A Century of Energy*, NATIONAL ENERGY TECHNOLOGY LABORATORY (FORMERLY US BUREAU OF MINES), US DEPARTMENT OF ENERGY, 2010, P. 92

indigenous substances to acquire the know-how for private commercial production of synthetic liquid fuels.

The 1944 Act established an Office of Synthetic Liquid Fuels in the Fuels and Explosives Branch of the Bureau with Wilburn C. Schroeder (1906–91) as acting and later chief of the new office and its five divisions. The five divisions were: (1) Research and Development Division with Henry H. Storch, who spent the previous year in London studying synthetic fuel production, its chief and Lester L. Hirst his assistant; (2) Synthesis Gas Production Division with William W. Odell its chief; (3) Hydrogenation Demonstration Plant Division; (4) Gas Synthesis Demonstration Plant Division; and (5) Oil Shale Section within the Bureau's Petroleum and Natural Gas Division with Albert J. Kraemer (1892–1973) its chief. The act provided for the construction of three experiment stations and three large-size demonstration plants to carry out the research. Two of the experiment stations and plants conducted research on coal, the third experiment station and plant conducted oil shale research. With the passage of the 1944 Act, the Bureau transferred its ongoing coal hydrogenation

FIGURE 7.6 Fischer-Tropsch Laboratory, Bureau of Mines, Pittsburgh Testing Station 1946
SOURCE: SHERIE MERSON AND TIM PALUCKA, *A Century of Energy*, NATIONAL ENERGY TECHNOLOGY LABORATORY (FORMERLY US BUREAU OF MINES), US DEPARTMENT OF ENERGY, 2010, P. 110

and F-T research from the Pittsburgh Central Experiment Station to Bruceton, the first of the coal research experiment stations where from 1945 to 1948 it constructed the Research and Development Laboratory for Synthetic Liquid Fuels.

The second coal research experiment station, the Appalachian Experiment Station constructed from 1952 to 1954 at a site five miles north of Morgantown, West Virginia, continued the earlier gasification research at West Virginia University. It complemented the Underground Gasification Project in Gorgas, Alabama, that the Bureau had underway since 1946 on land the Alabama Power Company donated for investigations on the underground conversion of coal into gas. The Bureau constructed the third experiment station, the Petroleum and Oil Shale Experiment Station, from 1945 to 1947 at a location next to the University of Wyoming in Laramie. Harold P. Rue, supervising engineer of the Bureau's Petroleum Experiment Station in Laramie, served as director. Prior to operating its Laramie Experiment Station, the Bureau in 1944 had conducted laboratory and pilot plant oil shale research in the University of Wyoming's laboratories.

FIGURE 7.7 Bruceton Synthetic Fuels Laboratories. Left to right on hilltop: Gas synthesis building, warehouse and administration, coal hydrogenation buildings
SOURCE: SHERIE MERSON AND TIM PALUCKA, *A Century of Energy*, NATIONAL ENERGY TECHNOLOGY LABORATORY (FORMERLY US BUREAU OF MINES), US DEPARTMENT OF ENERGY, 2010, P. 173

The Bureau constructed the oil shale demonstration plant, from 1945 to 1949 on the government's Naval Oil Shale Reserve in Rifle, Colorado. The shale plant studied methods for separating and upgrading shale oil. Large-scale mining also began in 1945 at the Bureau's Experimental Mine in Rifle to provide oil shale for the demonstration plant and to develop improved mining techniques. It constructed the two coal-to-oil demonstration plants in Louisiana, Missouri, the coal hydrogenation plant from 1947 to 1949, the F-T plant from 1948 to 1951.

In addition to the oil shale and coal research the Synthetic Liquid Fuels Act included funds for a six-year agricultural residues research program (1944–50) at the Department of Agriculture's Northern Regional Research Laboratory in Peoria, Illinois, and for construction in 1945–46 of a demonstration plant in Peoria. The agricultural program investigated the conversion of the cellulosic residue in corn cobs and in the hulls of cottonseed, oats, and rice, to ethanol for use as a natural fuel or fuel additive. A final provision of the act allocated funds for the years 1948–50 to expand current research on petroleum secondary recovery from stripper oilfields and oil refining. The Petroleum Experiment Station in Bartlesville, Oklahoma, and the Petroleum and Oil Shale Experiment Station in Laramie carried out much of the secondary recovery and refining research.

During the immediate postwar years when the Bureau initiated its synthetic liquid fuels program, domestic demand for petroleum had increased steadily. In spite of more exploration and drilling, imports continued to rise so that in 1948 the United States became a net petroleum importer. Amendments to the Synthetic Liquid Fuels Act of 15 March 1948 and of 22 September 1950 extended the Bureau's program to 1955 and raised the authorization to $60 million and finally to $87.6 million.[70] Representative Randolph of West Virginia, a co-sponsor of the act and a firm supporter of the Bureau's program, wrote in 1948, a year in which the United States experienced an energy crisis:

"We cannot survive a prolonged famine in liquid fuels. We must not rely on uncertain foreign sources. In the interest of national security, it is imperative that an American synthetic-liquid-fuels industry be established as soon as possible, before our petroleum is gone, before another national emergency."[71]

Randolph had given the Bureau's assessment of the United States' postwar energy situation. The Bureau wanted to construct the three demonstration plants in order to apply the results of its laboratory research and development on a scale large enough to provide information useful to industries planning industrial production. It believed that the discovery and perfection of basic reactions and processes occurred in research, the evaluation of the reactions and processes followed in demonstration plants. By acquiring plant engineering, operating, and cost information, the Bureau hoped to offer a pattern of experience to American industry and pointed out that benefits applicable to industrial plants, including a more accurate estimate of investment and operating costs and a more precise evaluation of synthetic fuel processes, already

70 US Statutes at Large, 78th Congress 1944, Vol. 59, Part I, pp. 190–91; "The Bureau of Mines Synthetic Liquid Fuels Program 1944–55," Part 1-Oil from Coal, *Bureau of Mines Report of Investigations 5506* (Washington, DC: Government Printing Office, 1959), pp. 1–306 (pp. 2–10). The other locations for petroleum secondary recovery research were the Bureau's Petroleum Field Offices in San Francisco, Dallas, and Franklin, PA, and its sub-offices in Wichita Falls, TX, and Bradford, PA. For the agricultural residues program see "Synthetic Liquid Fuels, Annual Report of the Secretary of the Interior for 1950," Part III – Liquid Fuels from Agricultural Residues, *Bureau of Mines Report of Investigations 4772* (Washington, DC: Government Printing Office, 1954), pp. 1–82; see Marchese (note 10), Chapter 3 for the oil shale program.

71 Jennings Randolph, "Oil from Coal Nearer," *Coal Age*, 51, no. 8 (August 1946): 86–91.

had accrued from improvements in plant design and construction. Many manufacturing firms, the Bureau said, wanted their products tested in the demonstration plants to obtain valuable information on performance under new and specialized conditions. Because of the shortage of American scientists and engineers with experience in either the F-T or coal hydrogenation process, the Bureau also hoped that by cooperating closely with industry it could prepare today for tomorrow's era of synthetic liquid fuel.[72]

7 The Demonstration Synthetic Fuel Plants in Louisiana, Missouri

With the end of World War II in August 1945, the United States War Department had available at Louisiana, Missouri, a $17.5 million high-pressure synthetic ammonia plant that Hercules Powder Company constructed and formerly operated since 1941–42 for the army's Missouri Ordnance Works. The Bureau acquired this 390-acre site located on the Mississippi River about 75 miles north of St. Louis on 1 February 1946, intending to convert the synthetic ammonia plant to a high-pressure coal hydrogenation demonstration plant. It selected the Missouri location mainly because of the plant's readily adaptable services and facilities after making a nationwide survey of 206 proposed sites in 21 coal-producing states. Missouri's advantages included existing power and water-treating units, heavy duty high-pressure gas compression equipment, hydrogen manufacturing apparatus, ready-to-occupy administrative offices, and adjacent government-owned land.

Selecting the Missouri Ordnance plant saved the Bureau several million dollars in construction costs and considerable time and was not without precedent. Both ICI's synthetic fuel plant at Billingham and IG Farben's plant at Leuna operated in conjunction with synthetic ammonia plants to make the best use of raw materials, plant facilities, and personnel. Construction on the $10 million, 200–300 bbl/d, two-stage coal hydrogenation plant started in May 1947, and its operation began two years later in 1949. The following year, the Bureau completed a smaller $5 million, 50 bbl/d F-T plant that became fully operational in 1951.

The Bureau decided to start with coal hydrogenation and to construct a larger hydrogenation plant for several reasons. In the first place the Bureau had long time but quite limited experience with hydrogenation rather than

72 *Coal to Oil Demonstration Plants, Dedication*, Office of Synthetic Liquid Fuels, Bureau of Mines, Louisiana, Missouri (8 May 1949), p. 24.

FIGURE 7.8 Bureau of Mines, Louisiana, Missouri, demonstration plants. Center: Original synthetic ammonia plant. Left: Coal hydrogenation and F-T demonstration plants
SOURCE: SHERIE MERSON AND TIM PALUCKA, *A Century of Energy*, NATIONAL ENERGY TECHNOLOGY LABORATORY (FORMERLY US BUREAU OF MINES), US DEPARTMENT OF ENERGY, 2010, P. 188

F-T. Secondly, the Germans and the British had concentrated more on hydrogenation, and the Bureau expected to benefit from their experience. The first German coal hydrogenation plant operated for eight years before the construction of a F-T plant, the Germans had more and bigger hydrogenation plants in operation, and the hydrogenation plants produced more synthetic liquid fuel than the F-T plants. Compared to F-T, coal hydrogenation had become a well-established industry in Germany, and the Bureau now had its complete record. The British had not constructed any industrial-size F-T plants, but the Bureau had available ICI's experience in operating its Billingham hydrogenation plant. Finally, the Bureau intended to synthesize high quality motor and aviation fuels and only coal hydrogenation could provide these fuels. In the Missouri research and development program, the Bureau utilized its almost 25 years' experience with synthetic petroleum production, ICI's operating data, and the wealth of information it had obtained from the German synthetic fuel documents that the TOM captured in World War II's closing months.[73] In addition to the 400 Bureau employees initially involved in the plants' design and construction, seven German synthetic fuel scientists that the War Department through Project Paperclip quietly brought to the United States assisted in the program. The German scientists led the world in synthetic fuel research, development, and production, and the Bureau intended to make full use of their experience and expertise.[74]

Plans for using the German scientists in the Bureau's demonstration program had begun in December 1945 only a few months before the Bureau acquired the Hercules synthetic ammonia plant. The TOM, many of whom the Bureau employed, invited Colonel P. W. Melton, US army, to a meeting in the Bureau's Washington office to explain the government's policy on the use of enemy aliens. At the meeting the TOM learned that only the State Department had the authority to issue a visa to an enemy alien, but it had not yet adopted any specific visa granting policy. The Treasury Department's ruling that made it illegal for a civilian agency to pay for the service of enemy aliens confronted the Bureau with a more complicated issue to resolve. The proper procedure, Melton said, required the Bureau to work through the War Department and the army's Quartermaster Corps. The War Department had the authority to admit an enemy alien to aid in defending the United States and in other situations involving the nation's well-being. In such cases the aliens had to work on a military installation and under the supervision of military personnel. The Quartermaster Corps therefore would employ them for research at the Missouri Ordnance Works. Melton advised the TOM to make requests for particular scientists through the Joint Intelligence Objectives Agency (JIOA), an agency of the Joint Chiefs of Staff.

Two other issues for which the government had not yet established a clear policy arose at this meeting. The first inquired whether any of the German

73 Note 72, pp. 1–24; L. J. Grunder, "Report of the Hydrogenation Demonstration Plant Division, Louisiana, Missouri," unpublished report of 15 March 1946, Synthetic Fuel Collection, Texas A&M University Archives.

74 The Bureau of Mines arranged for sixteen German scientists to assist with its fuel projects; five on coal hydrogenation, two on Fischer-Tropsch, eight on oil shale, one on acetylene production. Two of them were experts in more than one field, given the above categorization. See Grunder (note 73). Members of the British postwar investigating team were not impressed upon visiting the German plants. Kenneth Gordon, British team leader and ICI's research manager and synthetic fuels specialist, maintained that the Germans had not made any advances during the war years and that the British had little to learn from them. See *Report on the Petroleum* (note 45), p. 2.

scientists, once in the United States and having fulfilled the terms of their contracts, could obtain clearance to remain in the United States and work in a university or in private industry. The second dealt with the release of information contained in the captured documents. Melton informed the TOM that the State Department had not acted on the first issue, and regarding the second issue he replied that Military Intelligence would make the decisions and then authorize the Department of Commerce to publish any information released. Melton's presentation resulted in the immediate formation of a three-man committee to implement the procedure he had outlined for bringing the German synthetic fuel scientists to the United States.

How did the Bureau decide which German scientists it would ask to join its newly established synthetic liquid fuel program? The Bureau had no difficulty making a decision. Several TOM members, such as Schroeder, Hirst, William Appler Horne (b. 1912), Alfred R. Powell (1891–1980), and Vladimir Haensel, had done synthetic fuel research and studied Germany's prewar coal hydrogenation and F-T industries. TOM had interrogated many of the scientists who worked at the German plants, knew their areas of expertise, and obtained additional plant information from the captured documents.[75]

Ernst Donath was one of the German scientists interrogated at IG Farben's high-pressure laboratory in Ludwigshafen in July 1945 and later came to the United States. His personal account reveals how the seven German synthetic fuel scientists arrived in the United States as part of Project Paperclip. Donath had obtained his doctorate of engineering at the Technical University in Breslau in 1926. That same year he joined Pier's recently established high-pressure experiment group in Ludwigshafen, eventually becoming chief chemist for the Research High-Pressure Department. Donath's specialty was catalysis, and he held over 100 patents on high-pressure catalytic research. Donath recalled that in June or July 1946 Captain C. Tripp, of the War Department and a Paperclip member, came to his home in nearby Mannheim and asked if he had an interest in working for the United States government, specifically for the Bureau of Mines. Donath did not know that his intended destination was Louisiana, Missouri, and he also thought his stay in the United States might be short rather than last a lifetime. Germany's synthesis of oil from coal (soon forbidden by law), had ended with the war, and Donath knew that IG Farben, left with a surplus of scientists and engineers, would not oppose his leaving. Pier did not want to lose Donath, however, and he tried unsuccessfully to convince Donath

75 Note 3; *Proceedings of the Technical Oil Mission Meeting Held under the Auspices of the Petroleum Administration for War and the Bureau of Mines* (Washington, DC, 13–14 December 1945), 138 pp., on pp. 89–97.

to remain despite the employment problems that existed at IG Farben and throughout Germany. During the Nazi years Pier had protected Donath, who had a Jewish-Roman Catholic parenthood, and said he would help him now.

Donath arrived in Washington on 20 January 1947, and only at that time learned of his destination: the Bureau's demonstration plant in Missouri. Meanwhile in Germany several other synthetic fuel scientists, seven total, had accepted the War Department's invitation and in the years 1947–49 came to the United States. The coal hydrogenation specialists were Donath's friend, Hans M. Schappert (1901–78), a chemical engineer from IG Farben, Ludwigshafen; Max Josenhaus, director of the Blechhammer plant in Upper Silesia (East Germany); and Kurt Bretschneider, a mechanical engineer, and Erich Frese from Ruhröl. Helmut Pichler (1904–74), Franz Fischer's assistant at the Kaiser Wilhelm Institute in Mülheim and an authority on the F-T synthesis, and Leonard Alberts from the Ruhrchemie F-T plant in Sterkrade-Holten (West Germany) were the F-T specialists.[76] Establishing the identities of the German scientists who participated in the Missouri demonstration program has been a problem because of the War Department's secrecy, but the War Department acknowledged that Frese and Alberts spent time at the War College in Washington, DC. The War Department also covered up any political activity of the scientists, allowing Alberts' Nazi affiliation to remain unknown until the 1980s and after Congress passed the Freedom of Information Act (1966–74). Alberts returned to Kronberg near Frankfurt in 1973. Pichler, after leaving the Bureau in 1947 went to Hydrocarbon Research Inc. in Lawrenceville, NJ, and in 1967–68 to the Kaiser Wilhelm Institute in Mülheim and the University of Karlsruhe. Bretschneider returned to Altkönigstift near Kronberg. No other information exists on the status of the other German scientists.[77]

76 Ernst Donath, Interview by Anthony N. Stranges, 9–12 August 1982.

77 No agreement exists on the number of German synthetic fuel scientists who came to the United States. Ezekial Clark, head of gasification, Bureau of Mines, Washington, DC, who had the responsibility of bringing the German scientists from Washington to the Bureau of Mines station in Pittsburgh, believed 10 or 12 had come. I have discovered 13 German scientists who apparently had some connection with the Bureau of Mines for varying time periods. Alberts, Bretschneider, Otto Hubmann (coal gasification), Wilhelm Gunz returned to Essen in 1953; Ernst Graf to Vienna. Donath (later in the Virgin Islands); Schappert; Fred Kloiber a chemical engineer later in Sarasota, Florida; and Walther Oppelt with the Bureau of Mines in North Dakota and later in Pittsburgh, remained in the United States. Fritz Ringer and Alexander Trofimow who had some connection with the German synthetic fuel program, were later in Massachusetts. Pichler, Oppelt, Hubmann, Frese, and Gunz died by late 1970s. No information on H. A. Wahl and Hans Hartmann who allegedly came to the United States from Pölitz. See also Bar-Zohar (note 1), pp. 129, 294; note 9.

To illustrate the secrecy, Donath remembered that, while still in Germany, on 27 November 1946 he signed a War Department contract for the employment of German nationals. But he never again saw the contract. Under its terms the German scientists agreed to work for the army's Quartermaster Corps for six months for $6 a month plus room and board. After six months they became Bureau employees and received monthly salaries up to $500 depending on their education and experience. While in Missouri, they had no immigration visas and therefore no legal residence in the United States. The inhabitants of tiny Louisiana knew of the Bureau's German colony, however, and, according to Donath, showed no hostility to the transplanted scientists and their families, who only a short time ago were wartime enemies. The Bureau appointed Hirst, acting head of the Bureau's hydrogenation plant, as the German scientists' contact when they traveled outside the town and an emergency arose.

Bureau and German scientists worked together on the design and operation of the two demonstration plants. The Bureau wanted to determine the suitability of American coals and the plant equipment used for converting them to liquids and gases. While it had easy access to the latest American research on the industrial-size high-pressure synthesis of ammonia or on refining petroleum, nothing comparable existed for synthetic fuel. Only Germany, and to a far lesser extent Britain, had established a large and a successful synthetic fuel industry in World War II and had the skilled scientists and engineers to operate it. Unlike the German nuclear scientists who lagged behind their American counterparts, Germany's synthetic fuel scientists led the world. In high-pressure coal hydrogenation, Donath was the authority on catalysts, Schappert focused on thermodynamics and coal-oil paste formation, Frese specialized in plant construction and costs. Schappert and Frese worked closely with the plant's daily operation while Donath took charge of preparing technical reports and assisted in planning the construction of an industrial-size plant. In the F-T synthesis, Pichler co-invented a medium-pressure process for liquid fuel production; Alberts specialized in catalysis. All the German scientists shared responsibility for translating numerous documents that the Bureau required for the research. Although German and Bureau scientists including Hirst, C. C. Chaffee, L. C. Skinner, and J. A. Markavits worked closely together, Donath pointed out that "the Americans had their own ideas about designing a plant and expected the Germans to supply data about the size of converters, hydrogen consumption, catalysts, etc."[78]

Bechtel Corporation of California constructed the $10 million hydrogenation demonstration plant for operation at 900° F (477° C) and pressures up to

78 Note 76.

10,300 psi (700 atm). With a 200 bbl/d output, the plant produced only a fraction of the 10,000 bbl/d expected from an industrial-size plant. The Bureau's objective, however, was not size but construction of a plant that incorporated innovations and improvements not found in prewar and World War II German plants. Merely duplicating a German plant would accomplish little because the Germans had operated their plants under wartime conditions and despite the inefficiency of their plants could not shut them down even briefly for improvement. The Bureau's plant was, therefore, in many ways newly designed and engineered, and subject to new problems even though the Bureau had available a considerable amount of published research on coal hydrogenation in German and British journals and reports in TOM's extensive captured German experimental data.

The hydrogenation plant featured automation and used it as much as possible to replace manual control of temperature, pressure, and reactant flow found in German plants. The Bureau expected automation to permit accurate reaction conditions, to increase plant capacity and efficiency, and to lower personnel requirements and production costs. It intended to supply all of this information to industry. Other innovations that simplified the operation and reduced investment costs included flash distillation of the heavy oil, a single-stage vapor phase hydrogenation instead of German plants' two stages, and a gas-fired radiation-type preheater in both the liquid and vapor phases versus convection-type in the German plants. Because hydrogenation required high pressures and temperatures the Bureau had to design the large steel alloy converters, pipes, valves, compressors, and plunger-type injection pumps for both the coal-oil paste and hydrogen gas. Arranging for the construction of equipment, much of it new to American manufacturers, and securing the few experienced engineers, scientists, and technicians available in the United States, presented additional problems.

When the hydrogenation plant began operating in February 1949, the Bureau followed essentially the same liquid phase-vapor phase sequence used for over twenty years in the German and British industrial-size plants. This consisted of crushing, drying, and pulverizing the raw coal in the coal preparation area; transferring it to a 60-ton bin in the pasting area for mixing with an equal weight of a heavy oil obtained from the process or from another source; and adding small amounts of an iron (II) oxide or tin (II) chloride catalyst. Pumping the heavy viscous paste under pressure and injecting it by steam-driven, plunger-type pumps into the preheater at 477° C changed the paste to an easier flowing fluid. Upon leaving the preheater, the fluid passed into the first of two identical series-arranged liquid phase converters where it reacted at 700 atm with hydrogen gas and then into the second converter for further hydrogenation. The continuous mixing of the reactants in the converters improved heat transfer and temperature control; the high reaction pressure reduced the amount of unconverted reactant and gaseous hydrocarbon formation and permitted a single-stage vapor phase hydrogenation.

Each cylindrically shaped stainless steel converter had the same dimensions: 32 inches (81.3 cm) inner diameter, 39 feet high (12.1 m) inner measure, and contained a 24-inch (61 cm) diameter reaction "basket." Insulation filled the 8-inch (20.3 cm) space between the converter's shell and basket, keeping the shell cool and maintaining the alloy's full tensile strength. Injecting cooling hydrogen gas into the space also prevented the temperature of the highly exothermic reaction from rising above the desired value of 477° C. A single converter weighed about 100 tons, and to protect the operators and equipment in case of fire or explosion, reinforced concrete walls enclosed both converters, heat exchanger and hot catchpot, on three sides, leaving an open side, and giving a stall-like arrangement. A concrete stall adjacent to the converter stall housed the preheater. A reaction time of about 1 hour converted 95% of the coal to heavy, middle (kerosene, diesel oil), and light oils (low grade gasoline), and small amounts of hydrocarbon gases.

Separation of the oils and hydrocarbon gases, and pressure reduction to about 34 atm, occurred in the hot and cold catchpots. The catchpots were cylindrical containers, similar in construction and size to the liquid phase converters, into which the products passed after leaving the second converter. The hot (first) catchpot separated the high-boiling heavy oil and residual solids (ash, catalyst, and unreacted coal) from the middle and light oil vapors and hydrocarbon gases. After cooling, the heavy oil-residual solids, called heavy oil letdown (HOLD), underwent either flash distillation or passed through the HOLD and then to a centrifuge. These processes removed the solids from the heavy oil, which after recycling to the paste preparation area for mixing with incoming coal to form more paste, returned to the liquid phase converter for hydrogenation. The light and middle oil vapors and hydrocarbon gases, after leaving the hot catchpot and cooling in heat exchangers to less than 200° C, entered the cold (second) catchpot. There the oil vapors condensed and passed into the letdown container for pressure reduction to 1 atm, before moving to the liquid phase distillation unit to remove tar acids and bases and heavier oils. The hydrocarbon gases passed out the top of the cold catchpot to the gas recycle apparatus.

The vapor phase unit, consisting of a concrete stall with a single converter (similar in size to the liquid phase converters), heat exchanger, and hot catchpot, and a second stall containing the preheater, sat adjacent to the other side of the liquid phase stall. Middle and light oils coming from the liquid

FIGURE 7.9 Coal hydrogenation converter stalls bracketed by thick concrete walls
SOURCE: SHERIE MERSON AND TIM PALUCKA, *A Century of Energy*, NATIONAL ENERGY TECHNOLOGY LABORATORY (FORMERLY US BUREAU OF MINES), US DEPARTMENT OF ENERGY, 2010, P. 190

phase distillation unit, after mixing with hydrogen gas at 700 atm, passed into the vapor phase preheater and then into the vapor phase converter, flowing over a fixed bed of metal trays containing 100 ml of catalyst in a fixed bed. The catalyst (K-536) was identical to the catalyst Ruhröl had developed near the war's end for the production of aviation gasoline. It consisted of sulfur, zinc, and molybdenum supported on terrana earth and gave a 46% conversion to unleaded 78-octane number gasoline that performed as well or better than petroleum gasoline. In the final operation the gasoline and other liquids passed to a small refinery in the distillation area for separation by fractionation and conversion to aviation and other high-quality fuels.

The Bureau succeeded in liquefying bituminous and subbituminous coals from Wyoming, Pennsylvania, Illinois, and Kentucky and a lignite from North Dakota. Natural gas cracked (reacted) with steam in the cracking furnaces of the Missouri plant's Ordnance Works provided sufficient hydrogen gas. The liquefaction process consumed 120,000–150,000 ft^3/hr of hydrogen gas, and hydrogen's production, the major contributor to the liquefaction cost, accounted for 40% of the total cost. But the Bureau pointed out that an industrial-size plant would obtain hydrogen much more economically by removing it from the plant's waste gases. In addition to producing 1.5 million gallons of gasoline that tested well in a fleet of military automotive vehicles, the plant produced heavy Bunker C fuels for ships and industry, hydrocarbon gases for heating, and valuable by-products such as phenols for plastics, creosote for wood preservation, and solvents for paints. Coal hydrogenation's flexibility, requiring only a few raw materials and little or no plant modification to produce a variety of products, was its greatest strength. According to the Bureau, a plant could change its production from a full range of civilian products to military products almost overnight.[79]

The Bureau awarded a $5 million contract for the design and construction of the F-T demonstration plant to the Koppers Company Inc. of Pittsburgh, Pennsylvania, in March 1948. Construction began shortly afterward. Ruhrchemie and other German firms had constructed nine F-T plants in the 1930s, but the Bureau's program was very much a learning-on-the-job experience. Like coal hydrogenation, no precedent for the F-T synthesis existed in American industry.

As in the German plants the Bureau's demonstration plant contained four main units: coal gasification, gas purification, hydrocarbon synthesis, and product refining. Gasifying coal to give synthesis gas (carbon monoxide-hydrogen) was, as in hydrogenation, the most expensive part of the F-T process, amounting to 60% of the production cost for gasoline and oil. It required crushing and pulverizing the coal and then drying the resulting particles; suspending the coal particles in a stream of oxygen gas and adding superheated steam; and feeding the suspended coal particles and gaseous

---

79 *Coal to Oil Demonstration Plants* (note 72); *Bureau of Mines Report of Investigations 5506* (note 70); Wu and Storch (note 63), pp. 85, 88, 187; Markovits, Braun, Donovan, and Sandaker (note 69). E. E. Donath, "Hydrogenation of Coal and Tar," in *Chemistry of Coal Utilization*, ed. H. H. Lowry, 3 Vols. (New York: John Wiley & Sons, Inc., 1945 and 1963), Supplementary Volume, pp. 1041–80, contains considerable useful information on the Bureau's program. In a fixed bed catalyst the gas or liquid flow does not move the particles of the catalyst; in a fluidized bed the flow moves the particles. The composition of K-536 is 0.6% molybdenum-2% chromium-5% zinc-5% sulfur, supported on terrana (a Bavarian fuller's earth) activated with hydrofluoric acid.

Dimensions of the reinforced concrete enclosure that held all the high-pressure containers were 59 m long, 8.5 m wide, 17.7 m high, with walls 35.6 cm thick. The two cold stalls with catchpots, coolers, and condensers were in the center. Next to the cold stalls, one on each side, were the liquid hot stall and the vapor hot stall, each containing converter(s), high-temperature heat exchanger(s), and hot catchpot(s). The two end compartments, each next to a hot stall, held the preheaters, giving the arrangement, preheater-hot stall-cold stall-cold stall-hot stall-preheater.

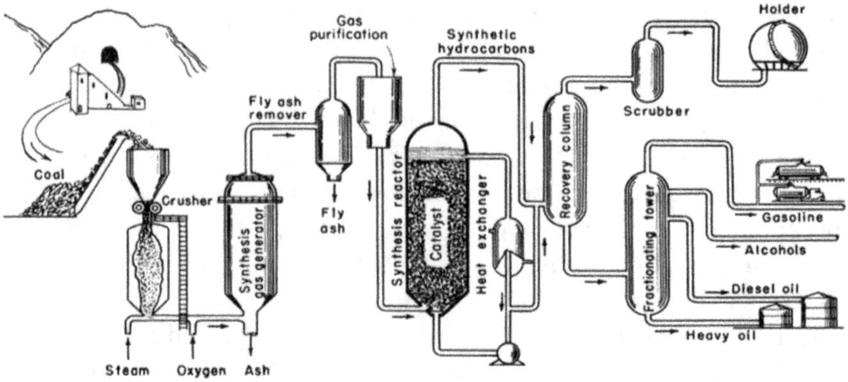

FIGURE 7.10 Fischer-Tropsch synthesis
SOURCE: SHERIE MERSON AND TIM PALUCKA, *A Century of Energy*, NATIONAL ENERGY TECHNOLOGY LABORATORY (FORMERLY US BUREAU OF MINES), US DEPARTMENT OF ENERGY, 2010, P. 151

FIGURE 7.11 F-T demonstration plant. Left: The two tall cylinders are the vertical gasifiers
SOURCE: SHERIE MERSON AND TIM PALUCKA, *A Century of Energy*, NATIONAL ENERGY TECHNOLOGY LABORATORY (FORMERLY US BUREAU OF MINES), US DEPARTMENT OF ENERGY, 2010, P. 192

reactants into a continuous gasifier for conversion to synthesis gas. A natural gas-fired pebble bed heater superheated the steam; a Linde-Fränkl oxygen producer, imported from Hoechst Chemical Company in Germany and completely reconditioned before use, extracted oxygen from air at –185° C (–300° F). It had a capacity of 23,000 ft^3/hr or 1 ton/hr of 98% oxygen. The Koppers-designed gasifier, consisting of a refractory-lined steel shell 6.5 feet (1.98 m) by 9 feet (2.7 m) internal measurement, operated at 1095° C (2000° F) and did not require the use of coking coals. According to the Bureau, its F-T plant contained the largest gasifier ever installed for the direct production of synthesis gas from pulverized coal, superheated steam, and oxygen. By feeding at both ends, it converted daily 28 tons of coal, 24 tons of oxygen, and 35 tons of superheated steam to 2 million ft^3 of synthesis gas.

After leaving the gasifier, cooling, removing dust and ash particles, and compressing, the synthesis gas went to the purification unit where precipitators and separators, catalytic converters, and activated carbon absorbers removed the impurities. The gas required purification because of the reaction's sensitivity to minute amounts of sulfur, tar, gums, resins, and hydrogen cyanide. Sulfur, almost always a catalyst poison, had an allowable content 1/100 the amount in gas purified for domestic use.

Conversion of the purified synthesis gas to petroleum liquids occurred in the hydrocarbon synthesis unit and enabled the Bureau to test two types of internally cooled converters it had developed. The first, an oil circulation converter, effectively removed the heat of reaction by circulating a nonvolatile coolant oil through a fixed catalyst bed of fused iron granules as the synthesis gas passed over the bed. It provided good temperature control, a wide choice of products, and long catalyst life. In the second type, an oil catalyst slurry converter, the synthesis gas bubbled through a rapidly circulating slurry of powdered catalyst and nonvolatile coolant oil, which removed the heat of the reaction. Both types operated at 20–40 atm (300–600 psi) and a maximum temperature of 315° C (600° F). The Bureau claimed that its converter design allowed for production of 500 bbl/d, but the two designs it constructed had much lower production capacities of 80 bbl/d for the internally cooled and 40 bbl/d for the slurry converter. Its converters were nevertheless a vast improvement over the cumbersome and expensive German externally cooled, fixed bed converters in which the indirect removal of heat led to severe overheating problems and limited German converter production to 18 bbl/d.

The refinery, the plant's fourth unit, allowed for maximum cooling and condensing of the gases and vapors leaving the converters and for their separation (distillation) and upgrading (refining) to useful liquid products. An oil-absorption system within the refinery recovered any uncondensed gases and vapors for further treatment; separation and recovery of the coolant oil, light products, and distillate fuels occurred in the distillation section. Distillation also removed wax from the circulating coolant oil thereby eliminating deposits that clogged the catalysts' pores and increased the oil's viscosity. The plant produced about 55 bbl/d of finished gasoline, 10 bbl/d of diesel oil, 12 bbl/d of heavy oils and waxes suitable for cracking, and 5–10 bbl/d of propane.[80]

By the early fifties the Bureau considered the coal hydrogenation and F-T demonstration plants a success. It had tested many ranks of coal, experimented with various temperatures and pressures, shown the effectiveness of different catalysts, and produced gasoline from coal hydrogenation with octane numbers as high as 89 after adding 3 ml lead tetraethyl. At about the same time, most of the German scientists had obtained immigration visas, and if they chose, could remain in the United States as legal residents. Donath accepted a position with Koppers Chemical Company in Pittsburgh as manager of the Research Department, Fuels Processing Section. Schappert went to Koppers Chemical Division, while Josenhans and Frese joined Koppers Engineering and Construction Division. Alberts worked for a time at Bechtel. The others stayed for different periods of time before returning to Germany.[81]

8 Industrial Development of F-T and Coal Hydrogenation

Outside of the Bureau's work only a few synthetic fuel projects were underway in the late 1940s and mid-1950s. Three were F-T plants, none of which achieved full operating capacity before shutdown for both economic and technical reasons in 1953–54, and one was a coal hydrogenation plant. Two of the F-T plants, the Stanolind Oil and Gas Company plant at Garden City, Kansas, and the Carthage Hydrocol Inc. plant at Brownsville, Texas, were $20 million industrial-size plants. They used natural gas to produce synthesis gas and had a rated capacity of 7,000 bbl/d (ca. 1,000 tons/d) of liquid fuel and chemical by-products. The third F-T plant under construction at Liberty, Pennsylvania, near Pittsburgh, was a joint operation of Pittsburgh Consolidated Coal Company and Standard Oil Development Company. The 50 ton/d pilot plant cost $300,000 and unlike the other two plants used pulverized coal to produce

80 *Coal to Oil Demonstration Plants* (note 72); *Bureau of Mines Report of Investigations 5506* (note 70); Storch, Golumbic, and Anderson (note 3).
81 Donath interview (note 76).

synthesis gas. The three plants adopted the fluidization process that Warren K. Lewis (1882–1975) and Edwin Gilland (1909–73) at MIT invented in late 1939 for the catalytic cracking of petroleum. In fluidization, which the F-T plants used for both preparing synthesis gas and converting it to liquid hydrocarbons, the preheated gaseous reactants or the coal particles suspended in a stream of oxygen and steam, when blown through a bed of powdered catalyst in the converter, surround the particles of the finely divided catalyst, transforming the bed into a mobile fluid that resembled a boiling liquid. The motion significantly increased contact between catalyst and reactants, greatly improving temperature control and the reaction rates obtained with the German and the Bureau's laboratory developed converters.

An older German F-T plant of 10,000 bbl/d capacity required 128 small, complicated, and costly fixed-bed converters to provide the large steel cooling surface areas that removed the heat of reaction and maintained the catalyst's activity within the narrow temperature range of synthesis. A similar-sized plant using the fluidized catalytic process had only four converters and 1% of the German fixed-bed converters' cooling surface. Fluidization simplified the plant's engineering design and resulted in a considerable reduction in plant investment cost. It enabled the F-T synthesis to produce gasoline and diesel oil at a price close to that of the petroleum derived products.[82]

In Germany the F-T and coal hydrogenation plants were complementary rather than competitive: F-T produced mainly diesel oil whereas coal hydro-

82 Stanolid Oil and Gas Company was a wholly-owned subsidiary of Standard Oil Company of Indiana; see William F. Bland, "While Government Talks About What to do on Synthetics, the Oil Industry Does It," *National Petroleum News*, 40 (18 February 1948): 22–3. Carthage Hydrocol, Inc. consisted of the following companies: Forest Oil Corporation, Chicago Corporation, La Gloria Corporation, Niagara Shore Corporation, Stone and Webster, Inc., United Gas Corporation, Western Natural Gas Company, Hydrocarbon Research, Inc., and the Texas Company.
In addition to the large F-T projects, a significant number of smaller operations were ongoing in the 1950s. The Texas Company was spending $1 million a year on equipment design and catalyst research at its laboratories in Beacon, NY, and Montebello, CA; Hydrocarbon Research, Inc. carried on investigations at its pilot plant in Olean, NY, and its laboratories in Trenton, NJ; M. W. Kellogg Co. operated pilot plants in Jersey City, NJ, and Lake Charles, LA, and since 1946 had worked with Cities Service Co. on synthesizing liquid fuels from natural gas. The only commercial-size F-T plant operating in the 1950s was the government owned 5,000 bbl/d Sasol plant in Sasolburg, South Africa, about 50 miles south of Johannesburg. It used both fixed bed and fluidized bed catalysts. Esso Standard Oil Company tested a fluidized rather than a fixed bed catalyst at its experimental oil shale retorting plant in Baton Rouge, LA. See "Synthetic Fuels," *Chemical Engineering*, 55, no. 6 (June 1948): 131–38 and Fieldner (note 44).

genation produced high quality gasoline. With the introduction of the Lewis-Gilland fluidized bed and engineering improvements such as automation the F-T synthesis now gave a high-quality gasoline and according to its supporters at a cost less than that of coal hydrogenation and comparable to that of petroleum derived gasoline. The Bureau welcomed the advances industry made to the F-T synthesis. At the same time it pointed out the relevance of its demonstration program which tested different coals as feedstocks and avoided duplicating much of private industry's F-T research using a natural gas feedstock. The Bureau noted that the improvements incorporated in its hydrogenation plant continued to make coal hydrogenation economically competitive and provided figures showing the same cost for either process.

Despite the Bureau's favorable economics, American industrialists constructed only one coal hydrogenation plant, Union Carbide's 300 ton/d pilot plant at Institute, West Virginia. The $11 million, 200–400 atm plant, which Carbide operated from 1952 to 1956, incorporated some of the older German technology as well as the latest advances in coal conversion, particularly accurate temperature control, and enabled Carbide's scientists to decrease reaction time from 45 minutes to 3 minutes. Unlike the German hydrogenation plants, the Carbide plant produced primarily chemicals such as highly demanded benzene, phenols for use in plastics, pharmaceuticals derived from quinoline (nicotinic acid), picoline for tuberculosis drugs, and new rocket fuels. Carbide wanted to develop new products from coal and thereby permit a gradual return to coal as a primary chemical feedstock rather than natural gas. The Bureau had taken an identical position regarding its choice of a coal feedstock. Carbide later planned to build a full-sized 1000–4000 ton/d plant to hydrogenate West Virginia's coal deposits, but for economic reasons it closed the Institute plant in 1956.[83]

9 Concluding the Bureau's Synthetic Fuel Program

What happened to the Bureau's two Missouri coal conversion plants? During their years of operation the two plants demonstrated clearly the technological feasibility of coal hydrogenation and the F-T synthesis. But economically,

83 John R. Callaham, "Coal Hydrogenation Process Unlocks Vast Aromatics Field," *Chemical Engineering*, 59, no. 6 (June 1952): 152–58; "Carbide Takes the Hex out of Coal," *Chemical and Engineering News*, 30, no. 19 (12 May 1952): 1954–59; "New Emphasis on Coal Chemicals," *Chemical and Engineering News*, 34, no. 34 (20 August 1956): 4009; Lawrence P. Lessing, "Coal," *Scientific American*, 193 (July 1955): 38–67.

coal-derived petroleum could not compete with natural petroleum. The cost of synthesizing gasoline in a projected 30,000 bbl/d coal hydrogenation plant in Rock Springs, Wyoming, according to the Bureau's figures of January 1952, was 19.1¢/gal (after 6% profit and 50% income tax) or almost double the 10.6¢/gal cost of regular grade gasoline at petroleum refineries. Two cost studies on synthetic gasoline from coal hydrogenation reported even higher figures. Ebasco Services, a New York City engineering construction and consulting firm that Oscar Chapman (1896–1978), Secretary of the Interior, authorized to review the Bureau's fourth and final cost revision, estimated 21.8¢/gal; while the National Petroleum Council (NPC), whose members mostly opposed the Bureau's findings, arrived at 34.8¢/gal. For F-T synthesized gasoline the Bureau had no detailed cost analysis, estimating a cost in 1952 of about 19¢/gal, while the NPC calculated 29.2¢/gal, assuming operation of a 30,000 bbl/d plant.[84]

Several factors contributed to the great variance in cost studies, especially for hydrogenation, the most important of which was the widely different estimates of profits earned annually from sales of the chemical by-products. NPC projected $45,000; Ebasco $122,000. NPC also estimated much higher

84 *Bureau of Mines Report of Investigations 5506* (note 70), pp. 210–15; "Developments in Synthetic Liquid Fuels," *Chemical and Engineering News*, 30, no. 32 (11 August 1952): 3248–56. President Harry Truman established the NPC in June 1946 in order that the wartime government-industry cooperation would continue in peacetime. Julius Krug (1907–70), Secretary of the Interior at that time, assisted by Ralph Davies (1897–1971), a Standard Oil of California vice-president, selected 85 industrial leaders, mainly from the petroleum industry, to form the council.
The cost of the Rock Springs hydrogenation plant, according to Ebasco and the Bureau's revised estimate, was $351 million, or $404 million including the coal mine and all other expenditures. NPC's figures were $396 million for the plant and $533 million total cost. Two earlier studies had given lower costs. Percival C. Keith (1901–76), president of Hydrocarbon Research Inc., in 1946 reported that for a F-T plant using natural gas costing 5¢/1,000 cubic feet could produce 80-octane gasoline for 6.5¢/gal; for a hydrogenation plant using coal as the raw material the cost was 8–9¢/gal of gasoline. See P. C. Keith, "Gasoline from Natural Gas," *Oil and Gas Journal*, 45, no. 6 (15 June 1946): 104–12. A 1949 study by George Roberts and Paul R. Schultz of the Stanolid Oil and Gas Company gave significantly higher figures. Their estimates of investment and operating costs for new plants producing 10,000 bbl/d of 80-octane gasoline from crude oil gave a per gallon price of 12.2¢; from natural gas 15¢; from oil shale 22.5¢; from bituminous coal 22.5¢; and from sub-bituminous coal 27.6¢. Roberts and Schultz concluded that production of F-T synthetic gasoline from natural gas was competitive with gasoline derived from crude petroleum, but its production from coal and oil shale hydrogenation was uneconomical. They believed synthetic gasoline obtained from coal using an improved F-T synthesis could lower the cost from 22.5¢/gal to 17¢/gal, George Roberts and Paul R. Schultz "Liquid Fuels from Coal and Shale," *Oil and Gas Journal*, 48, no. 19 (15 September 1949): 116–17, 120–21, 123–25.

plant and housing construction costs, general and administrative overhead, and required operating costs such as oxygen production, steam and electrical power. Not only did Ebasco, NPC, and the Bureau disagree on cost figures, but the Bureau's synthetic fuel scientists differed on the priorities and even the necessity of the demonstration program. Ezekial L. Clark, later head of the Bureau's Gasification Division in Washington, DC, noted the high cost of hydrogenating coal, and unlike Hirst and Schroeder, he favored F-T development. Storch believed the Bureau had gone beyond its original objectives and should continue only with basic laboratory and pilot plant research.[85]

Spokesmen for both the petroleum and chemical industries questioned the wisdom of the government's support of synthetic fuel development. A. L. Solliday (1898–1982), executive vice-president of Stanolind and a NPC member highly critical of the Bureau's program, argued that because of fuel technology's rapid advances the plants would be obsolete before their completion. Another NPC member, Eger V. Murphee (1898–1962), president of Standard Oil Development Company, believed that the western coal bearing regions lacked sufficient water supplies to operate synthetic fuel plants. Walter J. Murphy (1899–1959), editor of *Chemical and Engineering News*, called for clarification of the Bureau's objectives. Was it to obtain technological and economic data for operation of an industrial-size plant not attainable in any other way? Was the construction of a 30,000 bbl/d hydrogenation plant or eight of these plants, as the Bureau had considered in January 1952, a matter of national security? Even eight synthetic fuel plants, producing 240,000 bbl/d, represented only 8% of the 3 million barrels of US daily petroleum consumption. Was this sufficient to justify the government's spending $3 billion for the construction of synthetic fuel plants? Murphy also asked whether the government should spend money earmarked for synthetic fuel research and plant construction on domestic petroleum exploration and cheaper imports from Canada.

The NPC raised two more issues about the Bureau's program. It pointed out that 53% of the plant's operating revenue would come from the production of chemicals, while simultaneously questioning whether the commercial market could absorb the additional products of coal conversion. NPC also viewed the Bureau's demonstration program as a drive to put the government into the synthetic fuel business under the guise of national security. Unlike the

85 Ezekial Clark, Telephone Interview by Anthony N. Stranges, 28 February 1978. For Storch see "Synthetic Liquid Fuels, Annual Report of the Secretary of the Interior for 1953," Part 1-Oil from Coal, *Bureau of Mines Report of Investigations 5043* (Washington, DC: Government Printing Office, April 1954), pp. 1–66, on p. 2.

aerospace and nuclear industries that used national security to their advantage and gained lucrative government contracts during the Cold War years, the petroleum industry largely resisted the Bureau's program. It maintained that the government was in direct competition with private industry. NPC reminded Bureau supporters that the petroleum industry had provided sufficient petroleum in past national emergencies and would continue to do so in the future. It dismissed a government supported synthetic fuel industry as not only uneconomic but unnecessary now and in the foreseeable future.[86] Despite the hostile environment the Bureau experienced, its demonstration program had succeeded technologically.

The economics of coal hydrogenation and F-T, however, ultimately doomed its survival in a time of abundant and cheap domestic and imported natural petroleum. Continued research on catalysts and hydrogen production, improvements in reaction conditions such as temperature control, and reduction of the high pressures used in hydrogenation would have improved the economics and possibly saved the demonstration program. But 1952 was a Republican year. With the election of Dwight Eisenhower (1890–1969), the first Republican president in twenty years, a Republican congress, eager to support Eisenhower's governmental hands-off approach and laissez-faire, test-of-the-market economic policy, and with additional opposition coming from oil-state Democrats, the Bureau's eleven-year program effectively had ended. In March 1953 the House Appropriations Committee cut $2 million from the Department of Interior's funds for the two coal conversion plants and another $9 million from the Bureau's budget. At about the same time, Douglas McKay (1893–1959), the new Secretary of the Interior (1953–56), appointed Storch, a critic of the Bureau's demonstration program, head of the Bureau's Office of Synthetic Liquid Fuels. On 29 May 1953, after having spent $82 million of its $87.6 million authorization, the Bureau shut down its Missouri demonstration program and dismantled the coal hydrogenation and F-T plants. Small-scale research continued at Bruceton. What the Truman administration originally had conceived as a strategic reserve and a national security measure during the closing World War II years and early Cold War years fell victim to petroleum economics, power, and politics.

86 "Synthetic Fuels" (note 93), 132; Walter J. Murphy, "Synthetic Fuels from Coal and Shale," *Chemical and Engineering News*, 30, no. 32 (11 August 1952): 3247. For an extended discussion on the controversy over the closing of the Bureau's demonstration plants see Richard H. K. Vietor, "The Synthetic Liquid Fuels Program: Energy Politics in the Truman Era," *Business History Review*, 54, no. 1 (1980): 1–34; Marchese (note 3), Chapters 7 and 8; and Arnold Krammer, "Technology Transfer as War Booty: The US Technical Oil Mission in Europe," *Technology and Culture*, 22 (1981): 68–103.

TABLE 7.1 Bureau of Mines synthetic fuel program, synthetic liquid fuels act

Administration	Government financial support
Roosevelt-Truman 1944–52	Act passed 5 April 1944, with later amendments totaled $87.6 million
Eisenhower 1952–60	Funding for Act ended 29 May 1953

SOURCE: COURTESY OF AUTHOR

10 Conclusion

Petroleum consumption and imports continued to increase in the 1960s–70s, and with the October 1973–March 1974 Arab oil embargo and the November 1979–January 1981 Iranian hostage problem further complicating the situation, the United States found itself experiencing an energy crisis. Petroleum consumption by this time had reached 16–18 million bbl/d, half of it imported. In response and in marked contrast to what occurred twenty-five years earlier, government and private corporations such as Gulf Oil, Hydrocarbon Research, and Exxon Corporation carried out joint synthetic fuel programs, intending through their cooperative effort to relieve the energy crisis. Indeed the events of the 1970s–early 1980s challenged the wisdom of the government's decision to terminate the Bureau's earlier synthetic fuel program.[87]

87 Stranges (note 2).

CHAPTER 8

Epilogue: Post-World War II Synthetic Fuel Development and the Emergence of Environmental Concerns

1 Introductory

The twentieth-century coal-to-petroleum, or synthetic fuel, industry evolved in three stages: (1) the invention and early development of the Bergius coal hydrogenation and Fischer-Tropsch (F-T) gas synthesis processes from 1910 to 1926; (2) Germany's industrialization, the expansion from laboratory research and development to Bergius and F-T pilot-demonstration plants to industrial-size Bergius and F-T plants, from 1927 to 1945; and (3) global transfer of the German synthetic fuel processes to Britain, France, Japan, Canada, the United States, South Africa, and other nations that occurred mainly in the 1930s–50s but has continued into the 1990s–2000s.

Of the three nations, Germany, Britain, and Japan, that industrialized or planned to industrialize their synthetic fuel programs in the prewar and World War II years, only the Japanese program failed. The Japanese, in their rush to have a domestic source of liquid fuel, bypassed the pilot plant stage. They had done excellent small-scale laboratory research but never tested their operational conditions (temperature, pressure, catalysts) in pilot plants before leaping to the construction of industrial-size plants. This lack of small-scale operational experience resulted in the large-scale industrial failure of the Japanese synthetic fuel program.

In addition to the large-scale German, British, and Japanese synthetic fuel programs, small-scale synthetic fuels programs were underway in prewar and World War II Italy and France. Both programs were war-related and ceased operation once the war ended. In the post-war years, the 1950s–80s, Sasol in South Africa constructed three F-T synthetic fuel plants that continue to operate in the 2000s whereas the synthetic fuel plants United States, German, and Japanese industrialists constructed in the 1970s–early 80s ceased operation by the late 1980s. The 1970s energy crises, the Arab oil embargo of October 1973–March 1974 and the Iranian hostage crisis of November 1979–January 1981, revitalized synthetic fuel research and development and led to joint government-industry programs, but the research-development and joint pro-

 | DOI:10.1163/9789004690912_009

TABLE 8.1 Government-industry scale-up of synthetic fuel programs

Country	Research and development	Pilot-demonstration plants	Industrial-size plants
Germany	yes	yes	yes
Great Britain	yes	yes	yes
Japan	yes	no	yes (failure)
Canada	yes	yes	yes
United States	yes	yes	yes
South Africa	yes	yes	yes

SOURCE: COURTESY OF AUTHOR

TABLE 8.2 Joint government-industry synthetic fuel research-development

Country	Research-development
Germany	Continuous until 1945, intermittent thereafter
Great Britain	Continuous until 1950s
Japan	Continuous until 1945
Canada	Continuous to 2000s
United States	Intermittent since 1950s
South Africa	Continuous to 2000s

SOURCE: COURTESY OF AUTHOR

grams quickly disappeared once the crises had passed and petroleum prices collapsed in 1982/83–2001. The discontinued programs included the F-T and coal hydrogenation pilot plants that Gulf Oil, Atlantic Richfield, and Exxon operated in the United States; the Saarbergwerke AG plant in Saarbrüken, Ruhrkohle AG's plant in Essen, and the 200 ton/d coal hydrogenation plant that VEBA-Chemie AG (in 1978 VEBA Oel AG in Gelsenkirchen) operated in Bottrop from 1981 to 1987; and the 50 tonne/d coal hydrogenation plant that Japan's NEDO Company, with headquarters in Kawasaki, constructed at Morwell, Australia, and operated from 1985 to 1990.

The cycle of fits and starts in the synthetic fuel industry has continued in the 1990s–2000s, and the only operational coal hydrogenation, F-T, and gas-to-liquid plants are Sasol's three F-T plants in South Africa and the six F-T and coal hydrogenation plants in China, Malaysia, Nigeria, and Qatar.[1] These nine operational synthetic fuel plants, in combination with the 1980s–90s hydraulic fracturing process that revolutionized the recovery of naturally occurring petroleum, and the scientific community's demonstration that global warming and environmental pollution are real provide an informative overview of the fossil fuel industry in the twentieth and twenty-first centuries.

2 Pre-World War II Small-Scale Synthetic Fuel Developments: Italy and France 1920s–40s

By the 1940s the Germans and British had developed the largest and most advanced synthetic fuel programs. Their success led other European nations, particularly Italy and France who like the British and Germans lacked sufficient domestic sources of liquid fuel, to establish their own synthetic fuel programs aimed at increasing their supplies of petroleum. Their synthetic fuel plants and programs, both hydrogenation and F-T, were smaller-size, war-related, and ceased once World War II ended in 1945 or shortly afterwards.

In Italy, having very little domestic oil, Benito Mussolini (1883–1945), Italy's Fascist prime minister, had shown considerable interest in hydrogenating coal and crude oil since 1933. The Italian government's Azienda Nazional Idrogenazione Combustibili (ANIC) reached an agreement with the International Hydrogenation Engineering Company (IHEC) in 1936 and in that same year decided to construct refineries for hydrogenating crude oil at Bari in southern Italy and Livorno (Leghorn) in Tuscany. Bari hydrogenated 150,000 metric tons of Albanian crude oil to give gasoline, diesel oil, and lubricating oil; Leghorn hydrogenated 150,000 metric tons of Venezuelan crude oil and Romanian Pacura (pitch) giving the same products. The Bari plant started operation in 1938, Leghorn in 1939, both continued into the 1940s. The original plan of hydrogenating Italian coal to synthesize liquid fuel never materialized.[2]

1 I have compiled the information on the synthetic fuel plants from "Fuel Fix" which is the energy section that appears regularly in the *Houston Chronicle* and is available online at *houstonchronicle.com* and in print form.

2 *High-Pressure Hydrogenation at Ludwigshafen-Heidelberg*, FIAT Final Report No. 1317, 9 Vols. (Dayton, Ohio: Joint Intelligence Objectives Agency, 1951), Vol. 1, p. 98. IG Farben joined with Royal Dutch Shell in the Hague and London, Imperial Chemical Industries in London, and Standard Oil in the spring of 1931 to establish International Hydrogenation Patents (IHP) and International Hydrogenation Engineering Company (IHEC) in the Hague, thereby acquiring international control of high-pressure coal hydrogenation.

The French, who from 1920 to 1926 produced 50–70% of their domestic petroleum consumption, had small oil shale and coal carbonization industries. They focused on synthesizing methanol (methyl alcohol) from coal (coke) using the water gas reaction rather than synthesizing petroleum liquids from coal and constructed three synthetic fuel plants, two coal hydrogenation plants in 1935 and one F-T plant in 1937.

The three synthetic fuel plants French industrialists constructed were large demonstration or semicommercial plants located in northern France. The Société des Carburants Synthétique des Mines de Béthune in 1935 completed construction of a two-stage semicommercial coal hydrogenation plant in the Pas-de-Calais region. The Compagnie Francais des Essences Synthétiques constructed a similar size, 48 metric ton/d two-stage plant at Liévin. France's coal hydrogenation program suffered from high plant construction costs estimated at FF 200 million ($12 million US) in 1936 and from lignite price increases in the late 1930s. Both plants discontinued operation after World War II. The Compagnie Francais dismantled the Liévin plant in 1964. The Société Courriéres-Kuhlmann constructed France's single F-T plant at Harnes in the Pas-de-Calais region. It had an annual capacity of 20,000 metric tons of liquid hydrocarbons, including gasoline and diesel oil, and by June 1937 was running at 50% capacity. Plant operation ceased after World War II, the year not specified.[3]

3 Henrie de Peyerimhoff, "The Coal Situation in France," *Proceedings of the Third International Conference on Bituminous Coal*, 1930 (Pittsburgh: Carnegie Institute of Technology, 1931), Vol. I, pp. 48–52. Most of the French research on LTC, as in Germany, was for the production of gasoline, fuel oil substitutes, and dense metallurgical coke; A. C. Fieldner, "Low-Temperature Carbonization of Coal," *Bureau of Mines Technical Paper 396* (Washington, DC: Government Printing Office, 1926), pp. 1–46, on p. 3; E. Audibert, "Synthetic Motor Spirit from Mixtures of Carbon Monoxide and Hydrogen," *Fuel*, 5 (1926): 170–77; Charles Lormand, "Industrial Production of Synthetic Methanol," *Industrial and Engineering Chemistry*, 17 (1925): 430–32 (1 kL = 264 gal; corresponding cost figures are FF 105 /kL, FF 130 /kg); Georges Patart, French Patent 540,343 of 19 August 1921, published 12 July 1922; T. P. Hilditch, *Catalytic Processes in Applied Chemistry* (New York: D. Van Nostrand Company, 1931), p. 88; Georges Patart, "The Industrial Transformation of Bituminous Coal into Organic Technical Products," *Proceedings of the International Conference on Bituminous Coal*, 1926 (Pittsburgh: Carnegie Institute of Technology, 1927), pp. 132–60, on p. 137; Paul Dumanois, "Methyl Alcohol as Motor Fuel," *Proceedings of the International Conference on Bituminous Coal*, 1926, pp. 161–64. Patart's article is also in *Fuel*, 6 (1927): 160–70; Jean Bing, "The Gas and Tar Industries in France," *Proceedings of the Second International Conference on Bituminous Coal*, 1928, 2 Vols. (Pittsburgh: Carnegie Institute of Technology, 1929), Vol. 2, pp. 766–88, on p. 774; Bing, "Bituminous Coal as a Source of Chemical Products," *Proceedings of the International Conference on Bituminous Coal*, 1926, pp. 404–18, on pp. 409–12; Étienne Hirsch, "La Fabrication D'Essence Synthétique de la Société Courriéres-Kuhlmann," *Memories de la Société des Ingénieurs Civils de France*, 90 (1937): 756–67; H. H. Storch, "Synthesis of Hydrocarbons

3 Post-World War II Synthetic Fuel Developments: Sasol 1950s–2000s

Following the synthetic fuel plant shutdowns after World War II and until the global energy crises of October 1973–March 1974 and November 1979–January 1981, all major synthetic fuel research and development ceased, mainly because of economic factors, except for the construction in March 1955 of the South African Coal, Oil, and Gas Corporation's (Sasol) $145 million F-T plant (Sasol I) in Sasolburg, south of Johannesburg. South Africa's desire for energy independence and the low quality of its sub-bituminous coal, dictated the choice of F-T synthesis rather than coal hydrogenation. Sasol's Johannesburg plant remained the only operational industrial-size synthetic fuel plant until the 1970s energy crises and South Africa's concern about hostile world reaction to its apartheid policy prompted Sasol to construct two larger F-T plants, Sasol II in March 1980 and Sasol III in June 1982, in Secunda. At this time, in 1983, the smaller Sasol I plant stopped producing synthetic crude oil.

Sasol's three plants used similar F-T coal-to-oil conversion processes whereby coal passed into Lurgi gasifiers for conversion into crude synthesis gas. At Sasol I the crude synthesis gas reacted in a low-temperature slurry phase distillate reactor to produce linear-chained hydrocarbon waxes and paraffins. At the Secunda plants the crude synthesis gas passed into higher temperature reactors and in a single stage produced C_1–C_{20} hydrocarbons and synthetic crude oil for refining and fuels production. Additional products include propylene, ethylene, solvents, and alpha olefins. The two Secunda plants, Sasol II built at a cost of R 2.3 billion (R = Rand, $0.16 billion US) and Sasol III built at a cost of R 4.7 billion ($0.32 billion US) and now called Sasol Synthetic Fuels, use an iron-based catalyst and operated at T = 350° C, P = 23.8 atm. Total production of both plants is 160,000 bbl/d, about twenty times the synthetic crude oil production of Sasol I.[4]

from Water Gas," in *Chemistry of Coal Utilization*, ed. H. H. Lowry, 3 Vols. (New York: John Wiley & Sons 1945 and 1963) Vol. 2, p. 1800; W. R. K. Wu and H. H. Storch, "Hydrogenation of Coal and Tar," *Bureau of Mines Bulletin 633* (Washington, DC: Government Printing Office, 1968), pp. 1–195 (p. 4); "Calais Petrol-from-Coal Works to be Abandoned," *Chemical Age*, 91 (15 February 1964): 265; "Coal Hydrogenation-France," *Chem. Trade J.*, 102 (1938): 109.

4 Sasol 50 Years of Innovation, Sasol.com, 79 pp.; South African Rand (R) 14.26 = $1.00 US, 2021 rate. In 1955 Sasol I cost £40 million or $145 million US; using a 2021 conversion rate, Sasol I's cost of £40 million = $54 million US.

FIGURE 8.1 Synthol (synthetic oil) reactors at Sasol I
SOURCE: https://www.sasol.com/70-years/celebrating-our-heritage

4 Post-World War II Synthetic Fuel Developments: United States 1960s–80s

A major reason for the United States government's and industry's renewed interest in coal-derived petroleum was that in 1958 the United States went from a net petroleum exporter to a net importer. Domestic petroleum consumption continued to increase in the 1960s, and in 1964–65 it began to increase more rapidly, going from approximately 4 billion bbl/yr to almost 5.5 billion bbl/year in 1970. The United States' petroleum production increased in this period but more slowly than demand. Domestic production reached its peak in 1970 and began to decline. Within two years, in 1972, the United States experienced its first trade deficit since 1935 and only the second deficit since 1888.[5]

5 Joseph M. Duckert, *A Short Energy History of the United States* (Washington, DC: The Edison Electric Institute, 1980), p. 63.

Although the drive to develop an American synthetic fuel industry did not occur for another decade, until after the Arab oil embargo of 1973–74, the mid-1960s marked the beginning of small-scale industrial research on coal hydrogenation. In anticipation of future petroleum shortages, the Office of Coal Research (OCR), a branch of the United States Department of Interior, in 1962 awarded Willard Bull, a chemist with Spencer Chemical Company in Kansas City, Missouri, a contract to continue the coal hydrogenation research that he had begun in the early 1950s.

Spencer carried out research in its laboratory in Merriam, Kansas, converting up to 50 pounds of coal daily to a liquid product that solidified upon reducing the reaction temperature and pressure. Spencer called its coal hydrogenation process Solvent Refined Coal (SRC) because the process consisted of mixing finely crushed dried coal with a heavy oil solvent (1 part coal: 2 parts oil) and then adding hydrogen gas at high pressure (69–137 atm) and temperature (450° C) to the coal slurry, producing a black, clean-burning glassy solid boiler fuel.

During the early research, Spencer's chemists found that in lowering the ash content of the solid product to 0.5% by weight they had removed most of the sulfur originally present in the coal. They had thereby not only minimized the ash disposal problem at coal-burning power plants but had reduced substantially the sulfur oxides emissions (SO_2, SO_3), the primary contributors to air pollution.[6] By 1966, because of favorable technical and economic analyses and a continually deteriorating balance between domestic petroleum consumption and production, Spencer decided to expand its program.[7]

Gulf Oil Corporation had now acquired the Spencer Company, and in 1968 with support from a second OCR contract for $88 million, Gulf began the design and construction of a 50 ton/d pilot plant at Fort Lewis, Washington, south of Tacoma. Gulf completed construction of the Fort Lewis Plant in September 1974. The SRC pilot plant successfully hydrogenated high-sulfur bituminous coals from the eastern United States and bituminous coals from Pennsylvania, Kentucky, West Virginia, and Ohio. Its solid product, called SRC-I, had a heating value of 16,000 BTU/pound, a value 33% greater than that of the coal reactant. The process converted 65% of the coal to the clean-burning solid SRC-I, 15% to distillable liquids, and the remaining 20% to ash and gases.[8]

6 Gulf Oil Corporation, Information Brochure, *Solvent Refined Coal-II: Turning Coal into Clean Energy*, pp. 4–5; *Assessment of Technology for the Liquefaction of Coal* (Washington, DC: National Research Council, 1977), pp. 47–50. The Gulf Oil Information Brochures have no specific date, they date from the late 1970s–early 1980s.

7 Gulf Oil Corporation, Information Brochure, *SRC-II; Background and Schedule.*

8 Note 6. Gulf acquired Spencer in 1963 for $120 million.

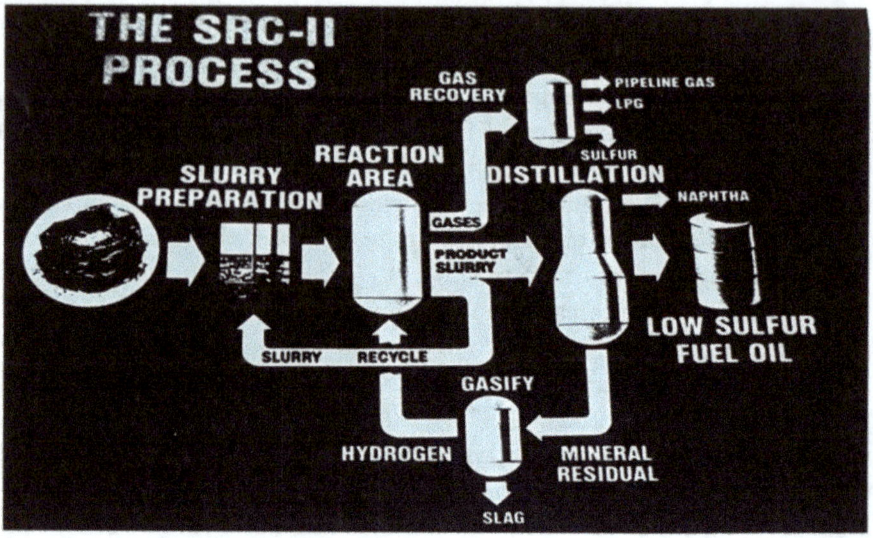

FIGURE 8.2 SRC-II process
SOURCE: *Solvent Refined Coal-II*, DIRECTOR OF PUBLIC AFFAIRS GULF OIL CORPORATION

Gulf's ongoing laboratory work early in the 1970s showed that modifying the SRC-I process to operate at a higher pressure (200 atm) and a slightly lower temperature (300–400° C) resulted in a greater hydrogen addition and gave a liquid rather than a solid product. The liquid resembled the petroleum-derived boiler fuels burned in oil-fired electric power plants. Gulf found that its new, improved version SRC process, SRC-II, also produced the heavy oil solvent used to dissolve the finely crushed dried coal, and thereby eliminated the need to add constantly external solvents.[9]

The Fort Lewis plant began intermittent SRC-II production in 1974 converting daily about 30 tons of high sulfur Eastern bituminous coals chiefly to 70 barrels of home heating oil and fuels for use in public utility power plants. In an August 1978 experiment, Consolidated Edison's 74th Street Station in New York City burned 4,500 barrels of SRC-II fuel oil from the Fort Lewis pilot plant to generate electrical energy. The coal-derived fuel oil performed as well technically and environmentally as petroleum-derived fuel oil.[10]

Because of the Fort Lewis pilot plant's success, DOE decided to take the next step toward developing an American synthetic petroleum industry. In

9 Gulf Oil Corporation, Information Brochure, *Solvent Refined Coal-II: Turning Coal into Clean Energy*, pp. 5–7; Gulf Oil Corporation, Information Brochure, *SRC-II: The Process*.

10 Gulf Oil Corporation, Information Brochure, *SRC-II: Background and Schedule*.

July 1978, it awarded Gulf a contract to design, construct, and operate a high-pressure coal hydrogenation demonstration plant. At the same time, in autumn 1978, DOE agreed to share the hydrogenation technology developed in the demonstration plant with Japan and West Germany, two countries almost totally dependent on imported petroleum for their liquid fuels. DOE investigated more than twenty coal-producing areas in the United States before selecting a 2,600-acre site north of Morgantown, West Virginia, on which to construct a $1.5 billion plant processing 6,000 tons of coal daily and produce about 18,000 bbl/d of synthetic petroleum. The plant's size, 120 times larger than Gulf's pilot plant, was still one-fifth the estimated size of a future industrial production plant, hydrogenating 30,000 tons of coal per day. Gulf scheduled major construction to begin in the spring of 1982 with operation set for 1985. The plant's projected cost, $1.4 billion, represented a major investment, but a cost equivalent to only six days' oil imports at the low price of $30/bbl.[11]

Gulf constructed its West Virginia demonstration plant to produce mainly distillate fuel oil, a low sulfur (less than 0.3%) nonpolluting fuel, for the generation of electrical power and steam in the Eastern United States, where utilities and industry used natural petroleum fuel oil. Gulf claimed that a larger industrial-size plant hydrogenating 30,000 tons of coal daily would yield 60,000 barrels of distillate fuel oil or enough to meet the electrical demands of a city with one million inhabitants. Naphtha was the second major product scheduled for production in the SRC-II demonstration plant. This lighter, low-boiling range hydrocarbon mixture (30–200° C) when upgraded, produces a high-octane unleaded gasoline. An industrial-size plant's naphtha production is about 20,000 barrels (800,000 gallons) of gasoline daily, an amount consumed in a city of 250,000 population.[12]

DOE and Gulf planned a two-year demonstration period, and if the plant's operation succeeded technically and economically, Gulf had the option to purchase the plant and proceed with industrial development. But in early 1981, one year before the start of construction on the West Virginia plant, DOE decided not to continue funding the project. The Reagan government's new policy, which called for private industry to play a larger role in the industrial development of synthetic fuels, dictated DOE's decision. Upon learning of DOE's decision, the German, Japanese, and American sponsors of the West Virginia plant met in Bonn, Germany, to discuss the fate of the proposed

11 *Oil and Gas Journal*, 78 (1980): 33.
12 Gulf Oil Corporation, Information Brochure, *SRC-II: Products*.

demonstration plant. Faced with the loss of DOE support the three sponsors terminated the entire project.[13]

Two other coal hydrogenation projects started operation in the early 1960s. Both were larger than Gulf's initial project and also seemed to reflect the United States' awareness of its increasing dependence on imported natural petroleum. The first, the H-Coal Process, a joint project, initially involved DOE and Hydrocarbon Research Inc. of Trenton and Lawrenceville, New Jersey. The H-Coal Process began in 1963 in Hydrocarbon Research's laboratory with a small unit hydrogenating 25 pounds of coal per day and then with a larger process development unit that hydrogenated three tons of coal daily.[14] By the early 1970s (1971–72), Ashland Synthetic Fuels, a subsidiary of Ashland Oil Inc. in Ashland, Kentucky, and Badger Plants Inc. in Cambridge, Massachusetts, joined the project.[15]

The H-Coal process required mixing crushed, dried coal with a heavy oil derived from the process to produce a slurry and pumping to a pressure of about 200 atm. After adding compressed hydrogen gas to the slurry and heating to 345–425° C, the slurry passed into an ebullated bed reactor containing a metallic catalyst (cobalt-molybdenum alumina extradate) where liquefaction occurred. All hydrogenation processes require large amounts of hydrogen gas. The H-Coal Process consumed 14,000–20,000 ft^3 of hydrogen for each ton of coal hydrogenated and converted 90% of moisture- and ash-free coal to synthetic petroleum. Each ton of coal produced three barrels of synthetic liquid.[16]

---

13 US, Germany, Japan to Discontinue SRC-II Project in West Virginia," *Energy Insider*, 4, no. 13 (6 July 1981): 2.

14 Seymour C. Schuman, Ronald H. Wolk, Michael C. Chervenak, "Hydrogenation of Oils," US Patent 3,183,180 of 11 May 1965; S. B. Alpert, E. S. Johanson, and S. C. Schuman, "Can coal compete for liquid fuels," *Hydrocarbon Processing, Petroleum Refiner*, 43, no. 11 (1964): 193–97. Hydrocarbon Research Incorporated, founded in 1943, changed its name to Hydrocarbon Technologies Incorporated in 1995. Headwaters in South Jordan, Utah, acquired Hydrocarbon Technologies Incorporated in 2001, changed its name to Headwaters Innovation Technology, and shortened its name to HTI. HTI developed coal hydrogenation (direct coal liquefaction DCL) and F-T synthesis and licenses its processes internationally.

15 Ashland Synthetic Fuels, Inc., Information Brochure, *H-Coal, A Future Energy Source*, 1979; "H-Coal Special Issue," Department of Energy, *Energy Countdown*, November 1979; Alfred G. Comoli, Carmine A. Battista, Edwin S. Johanson, Carol P. Laird, "Development and Demonstration of the H-Coal Process," paper presented at the Division of Petroleum Chemistry, American Chemical Society Meeting, Miami, 11–15 September 1978.

16 Seymour C. Schuman, Ronald H. Wolk, Michael C. Chervenak, "Hydrogenation of Coal," US Patent 3,321,393 of 23 May 1967; "The H-Coal Process," 9 January 1981, paper submitted by Hydrocarbon Research Incorporated to the Research and Development Council of

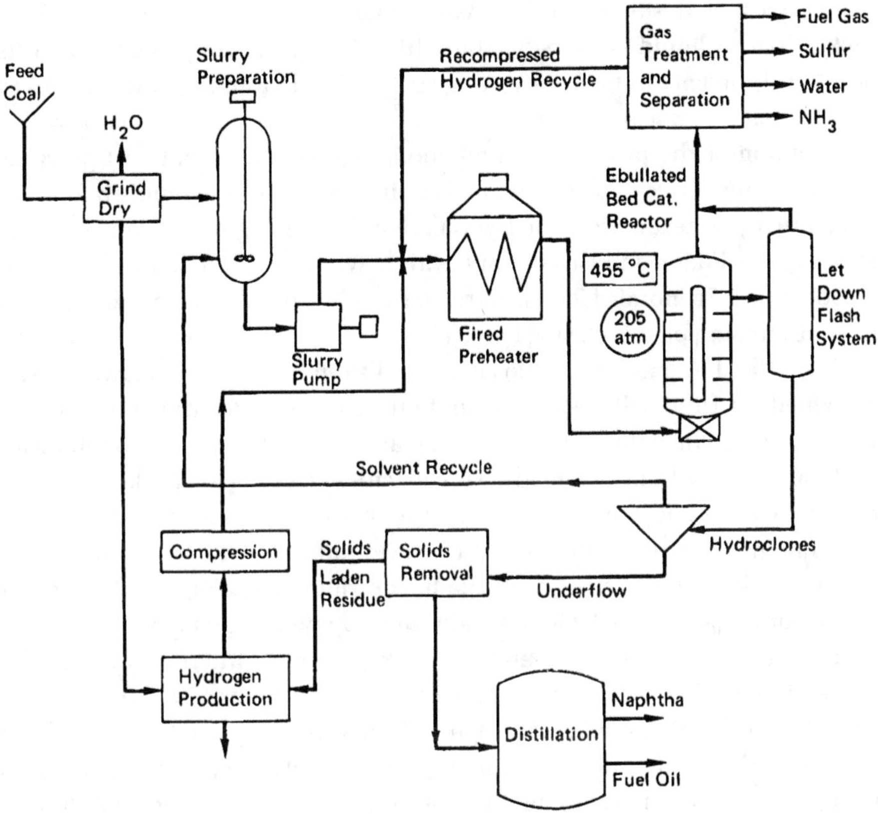

FIGURE 8.3 H-coal process
SOURCE: *Assessment of Technology for the Liquefaction of Coal.* US DEPARTMENT OF ENERGY, OFFICE OF SCIENTIFIC AND TECHNICAL INFORMATION, REPORT 7049743, DECEMBER 1977, P. 65. https://doi.org/10.2172/7049743

The Office of Coal Research, now a division of DOE, originally provided funds for Hydrocarbon Research's early work. By the mid-1970s, it decided that the H-Coal Process had progressed to the pilot plant stage, and in late 1976 Hydrocarbon Research and its partners began constructing a 200–600 ton/d

New Jersey for the Thomas A. Edison Award; Alfred G. Conoli, P. Ganguli, and Michael Merdinger, "H-Coal Processing of Kentucky No. 11 Coal and 1980 Status of H-Coal," 15th Intersociety Energy Conversion Engineering Conference, Seattle, WA, 18–22 August 1980, pp. 1818–23; Richard M. Eccles, and George R. DeVaux, "Current Status of H-Coal Commercialization," presented at the Syncrude Fuels Conference, San Francisco, CA, 13–16 October 1980; *Assessment of Technology for the Liquefaction of Coal* (note 6), p. 66.

plant on a 50-acre site adjacent to Ashland Oil's refinery in Catlettsburg, Kentucky. Coal to liquid conversion started in May 1980. The plant reached its 600-ton/d capacity in 1981, converting coals from all over the United States to liquid fuels for transportation, home, and industry. DOE originally provided $143 million of the project's $179 million anticipated cost, but cost overruns raised the final price to $296 million. Pending favorable economic evaluations after a two-year test period, the Hydrocarbon Research group planned to construct a $1.4 billion industrial plant with a synthetic petroleum capacity of 50,000 bbl/d. Economic factors, particularly falling oil prices in 1982–83, led to the Catlettsburg plant's closing in 1983.[17]

With the H-Coal plant's closing, the Exxon plant in Baytown, Texas, remained the United States' only operational coal hydrogenation pilot plant.[18] It represented the culmination of pilot plant studies that began in 1966 with units of 0.5 ton/d and 1 ton /d coal capacities. The Baytown plant, located on a 55-acre site adjacent to the Exxon refinery, was another joint industry-government operation with 50% of the $340 million cost coming from DOE. Exxon, the largest industrial participant, contributed 23.2%; two of the other participants, Japan Coal Liquefaction Development Company (8.3%) and Ruhrkohle AG of West Germany (2.7%) represented interests outside the United States. The plant, which opened in April 1980 and began operation on 24 June 1980, after a six-year design and construction period, converted 250 tons of coal per day into approximately 600 barrels of synthetic petroleum using a third coal conversion method called the Exxon Donor Solvent Process (EDS).[19]

Like the other coal hydrogenation processes, SRC-II and H-Coal, EDS required high pressures and temperatures, T = 425–470° C and P = 130–135 atm.

17 *H-Coal, a Future Energy Source*; "H-Coal Special Issue," *Energy Countdown* (both note 15); Eccles and DeVaux, "Current Status of H-Coal Commercialization," (note 16); "H-Coal Plant Sets February Restart Using Western, Sub-Bituminous Coal," *Energy Insider*, 5, no. 2 (February 1982): 1.

18 The Fluor Corporation reactivated and rebuilt the DOE test plant at Cresap, West Virginia. It conducted a coal liquefaction test project for DOE using Exxon's Donor Solvent Process. *Oil and Gas Journal*, 78 (1980): 71.

19 *Coal R&D*, 3 (February 15, 1980): 1; Robert Porter, "April startup set for synfuel plant in Texas," *Energy Insider*, 3, no. 4 (18 February 1980): 1, 5; Robert Porter, "Second synfuels plant now operating in Texas," *Energy Insider*, 3, no. 14 (7 July 1980): 1; L. E. Furlong, E. Effron, L. W. Vernon, and E. L. Wilson, "Coal Processing: the Exxon Donor Solvent Process," *Chemical Engineering Process*, 72, no. 8 (1 August 1976): 69–75; "Startup near for Exxon coal liquefaction plant," *Chemical and Engineering News*, 58 no. 5 (4 February 1980): 7–8; *Assessment of Technology for the Liquefaction of Coal* (note 6), pp. 55–57.

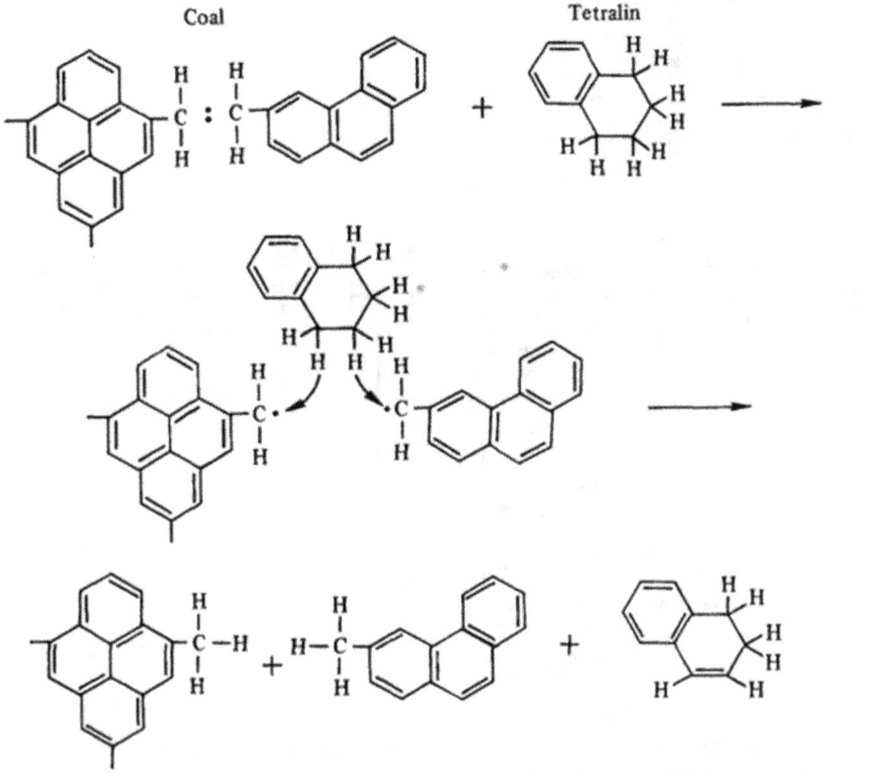

FIGURE 8.4 Donor action in coal structure
SOURCE: *Assessment of Technology for Liquefaction of Coal*, US DEPARTMENT OF ENERGY, OFFICE OF SCIENTIFIC AND TECHNICAL INFORMATION, REPORT 7049743, DECEMBER 1977, P. 19. https://doi.org/10.2172/7049743

It differed from the other processes because it used a previously catalytically hydrogenated recycle liquid such as a tetralin, $C_{10}H_{12}$, as the solvent oil added to the crushed dried coal to produce a paste or slurry. The liquefaction of coal therefore resulted not only from the addition of hydrogen gas to the slurry but because the solvent oil donated some of its hydrogen atoms to the coal. Exxon concluded that use of a hydrogenated rather than a non-hydrogenated solvent greatly improved the liquefaction process. The nickel-molybdenum metallic catalyst did not contact coal minerals or high-boiling liquids and had a longer, more active lifetime.[20]

20 John W. Taunton, K. L. Trachte, and R. D. Williams, "Coal feed flexibility in the Exxon Donor Solvent coal liquefaction process," *Fuel*, 60, no. 9 (September 1981): 788–94; Debra

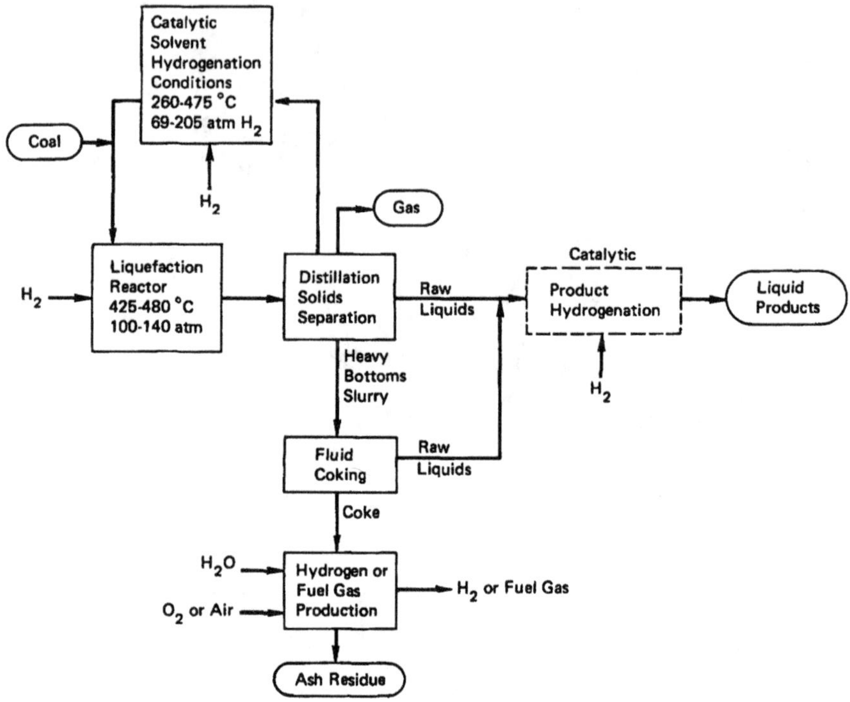

FIGURE 8.5 Exxon donor solvent process
SOURCE: *Assessment of Technology for the Liquefaction of Coal*, US DEPARTMENT OF ENERGY, OFFICE OF SCIENTIFIC AND TECHNICAL INFORMATION, REPORT 7049743, DECEMBER 1977, P. 56. https://doi.org/10.2172/7049743

The Baytown plant was self-sufficient, providing its own fuel and the hydrogen gas required for liquefaction. Gasifying the coke obtained in the hydrogenation process supplied the fuel, reforming the light gaseous products, CH_4 and C_2H_6, by reacting them with steam, produced hydrogen gas. EDS had about a 71% yield of dry coal to synthetic petroleum, converting one ton of bituminous coals, subbituminous coals, or lignites to 2.7–3.1 barrels of low-sulfur gasoline-size molecules (C_4–C_{15}, bp = 200–300° C) and lubricating oil (C_{15}–C_{25}, bp = 300–400° C). The pilot plant operated until 1982 when Exxon

---

Lewis, "Information an Important Product at DOE's Synfuel plant at Baytown, Texas," *Energy Insider*, 4, no. 16 (7 September 1981): 4–5; G. K. Vick 20. John W. Taunton, K. L. Trachte, and R. D. Williams, "Coal feed flexibility in the Exxon Donor Solvent coal at DOE's Synfuel plant at Baytown, Texas," *Energy Insider*, 4, no. 16 (7 September 1981): 4–5; G. K. Vick and W. R. Epperly, "Status of the Development of EDS Coal Liquefaction," *Science*, 217 (23 July 1982): 311–16.

closed and dismantled the plant because of economic issues. Exxon had planned to open its first industrial-size plant in 1997.[21]

DOE support for coal hydrogenation processes, such as SRC-II, H-Coal, and EDS, reached its maximum in the period 1975/76–2000, totaling $3.6 billion. Large-size demonstration programs received about 90% of the spending, but none survived beyond the early 1980s.

5 Fossil Fuel, Synthetic Fuel, the Environment

The 1960s–70s, particularly after the two 1970s oil crises, were also a time when the scientific community and the public became increasingly aware of the detrimental environmental impact fossil fuel combustion, because of the polluting gases coal, petroleum and natural gas released, had on the environment.

5.1 *Fossil Fuel, Synthetic Fuel Impact on Environment*

The US Geological Survey in 1976 established that the United States had huge quantities of coal to burn. It identified American coal resources at approximately 1,731 billion tons, with a demonstrated reserve base of 438 billion tons which existing technology could mine with at least 50% recovery. American coal reserves far exceeded the combined known reserves of all other fossil fuels (oil, natural gas, oil shale, and tar sands) and had a 200–300 years' estimated lifetime.[22] Other nations, Russia (Soviet Union), England, Germany, Australia, China, could make similar claims about their coal deposits.

Fossil fuel research and development programs in the oil and natural gas industries focused on discovering new sources of oil and natural gas and improving production yields. Synthetic fuel scientists directed their research to developing better catalysts and establishing optimum reaction temperature and pressure conditions that increased the synthesis' reaction rate and percent conversion to liquid fuels. Curbing atmospheric polluting gaseous emissions offered little incentive to many scientists within, and to some scientists outside, the fossil fuel energy sector.

The two major concerns of globally increasing fossil fuel consumption (combustion), which environmental scientists then and now recognized, were the increasing greenhouse gas emissions (carbon dioxide, water vapor, methane) that were warming the Earth's surface (global warming) and the

21 Robert Porter, "Second synfuels plant now operating in Texas" (note 19).

22 "Coal: Energy Source Past and Future," *Energy Insider* 3, no. 2 (8 December 1980): 2.

increasing release of the acidic polluting gases (sulfur dioxide, nitrogen dioxide) that were increasing the acidity of rainfall (acid rain). Their research showed that global warming caused a subsequent shift in Earth's agricultural zones, variance in rainfall, changes in marine production, glacier meltings, and a rise in sea levels.[23] Extensive acid rain studies showed that acid rain disrupted natural systems: producing waterways in which fish eggs do not hatch and mature fish die and degrading soil, plants, and forests; corroded human-constructed objects such as buildings and monuments; and impaired the health of humans (asthma, bronchitis).

Environmental scientists also expressed concerns about the synthetic fuel industry, noting that a geographic area, after having thousands of tons of coal removed for conversion to petroleum, would become a mined-out area, where no vegetation would grow. Water supply presented another problem. Coal-to-liquid fuel processes consumed large quantities of water for power plant cooling towers, for generating electricity, and for manufacturing the hydrogen gas essential for liquefaction. A 600 ton/d coal hydrogenation demonstration plant required 4,000 gal/min (1.1 ft^3/sec) or about four gallons of water for every gallon of synthetic petroleum. A limited water supply in coal-bearing regions therefore would hamper development of a coal hydrogenation industry.[24]

Nevertheless, until the 1960s–70s, environmental concerns did little to change the scientific community's and the public's longtime acceptance of

23 R. A. Scriven, "The carbon dioxide problem; a brief assessment of the present position," *J. Inst. Energy*, 53 (1980): 15–16; Roland A. Madden and V. Ramanathan, "Detecting Climate Change Due to Increasing Carbon Dioxide," *Science*, 209 (15 August 1980): 763–68; Gordon J. MacDonald, "The impact of carbon dioxide on climate," *The Physics Teacher*, 17 (September 1979): 367–73.

24 *Chemical Week*, 127 (9 July1980): 25; K. A. Rogers and R. F. Hill, "Coal Conversion Comparisons," The Engineering Societies Commission on Energy, Inc., Washington, DC, July 1979, pp. 69–7; *Assessment of Technology for the Liquefaction of Coal* (note 6), pp. 111–15; Gulf Oil Corporation, Information Brochure, *SRC-II: Environmental Issues and Protection Measures.*

The danger of carcinogenic hydrocarbon emission also arose. During the 1950s, workers at Union Carbide's coal liquefaction plant at Institute, West Virginia, developed skin cancer. A follow-up study completed during the late 1970s showed, however, that none of the workers developed systemic (internal) cancers twenty years later. Exposure to coal liquefaction chemicals appeared to cause no increase of death due to systemic cancers; many biomedical experts believe that because present worker protection procedures are so superior to the primitive techniques used at the Institute plant, the incidence of skin cancer will decrease. See Alan Palmer, *Journal of Organizational Medicine*, 21, 41 (January 1979); *Assessment of the Technology for the Liquefaction of Coal* (note 6), pp. 115–19; Gulf Oil Corporation, Information Brochure, *SRC-II: Health Issues and Protection Measures.*

industrial and domestic atmospheric pollution that resulted from fossil fuel combustion. Irrefutable experimental scientific evidence that Gilbert Plass, Roger Revelle, Charles Keeling, and other environmental scientists published at that time, however, led to an increasing awareness that the greenhouse gas pollutants resulting from fossil fuel combustion were warming the planet. Similarly, the publications of F. Hubbard Bormann, Noye Johnson, and Eugene Likens demonstrated clearly acid rain's environmental damage. The scientific evidence of climate change-global warming and acid rain that these and later environmental scientists amassed and published has remained largely unchallenged and dominant to this day.

5.2 *Environmental Impact: Climate Change-Global Warming and Acid Rain*

A brief discussion of the scientific thinking on climate change-global warming and acid rain follows. It provides a summary of the advances that occurred in the environmental sciences in the late 1800s–1900s and a comparison with the advances that occurred contemporaneously in fossil and synthetic fuel research and development.

Scientific thinking on climate change and global warming underwent a three-stage evolution. In the first stage, late-1800s–1900, a small group of scientists established a link between a changing atmosphere and climate change and proposed an atmosphere-climate change theory. They suspected that the increasing fossil fuel combustion was increasing the concentration of atmospheric carbon dioxide (human-produced or anthropogenic carbon dioxide, CO_2), changing (Δ) the atmosphere's composition, and that the changing atmosphere could change Earth's climate and temperature.

$$\Delta\, CO_2 \rightarrow \Delta \text{ Atmospheric Composition} \rightarrow \Delta \text{ Earth's Climate and Temperature}$$

The small number of nineteenth-century scientists who accepted a changing atmosphere-changing climate theory initially focused on accounting for ice ages and glacier formation and to a lesser extent on global warming. Despite the tremendous technological transformation occurring in the western world in the last half of the nineteenth century and the recognition that the transformation was spewing huge amounts of heat-absorbing gases such as CO_2 into the atmosphere, much of the scientific community remained convinced that little evidence supported a warming of Earth's surface temperature. If any increase of Earth's surface air temperature had occurred, the warming would be slow and insignificant. Scientific belief in a slow and insignificant warming continued into the 1940s.

The decades from 1900 to the 1940s represent the second stage, in which scientific interest in a changing atmosphere-climate change theory declined. As the twentieth century passed through its early decades, from 1900 through the 1940s, and convincing evidence of any global warming still lacking, research on the changing atmosphere-changing climate theory gradually lost the scientific community's interest and faded from discussion. The successive tragedies of World War I, the Great Depression, and World War II pushed climate science theories further into the background. During this second time period, the first applications of numerical methods to analyze weather patterns and other climate issues remained relevant research interests of atmospheric scientists. When linked to new evidence from infrared absorption spectra and gas chromatography published in the 1950s and to global atmospheric surface temperature measurements for the same time period, the scientific community looked anew at the forgotten nineteenth-century changing atmosphere-climate change theory. Absorption spectra and gas chromatography showed an increasing concentration of atmospheric CO_2 compared to pre-1800s–50s concentrations, the atmospheric temperature measurements revealed a corresponding global surface temperature rise.

The research that began in the 1950s and continues in the 2000s represents the third stage in the history of climate change. The introduction of high-speed computers and computer programs that performed the numerous, difficult, and tedious calculations that the laws of physics and chemistry required, such as the differential equations of fluid dynamics and thermodynamics, resulted in numerical models that much more comprehensively and accurately simulated the consequences of climate-related variables (temperature, pressure, concentration) on the atmosphere. At that time, in 1975, Wallace Broecker (1931–2019), a geoscientist at Columbia University, upon evaluating the recent evidence, introduced in a scientific context the expression *global warming*.[25] The scientific evidence supporting Earth's warming, obtained from surface air temperature measurements at thousands of land-based meteorological (monitoring) stations and sea surface measurements from ships, moored buoys, and since 1978 from satellites with remote sensing instruments; from weather balloons that provided vertical profiles of temperature, humidity, and winds; and instruments that measured subsurface ocean temperature, salinity, and currents, continued to accumulate, so that by the first decade of the twenty-first century, climate science researchers nearly unanimously accepted the reality of the globe's warming. The combined evidence showed that the average

25 Wallace Broecker, "Are We on the Brink of a Pronounced Global Warming?" *Science*, 189 (August 1975): 460–63.

global surface temperature, which for the period 1951–80 was 14° C (57° F), gradually had warmed over the twentieth century. It increased by more than 0.8° C (1.4° F) over the past century, much of the increase occurring in the last 35 years. For comparison, a decrease of 5° C (9° F) represents the difference between today's global temperature and an ice age.

Arvid Högbom (1857–1940), a chemist-geologist at Uppsala and Svante Arrhenius (1859–1927), a physical chemist at Stockholm, were among the small group of late nineteenth-century scientists (first stage) who recognized that the increasing industrial combustion of fossil fuels, especially coal, would increase the concentration of atmospheric CO_2. The increasing CO_2 concentration would change the atmosphere's composition, and they suspected that the changing atmospheric composition could change Earth's climate. The change in climate, resulting from the increasing concentration of atmospheric CO_2, a gas then known to absorb the Sun's heat rays that Earth reflected rather than letting the rays pass into space, would increase Earth's atmospheric temperature and produce a global warming. They shared an interest in environmental and climate science issues, such as the effect of heat-absorbing atmospheric gases (CO_2) and recognized the long-term consequences of human-caused warming of Earth's atmosphere.

Högbom's study of the carbon cycle led him in 1894 to calculate the amount of CO_2 natural sources emitted and to compare that amount with the amount of CO_2 emitted mainly by the increased industrial combustion of coal. He showed that at the current 1890s low rate of increasing CO_2 emissions, the amount of atmospheric CO_2 would double over a very long time, likely thousands of years. Arrhenius investigated the influence of heat-absorbing gases in the atmosphere, and in April 1896 he published his results in the *Philosophical Magazine*.[26] Arrhenius used Högbom's 1894 estimate of human-produced CO_2 and showed that doubling the amount of CO_2 in the atmosphere produced a global average temperature rise of 5–6° C. The mean global temperature calculated from measurements recorded at meteorological stations in the mid-to-late 1800s was about 13.5° C (56.3° F).

Arrhenius, like Högbom, did not seriously worry about European and American industrial expansion that was putting more CO_2 into the atmosphere. He recognized that CO_2's increased presence in the atmosphere would trap more of Earth's reflected heat and produce a global warming. But he believed that

26 Svante Arrhenius, "On the Influence of Carbonic Acid in the Air upon the Temperature of the Ground," *Philosophical Magazine and Journal of Science* (fifth series), 41 (April 1896): 237–75. Arrhenius discussed Högbom's results in his 1896 paper.

at the current pace of industrial expansion, a 5–6° C temperature rise from doubling CO_2's concentration would require a 3,000-year time span. Arrhenius discussed global warming again in his 1908 book, *Worlds in the Making: The Evolution of the Universe*.[27]

Guy Callendar (1898–1964), a Canadian-born steam engineer and amateur meteorologist in West Sussex, England, and Gilbert Plass (1920–2004), Canadian-born physicist who worked in both the academic and industrial worlds (Chicago's Metallurgical Laboratory 1942–45, Ford Motor Company 1950s–60s, from 1968 at Texas A&M University) represent the transition from the 1900–1940s denial period to the 1950s–2000s period that acknowledged the reality of global warming. Callendar linked the presence of CO_2, methane (CH_4), and water vapor (H_2O) in the atmosphere to climate change. He believed humans caused global warming and beginning in 1938 provided experimental evidence from global temperature readings that showed the connection between an increasing atmospheric temperature and the increasing consumption of coal. Callendar used the new instrumentation of infrared spectroscopy to establish the presence of methane and water vapor in the atmosphere and that they were greenhouse gases.[28]

In his 1956 paper published in *Tellus*, Plass calculated that doubling the Earth's CO_2 concentration would produce a 3.6°C increase in the Earth's surface temperature, whereas a 50% reduction of CO_2 would decrease the surface temperature by 3.8° C. He predicted that CO_2's atmospheric concentration would increase 30% and Earth's surface temperature would be 1° C higher in 2000 than in 1900.[29] Plass published remarkable predictions in the 1950s about future atmospheric temperature increases resulting from increasing CO_2 concentrations and showed that contrary to prevailing scientific opinion the oceans would not absorb all the human-generated fossil fuel emissions. Together Callendar's and Plass's revitalization of global warming research paved the way for the next group of investigators that included Roger Revelle, Charles Keeling, and James Hansen.

Roger Revelle (1909–91) at the Scripps Institute of Oceanography in La Jolla, California, provided the research that showed the oceans' important role on climate change and global warming. By the mid-1950s Revelle had

27 Svante Arrhenius, *Worlds in the Making: The Evolution of the Universe* (New York and London: Harper & Brothers Publishers), 1908.

28 Guy Callendar, "The Artificial Production of Carbon Dioxide and its Influence on Climate," *Quarterly Journal of the Royal Meteorology Society*, 64 (275): 223–40.

29 Gilbert Plass, "The Carbon Dioxide Theory of Climatic Change," *Tellus* VIII, 2 (1956): 140–54.

become aware of Callendar's and Plass's studies on global warming, and he began to speak publicly about the long-term consequences of global warming. He increasingly believed that Earth's temperature would rise in the next 100 years because of greenhouse gases but acknowledged his uncertainty about the number of degrees, possibly 2° C or 10° C. The uncertainty led Revelle, beginning in 1956, to continue making mass spectrometric measurements to determine how much CO_2 the oceans could absorb from the atmosphere and how long the absorption would take. His pioneering studies of the ocean waters' turnover rate and the oceans' buffering effect, which he published in *Tellus* in 1957, showed that, contrary to prevailing scientific opinion, the world's oceans could not absorb unlimited quantities of atmospheric CO_2.[30] His ocean studies and other studies that showed a steadily increasing concentration of human-produced CO_2 in the atmosphere left no doubt about Earth's future warming. Revelle provided the first quantitative measure of the oceans' ability to dissolve atmospheric CO_2. Later, in 1977, Revelle, with new data on projected CO_2 emissions, concluded that a 40% rise in atmospheric CO_2 concentration would occur over the next centuries.[31]

In 1956 Revelle brought the chemist Charles Keeling (1928–2005) to Scripps to head the Atmospheric Carbon Dioxide Program, and in 1957–58 Keeling established the Mauna Loa Observatory (MLO), a meteorological monitoring station in Hawaii, to provide a continuous record of atmospheric CO_2's daily concentration. Keeling located the observatory in Hawaii on the large (4,600-meter) active volcano Mauna Loa (long mountain) because of its undisturbed air, remote location, and minimal influence of vegetation and human activity. Measurements began in March 1958 and provide readings obtained at an altitude of 3,400 meters. Keeling remained at Scripps his entire career.

At that time, the 1950s, uncertainty existed about the oceans' buffering effect on atmospheric CO_2 concentrations. The daily CO_2 concentration record from MLO over a few years and the graph obtained from the record, called the Keeling Curve, ended the uncertainty. The global increase in atmospheric CO_2 concentration was a fact. On its first day of operation in March 1958, the Mauna Loa infrared gas analyzer recorded an atmospheric CO_2 concentration of 313 ppm. Measurements also showed that the annual average atmospheric CO_2 concentration increased from 1958 to 1959 and even more in

30 Roger Revelle and Hans E. Suess, "Carbon Dioxide Exchange Between Atmosphere and Ocean and the Question of an Increase of Atmospheric CO_2 during the Past Decades," *Tellus* IX, 1 (1957): 18–27.

31 Roger Revelle, Overview and Recommendations. *Energy and Climate*. Washington, DC: National Academy of Sciences, 1977.

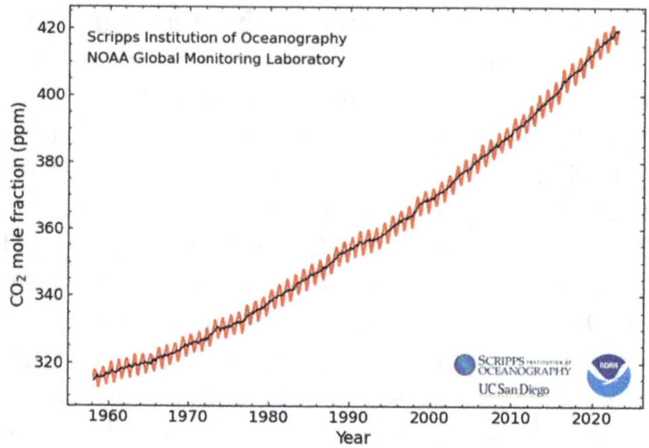

FIGURE 8.6 Keeling curve of CO_2 atmospheric concentration, Mauna Loa record
SOURCE: NOAA. https://gml.noaa.gov/ccgg/trends/

1960. Keeling published his results in his 1960 *Tellus* article "The Concentration and Isotopic Abundances of Carbon Dioxide in the Atmosphere."[32]

The Mauna Loa measurements, which Keeling recorded until April 1974 and the National Oceanic and Atmospheric Administration (NOAA) has continued in the early 2000s, showed that the increasing atmospheric CO_2 concentrations matched the increasing industrial and domestic fossil fuel consumption. Keeling calculated that if 57% of airborne fossil fuel emissions remained airborne, the increasing fossil fuel consumption matched very well the experimentally established CO_2 concentration curve over time. The slope of the curve was gradual in the first years and steeper in the ensuing years. Recent Mauna Loa readings were 315 ppm in 1958, 381 ppm in 2006, 389 ppm in 2010, 398.35 ppm in May 2013, 401.32 ppm in May 2014, 411.27 ppm in May 2017, 412.60 ppm in May 2018, and 424.0 ppm in May 2023.

James Hansen (b. 1941), a physicist at NASA's Goddard Institute for Space Studies in New York City since 1981 and professor at Columbia, is the last of the above mentioned twentieth-century atmospheric scientists whose research proved critical to establishing the link between global warming and fossil fuel combustion (greenhouse gas emissions). He remains a leading spokesman for climate scientists who support the CO_2 changing atmosphere-changing climate theory.

Beginning in 1982 Hansen introduced three-dimensional general circulation models in his climatology studies that aimed to simulate (reproduce)

32 David Keeling, "The Concentration and Isotopic Abundances of Carbon Dioxide in the Atmosphere," *Tellus* XII, 2 (1960): 200–03.

and predict accurately atmospheric conditions. The circulation models are mathematical equations of chemistry and physics that with aid of high-speed computers enabled him to include variables, such as air convection schemes and snow depths, in his calculations of global surface temperatures, and to account for factors that influence atmospheric conditions, such as variations in composition and temperature. The results he obtained, which were similar in the northern and southern hemispheres, indicated a warming of 0.5–0.7° C in the last century and revealed a strong warming trend between 1865 and 1880 that raised the global mean temperature in 1980 and 1981 to the highest-ever instrumentally recorded level. Hansen noted initially that the four warmest recorded years occurred in the 1980s with 1981 and 1987 the highest, and that 1998 was the warmest year since 1880. The rate of temperature change in the 1980s–90s exceeded all rates of change recorded since 1880.

Hansen commented that atmospheric CO_2 concentrations increased from 280–300 ppm in 1880 to 335–340 ppm in 1980 mainly because of fossil fuel combustion. From the mid-1960s to 1980 the global temperature rose by 0.2° C, resulting in a 0.4° C warming in the past century. The measured temperature increase agreed with the greenhouse effect calculations based on measured increases of atmospheric CO_2. By 2001 Hansen had improved the validity of his temperature measurements by including satellite measurements.[33]

Hansen's public statements have had a significant impact on alerting the public about the long-term consequences of human-caused, essentially industrial, CO_2 emissions. His appearance in June 1988 before the US Senate Committee on Energy and Natural Resources in which he summarized his years of research that showed Earth's warming, received wide coverage. Hansen testified that Earth had been warmer in the months from January to June 1988 than in any five-month period since temperature recording began in 1858. He believed with 99% certainty that the anticipated warming trend did not result from a natural variation or cycle of Earth's temperature. Increasing concentrations of CO_2 and other greenhouse gases caused the warming and made necessary a significant reduction of greenhouse gas emissions.

The Scottish chemist Robert Angus Smith (1817–74) in Manchester and London; and F. Hubbard Bormann (1922–2012), Noye Johnson (1930–87), and

---

33 James Hansen and S. Lebedeff, "Global Trends of Measured Surface Air Temperature," *Journal of Geophysical Research*, 92, D11 (20 November 1987): 13345–72; James Hansen and S. Lebedeff, "Global Surface Air Temperatures: Update through 1987," *Geophysical Research Letters*, 15 (1988): 323–326; James Hansen, I. Fung, A. Lacis, D. Rind, S. Lebedeff, R. Ruedy, G. Russell, and P. Stone, "Global Climate Changes as Forecast by Goddard Institute for Space Studies Three-Dimensional Model," *Journal of Geophysical Research*, 93, D8 (20 August 1988): 9341–64.

Eugene Likens (b. 1935) at the Hubbard Brook Experimental Forest in New Hampshire contributed significantly to our understanding of acid rain. Smith, whom Queen Victoria (1819–1901) in 1863 appointed the first inspector in the Alkali Acts Administration, accidentally observed an unusual acidity in Manchester's rains during the 1852 winter. He believed the soot particles released into the atmosphere from burning bituminous soft coal, and the source of the city's smog, also caused the acidity of the precipitated fog. His 1852 article "On the Air of Towns" reported cases of stunted tree growth and an acidity of the air that was at times high enough to turn litmus paper red (the red color indicates acidity) immediately during a rain. The acidity resulted from the conversion of sulfur in the coal to sulfur dioxide gas and its absorption by soot. Smith also estimated the amounts of carbonic acid and organic matter in the air and compiled figures showing a higher death rate in urban areas because of atmospheric pollution.

In a final set of experiments in which he bubbled polluted air through blood samples of inhabitants from different towns including Manchester, Smith concluded that the town's atmosphere had a peculiar effect on the blood's state. It caused a "remarkable reddening" "distinctly perceptible to the ordinary eye." Smith coined the phrase *acid rain* in his 1872 book, *Air and Rain: The Beginnings of a Chemical Climatology*.

Despite the significance of Smith's discovery of rain's acidity, his work failed to make an impact in its day. Why not? Smith consciously chose not to write *Air and Rain* as a popular account but as a compilation of the experimental evidence he had accumulated for years to highlight the damaging effects of pollution to air, land, and water. To quote from *Air and Rain*, "This volume could not have been made interesting to readers without sacrificing that which I suppose to be its best point – namely the large number of facts. Another, more popular in character, might perhaps be made to follow with propriety. We have been wearied with speculations about the Air, and it seemed to me advisable to begin building a systematic foundation with experiments which may be repeated frequently, or they have already been repeated."

During the next century industrializing nations such as the United States and Britain did little to limit or control the passage of sulfur dioxide and nitrogen oxides, particularly nitrogen dioxide, into the atmosphere. In the United States the earliest regulations, the Air Pollution Control Act of 1955 and the Clean Air Act of 1963, placed no restrictions on sulfur dioxide emissions, the main cause of acid rain, or on nitrogen oxides emissions. The year, 1963, however, marked a turning point in the history of acid rain.

F. Hubbard Bormann (1922–2012), Noye M. Johnson (1930–87), and Eugene Likens (b. 1935), ecologists at Dartmouth, with Robert S. Pierce of the US Forest Service, in 1963 began the long-term Hubbard Brook Ecosystem Study at the twelve-square mile Hubbard Brook Experimental Forest in New Hampshire's White Mountains. Hubbard Brook was the first study since Smith's linking air pollutants such as sulfur dioxide to the acidity of rain (acid rain). It became the model for ecological science studies. In the Hubbard Brook study, Bormann and Likens experimentally varied an entire ecosystem in order to establish its biological, chemical, and physical patterns and characteristics and the effects of acid rain on natural ecosystems. The results enabled them to evaluate human influence on the chemistry of soil, plants, trees, precipitation and clouds, and the ecology and biogeochemistry of aquatic ecosystems. Their work, particularly the 1974 paper in *Nature*, "Acid rain: A Serious Regional Environmental Problem," showed that the acidity of rain had increased over 20 years and remained acidic. Their paper initiated the later passage of national legislation that dealt with the acid rain problem.

The research of the Likens-Bormann group at Dartmouth and the Forest Service, and other ecologists triggered a debate on the cause of acid rain. Most ecologists agreed that acid rain resulted primarily from human causes, particularly the combustion of fossil fuels by utility companies. The main fossil fuels, coal, petroleum, and natural gas, contain varying amounts of sulfur and nitrogen and toxic metals such as mercury and lead that they release during their combustion. In the United States, about 66% of sulfur dioxide and 25–40% of the nitrogen oxides (NO, NO_2) come from the utilities companies' combustion of fossil fuels for electrical generation. Automobile exhaust contributes most of the remaining nitrogen oxides, about 50% in the atmosphere.

Manufacturers and industrialists, especially the major consumers of fossil fuels, opposed any regulations to reduce acid rain, mainly for economic reasons. They did not want to install any plant equipment, such as smokestack scrubbers designed to lessen and even eliminate the gaseous emissions responsible for acid rain. Political ideology, their opposition to *big activist* government, not peer-reviewed scientific evidence, directed their thinking.

The investigations of Likens, Bormann, Harvey, and others beginning in the 1960s and extensive studies by the EPA and the Canadian government (1996) had shown clearly that acid rain affected natural systems: waterways, soil, plants, and forests; human-constructed objects: buildings and monuments; and the health of humans. As a result of these investigations, in the United States the EPA in 1971 set a maximum limit on sulfur dioxide in the air, and in 1977 revisions to the 1970 Clean Air Act required smoke stack scrubbers to remove sulfur dioxide installed in all new power plants. Scrubbers remove about 90% of sulfur dioxide emissions.

Congress, in 1980, passed a second act to deal with the acid rain problem. The Acid Deposition Act of 1980 initiated a 10-year research program under the direction of the National Acidic Precipitation Assessment Program (NAPAP). NAPAP provided its first evaluation of acid rain in the United States in 1991. Congress passed additional amendments to the 1970 Clean Air Act in 1990. Title IV of the amendments established the Acid Rain Program. The Acid Rain Program introduced a Cap and Trade Program to control sulfur dioxide emissions. Title IV had two stages, the first beginning in 1995, the second in 2000, and proposed to reduce permanently power plant emissions of sulfur dioxide from their 1980 total of 18 million tons by 10 million tons, about a 55% reduction. Stage One limited sulfur dioxide emissions from the 110 largest power plants to 8.7 million tons of sulfur dioxide per year. Stage Two included all US power plants larger than 25-megawatts capacity. Total sulfur dioxide emissions in 2007 were 8.9 million tons sulfur dioxide, which met the 1990 requirement.

In the Cap and Trade Program that Congress established, each polluter receives or can purchase an emission allowance for each ton (or for any designated unit) of a specific pollutant it emits. If the plant emits fewer than its allotted tons of the pollutant, it can sell its extra tons and thereby reduce its operating cost or the cost of equipment it may have purchased to reduce its polluting emissions. Since establishing the Cap and Trade Program in 1990, sulfur dioxide emissions in the United States had decreased 40 percent by 2010 and 65% since 1976. During this same period the European Union, a 27-nation assembly whose roots date from 1952 and 1958 and includes France, Germany, and most of western Europe's nations with a population of nearly 450 million people, adopted effective acid rain reduction programs and achieved a 70% reduction in sulfur dioxide emissions.

The United States' 1990 Clean Air Act amendments placed no nationwide cap on nitrogen oxides emissions. The Act aimed to reduce nitrogen oxides emissions by 400,000 tons between 1996 and 1999 and allowed each plant to establish its own maximum. Total plant emissions, which were 375,000 tons in 1975, rose to about 600,000 tons in 1995 but since that time have fallen to 300,000 tons in 2010. Total nitrogen oxides' emissions from all sources decreased from 25 million tons in 1990 to 14.1 million tons in 2011 to 10.5 million tons in 2014. Nitrogen oxides' emissions in China, England, and in Europe since 2005, have fallen but remain high in Germany and Poland, and in India, South Africa, and Australia.

In 2005 the EPA announced the Clean Air Interstate Rule (CAIR) to deal with the problem of interstate pollution from power plant emissions of sulfur dioxide and nitrogen oxides. CAIR has a long-term reduction plan of 70% sul-

TABLE 8.3 US regulations to reduce atmospheric pollution (acid rain)

Regulation	Year
Air Pollution Control Act	1955
Clean Air Act	1963
Clean Air Act	1970
Clean Air Act revisions	1973
Acid Deposition Act	1980
Clean Air Act amendments (Acid Rain Program)	1990
Cap and Trade Program	1995 (stage 1), 2000 (stage 2)
Clean Air Interstate Rule (CAIRN)	2005

SOURCE: COURTESY OF AUTHOR

fur dioxide emissions in 28 eastern states and the District of Columbia from their 2003 level and a 60% reduction of nitrogen oxides emissions.

Little disagreement about the causes of global warming, acid rain, and other environmental issues exists within the international scientific community, especially among active climate science researchers. The challenges to the scientific evidence have come from the political side and the doubters in the coal and petroleum industries, who have an economic interest in denying the scientific evidence. The scientific debate, if ever one existed, has ended, but unfortunately the political debate has not. Climate politics has replaced climate science.

6 Oil Crises 1990s, Oil Shale Fracking Revolution and Its Consequences

Despite the growing awareness of environmental concerns, two developments in the mid-1990s triggered another synthetic fuel revival in the United States: (1) petroleum imports again reached 50% of total petroleum consumption, equaling the percentage of petroleum imported during the October 1973–March 1974 Arab petroleum embargo, and (2) an abundance of natural gas, equivalent to 800 billion barrels of petroleum, but largely inaccessible by pipeline, existed. Syntroleum (in 2007 Emerging Fuels Technology) in Tulsa; Rentech in Los Angeles; Exxon in Baytown, Texas; and Atlantic Richfield in Plano, Texas, recognized that the abundance of natural gas offered a way of reducing the United States' dependence on petroleum imports and in pilot plant studies at that time developed modified F-T syntheses that produced liquid fuels from natural gas (GTL).

DOE at its Pittsburgh Energy Technology Center through the 1980s–90s also continued small-size research on improved versions of coal hydrogenation. DOE pointed out that global coal reserves greatly exceeded petroleum reserves, anywhere from five to twenty-four times, and that it expected petroleum reserves to decline significantly in 2010–30. Energy forecasters likewise predicted a US petroleum shortfall of 1–3.5 million bbl/d in 2030.

The imbalance between the United States' domestic petroleum production and consumption had persisted since the 1960s, and the development of the coal hydrogenation and F-T synthetic fuel processes that followed the oil crises of 1973–74 and 1979–1981 did not eliminate the imbalance. But a 1980s–90s advance in petroleum recovery, combining horizontal drilling and hydraulic fracturing (fracking), revolutionized petroleum and natural gas production, and in the United States the combined drilling-fracking process significantly reduced, nearly eliminating, the gap between production and consumption.

Combined horizontal drilling and hydraulic fracturing began in 1986–91 in the Barnett Shale, located 5,000–8,000 ft deep in the Dallas-Fort Worth region of North Texas. It enabled American oil producers to recover petroleum and natural gas from the extensive, deeply buried (1,000–14,000 ft), dense oil shale deposits. The fast growing application of the combined process to other oil shale deposits in North Dakota (Bakken Shale 7.4 billion barrels estimated); Pennsylvania (Marcellus Shale 47 billion barrels estimated and Utica Shale 7.9 billion barrels estimated); New York (Marcellus and Utica Shale), West Virginia (Marcellus Shale); Texas (Barnett-North Texas, Eagle Ford-South Texas, Haynesville-East Texas 7.9 billion barrels total estimated and Permian Basin-West Texas 15 billion barrels estimated); and other states, significantly reduced the United States' reliance on petroleum imports to meet its domestic demand. At the same time the fracking revolution has contributed to growing environmental concerns that the process has increased earthquake activity because of its wastewater injected underground, and increased pollution of drinking water resources and methane gas emissions.[34]

The 1990s fracking revolution that followed the 1970s oil crises also shifted the construction of any potential industrial-size Bergius coal hydrogenation, F-T synthesis, and gas-to-liquid (GTL) plants from the United States and Europe to Asia, the Middle East, and Africa. Synthetic fuel research and development

34 See note 1, articles on fracking in "Fuel Fix."

in the United States experienced another setback in 1999 when DOE switched its coal hydrogenation research to *standby* and ended its coal hydrogenation funding, leaving Hydrocarbon Technologies Incorporated's (HTI) 1990s–2000s pilot plant in Lawrenceville, New Jersey, the only Bergius coal hydrogenation research project operating in the United States. HTI in 2002 received a $100,000 grant from DOE to continue its synthetic fuel research and has licensed its coal hydrogenation and F-T processes internationally, operating a coal hydrogenation plant in Mongolia, China, in 2009.[35]

With the exception of Sasol's three industrial-size F-T plants in South Africa constructed in 1955, 1980, and 1982; the few pilot plant and semiindustrial coal hydrogenation plants constructed in the United States and Europe in the 1970s–80s; and the six industrial-size F-T, gas-to-liquid, and coal hydrogenation plants constructed in the 1990s–early 2000s in China which has large coal fields, in Qatar with its abundant natural gas, and in Malaysia, and Nigeria, the synthetic fuel industry, like most alternative energies, has endured a series of fits and starts that plagued its history.

7 Conclusion

A combination of four factors led industrialized nations at various times during the twentieth century to conclude that synthetic liquid fuel could contribute to their growing liquid fuel requirements: (1) the shift from lower-energy solid fuels to higher-energy liquid fuels as the major energy source, (2) the invention of the Bergius and F-T coal-to-petroleum conversion and gas-to-liquid synthetic fuel processes, (3) recognition that global petroleum reserves were finite and much less than global coal reserves and limited petroleum's days as a plentiful energy source, and (4) the desire for energy independence.

The history of synthetic fuel has shown that shortly after Bergius invented high-pressure coal hydrogenation in 1913 the enthusiasm his process generated in petroleum-poor Germany during the early World War I years largely disappeared once Germany acquired Romania's oil fields. Similarly, in the United States, the East Texas petroleum discoveries that began on 5 October 1930 and continued throughout the decade, ended the Bureau of Mines Pittsburgh's 1920s–Depression-era F-T and coal hydrogenation pilot plant programs. The Bureau of Mines' promising resumption of research shortly after World War II

35 https://www.sbir.gov/sbc/hydrocarbon-technologies-inc.

TABLE 8.4 United States government synthetic fuel financial support

Administration	Financial support
Roosevelt-Truman (Democratic) 1944–52	Synthetic Liquid Fuels Act passed 5 April 1944, with later amendments totaled \$87.6 million
Eisenhower (Republican) 1952–60	Funding for 1944 Act ended 29 May 1953
Carter (Democratic) 1976–80	Synfuels Corporation \$20 billion initial budget, 2 April 1980
Reagan (Republican) 1980–88	Funding for Synfuels Corporation ended 19 April 1986

SOURCE: COURTESY OF AUTHOR

also ceased within a decade with the discovery of new petroleum fields, which allayed all fears of an impending energy crisis, and the emphasis the United States placed on developing nuclear energy. Despite the technological success of coal hydrogenation, F-T, and gas-to-liquid synthetic fuel processes, economics, politics, or a combination of economics and politics, significantly curtailed and sometimes halted the earlier 1920s and 1950s, and the later 1970s–2000s synthetic fuel research and development in the United States and Europe.

Two examples taken from the United States' synthetic fuel program in the post-World War II years, 1950s–70s, exemplify the politics of synthetic fuel research and development. Truman's Democratic administration (1948–52) financed the Bureau of Mines \$87.6 million 1949–53 synthetic fuel program, whereas the incoming Eisenhower Republican administration (1952–60) backed by the petroleum industry ended the government's financial assistance. Twenty years later, the sequence repeated itself. The Democratic Carter administration's (1976–80) \$20 billion 1980 Synfuels Act promoted development of a joint government-industry synthetic fuel program, Reagan's Republican administration (1980–88) bowing to the wishes of the petroleum industry defunded the Synfuels Act in 1986 and thereby dismantled the government-industry synthetic fuel program.

The historical record has demonstrated further that after nearly a century of research and development synthetic liquid fuel has not emerged as an important alternative energy source. Despite the technological success of synthesizing petroleum from coal, its lack of progress and cyclical history are largely the result of government and industry uninterest in committing firmly and long-term to synthetic fuel research and development. The synthetic fuel industry experienced intermittent periods of intense activity internationally in times of crises, only to face quick dismissal as unnecessary or uneconomical upon disappearance of the crises. Even the industry's argument that synthetic liquid fuels burn much cleaner than coal, and if substituted for coal they would reduce the emissions that have contributed to acid rain formation (NO_2, SO_2, SO_3), global warming (CO_2, CH_4), and to an overall deterioration of air quality, has failed to silence its critics. The synthetic fuel industry's hope of transforming its science and engineering research and development technological successes into large-scale industrial success failed to materialize.

Bibliography

A Note on the Sources. Much of the information I have used in my research on the German, Japanese, British, Canadian, and American synthetic fuel programs has come from the vast collection of documents and the 305 microfilm reels on the German synthetic fuel program, and the 185 microfilmed reports that the US Naval Technical Mission to Japan compiled. The reports on the Japanese synthetic fuel industry are the same as those in the National Diet Library in Tokyo and in the National Archives in Washington. Complete sets of both microfilm collections and documents are in the Synthetic Fuels Collection, Texas A&M University Archives. I have not listed each report in the bibliography but have included instead the articles I wrote based on these sources.

Abelshauser, Werner, Wolfgang von Hippel, Jeffrey Allan Johnson, and Raymond G. Stokes. *German Industry and Global Enterprise. BASF: The History of a Company.* New York: Cambridge University Press, 2004.

Alderson, Victor C. *The Oil Shale Industry.* New York: Frederick Stokes Company, 1920.

Allen, G. C. *A Short Economic History of Modern Japan 1867–1937.* London: George Allen & Unwin Ltd., 1946.

Ansell, L. L. et al. "Advanced Gas-to-Liquids Conversion Technology for Natural Gas Development." San Antonio, TX: World Conference on Refining Technology and Reformulated Fuels, 18–20 March 1997. Preprint.

Arrhenius, Svante. *Worlds in the Making: The Evolution of the Universe.* New York and London: Harper & Brothers Publishers, 1908.

Ashland Synthetic Fuels, Inc., Information Brochure, *H-Coal, A Future Energy Source,* 1979.

Assessment of Technology for the Liquefaction of Coal. Washington, DC: National Research Council, 1977.

Athabasca Oil Sands Conference, Proceedings. Edited by K. A. Clark et al. Edmonton, Alberta, September 1951.

Bailyn, Bernard et al., *The Great Republic: A History of the American People,* Vol. 1. Boston: D. C. Heath and Co., 1977.

Bar-Zohar, Michel. *The Hunt for German Scientists 1944–60.* New York: Hawthorn Books, Inc, 1967.

Barnhart, Michael A. *Japan Prepares for Total War.* Ithaca, New York: Cornell University Press, 1987.

Beck, Harald. *Friedrich Bergius, ein Erfinderschicksal.* Munchen: R. Oldenbourg, 1982.

Berghahn, Volker R. *The Americanisation of West German Industry 1945–1973*. New York: Cambridge University Press, 1986.

Bergius, Friedrich. *Die Anwendung hoher Drucke bei chemische Vorgangen und eine Nachbildung des Entstehungsprozesses der Stienkohle*. Halle: Wilhelm Knapp, 1913.

Bergius, Friedrich. "An Historical Account of Hydrogenation." Vol. 2. *Proceedings of the World Petroleum Congress*, 2 Vols. London: Office of the Congress, 1934.

Bergius, Friedrich. "The Transformation of Coal into Oil." *Proceedings of the International Conference on Bituminous Coal*, 1926. Pittsburgh: Carnegie Institute of Technology, 1927.

Bergius, Johannes (son of Friedrich Bergius), interview 4 July 1981.

Bergius-Verträge, Summary Survey, pp. 1–8, Bergius Papers, BASF Archives, Ludwigshafen, Germany.

BIOS Reports (British Intelligence Objectives Subcommittee). Synthetic Fuels Collection, Texas A&M University Archives.

Birkenfeld, Wolfgang. *Der synthetische Treibstoff, 1933–1945*. Göttingen: Musterschmidt-Verlag, 1964.

Blair, John M. *The Control of Oil*. New York: Vintage Books, 1978.

Blair, Sydney M. *Report on the Alberta Bituminous Sands, Government of the Province of Alberta*. Edmonton, Alberta: A. Shnitka, King's Printer, 1951.

Bone, William A. and Godrey W. Himus. *Coal: Its Constitution and Uses*. London: Longmans, Green & Co., 1936.

Borkin, Joseph. *The Crime and Punishment of I.G. Farben*. New York: The Free Press, 1978.

Borkin, Joseph and Charles A. Welsh. *Germany's Master Plan: The Story of Industrial Offensive*. New York: Duell, Sloan and Pearce, 1943.

Bosch, Carl. "The Development of the Chemical High Pressure Method During the Establishment of the New Ammonia Industry." Nobel Lecture, 21 May 1932, in *Nobel Lectures: Chemistry 1922–1941*. New York: Elsevier Inc., 1966.

British Fuel and Power Industries, P.E.P. (Political and Economic Planning. London: The Stanhope Press, October, 1947.

Brückmann, Heinrich. "Coal Liquefaction and its Importance in World Economy." *The Mining Journal*, 160 (28 January 1928): 69–70; 160 (4 February 1928): 94, and 160 (11 February 1928): 116.

Burgess, Renate (daughter of Friedrich Bergius), correspondence and interviews 1970s–80s (June 1981).

Bywater, Hector C. *Sea-Power in the Pacific* (reprint of 1921 ed.); New York: Arno Press and the New York Times, 1970.

Cahan, David L. "The physikalisch-technische Reichsanstalt: A Study in the Relations of Science, Technology, and Industry in Imperial Germany." PhD dissertation, Johns Hopkins University, 1980.

Campbell, Colin and Jean H. Leherrère. "The End of Cheap Oil." *Scientific American*, 287 (March 1998): 78–83.

Carroll, Berenice. *Design for Total War: Arms and Economics in the Third Reich*. The Hague, 1968.

CIOS Reports (Combined Intelligence Objectives Subcommittee). Synthetic Fuels Collection, Texas A&M University Archives.

Donath, Ernst. "Hydrogenation of Coal and Tar." Chemistry of Coal Utilization. Supplementary Volume. Edited by H. H. Lowry. New York: John Wiley & Sons, 1963, 1041–80.

Donath, Ernst, interview 9–12 August 1982.

Donath, Ernst and Maria Höring. "Early Coal Hydrogenation." *Fuel Processing Technology*, 1 (1977): 3–20.

Donath, Ernst and Maria Höring. "Kohle- und Ölhydrierung." *Ullman's Enzyklopädie der Technischen Chemie*, 3rd ed., Vol. 10. Munich: Urban & Schwarzenberg, 1958, 483–569.

Dubois, Josiah E., Jr. *The Devil's Chemists*. Boston: Beacon Press, 1952.

Duckert, Joseph M. *A Short Energy History of the United States*. Washington, DC: The Edison Electric Institute, 1980.

Edgerton, David. *Britain's War Machine: Weapons, Resources, and Experts in the Second World War*. New York: Oxford University Press, 2011.

Edwards, Tony. *The New Republic, History Broadsheet, No. 4*. London: Heinemann Educational Books, Ltd., 1972.

Eidt, Clarence M. et al. "Current Developments in Natural Gas Conversion Technology." In *Proceedings of Fourteenth World Petroleum Congress*, 1994: 249–57.

Ellis, Carleton. *Hydrogenation of Organic Substances*, 3rd ed. rev. New York: D. Van Nostrand Company, Inc., 1930.

Enos, John Lawrence. *Petroleum Progress and Profits*. Cambridge, MA: MIT Press, 1962.

Evonik.com, History.

Federal Oil Conservation Board, *Complete Record of Public Hearings*, 10 and 11 February 1926. Washington, DC: Government Printing Office, 1926.

Federal Oil Conservation Board, *Public Hearing*, 27 May 1926. Washington, DC: Government Printing Office, 1926.

Federal Oil Conservation Board, *Report*, 6 September 1926. Washington, DC: Government Printing Office, 1926.

Ferencz, Benjamin B. *Less Than Slaves*. Cambridge, MA: Harvard University Press, 1979.

FIAT (Field Information Agency Technical) Final Report No. 426. *Interrogation of Dr. Pier and Staff, IG Farbenindustrie AG, Ludwigshafen-Oppau*. Washington, DC: Joint Intelligence Agency, 14 November 1945.

FIAT Reports. Synthetic Fuels Collection, Texas A&M University Archives.

Fischer, Franz. *Die Umwandlung der Kohle in Öle*. Berlin: Verlag von Gebrüder Borntraeger, 1924. Translated by Richard Lessing. *The Conversion of Coal into Oils*. London: Ernst Benn Limited, 1925.

Fuel Research Technical Paper. London: His Majesty's Stationery Office, 1930s–40s.

Gibb, George S. and Evelyn H. Knowlton. *The Resurgence Years, 1911–1927: History of Standard Oil Company (New Jersey)*, Vol. 2. New York: Harper and Brothers, 1956. Four volumes written by different authors and published at different times (1956, 1958, 1971, 1988) give the history of Standard Oil.

Gillingham, John R. *Industry and Politics in the Third Reich*. New York: Columbia University Press, 1985.

Gimbel, John. *Science. Technology, and Reparations: Exploitation and Plunder in Postwar Germany*. Stanford, CA: Stanford University Press, 1990.

Gordon, Kenneth. "The Development of Coal Hydrogenation by Imperial Chemical Industries, Ltd." *Journal of the Institute of Fuel*, 9 (1935): 69–84.

Goudsmit, Samuel. *Alsos*. New York: H. H. Schuman, 1947.

Gould, Ed. *The History of Canada's Oil and Gas Industry*. Saanichton, BC: Hancock House Publishers, 1976.

Gulf Oil Corporation, Information Brochure, 1970s–80s.

Haber, Fritz. *Thermodynamik technischer Gasreaktionen*. Munich: Oldenbourg, 1905.

Haber, Ludwig. *The Chemical Industry, 1900–1930*. Oxford: Clarendon Press, 1971.

Haslam, R. T., R. P. Russell, and W. C. Asbury. "Comparison of Cracking and Hydrogenation and Methods of Producing Gasoline." *Proceedings of the World Petroleum Congress*, 2 Vols. London: Office of the Congress, 1934, Vol. 2, pp. 309–16.

Hayes, Peter. *Industry and Ideology*. New York: Cambridge University Press, 1989.

Hiebert, Erwin. "Walther Nernst," *Dictionary of Scientific Biography*, 16 Vols. New York: Charles Scribner's Sons, 1970–1980. Vol. 15, Supplement 1, pp. 432–53.

High-Pressure Hydrogenation at Ludwigshafen-Heidelberg, FIAT Final Report No. 1317, 9 Vols. Dayton, Ohio: Joint Intelligence Objectives Agency, 1951, Vol. 1.

Higham, Charles. *Trading with the Enemy*. New York: Delacorte Press, 1983.

Hilditch, T. P. *Catalytic Processes in Applied Chemistry*. New York: D. Van Nostrand, 1931.

Holroyd, Ronald. *IG Farbenindustrie AG Works, Ludwigshafen and Leuna. XIII. Hydrogen Resistant Steels, etc.*, CIOS XXX-103, Item No. 30 (1945).

Holroyd, Ronald. *IG Farbenindustrie AG Works*, US Bureau of Mines, Information Circular 7375. Washington, DC: Bureau of Mines, August 1946.

Homze, Edward. *Foreign Labor in Nazi Germany*. Princeton: Princeton University Press, 1967.

Höring Maria and Ludwig Raichle. "Zur Technologie der Kohle und Ölhydrierung, Matthias Pier zum Gedachtnis." *Chemie Ingenieur Technik*, 38, 3 (March 1966): 205–08.

Howard, Frank A. *Buna Rubber*. New York: D. Van Nostrand Company, Inc. 1947.

Howell, J. H. and R. M. Crawford. *The Production of Synthetic Fuels by the Hydrogenation of Solid and Liquid Carbonaceous Materials. XII. Materials of Construction.* US Naval Technical Mission in Europe, Technical Report 217-45. New York, New York, August 1945.

Hughes, Thomas Parke. "Technological Momentum in History: Hydrogenation in Germany 1898–1933." *Past and Present*, 44 (1969): 106–32.

Hunt, Linda. *Secret Agenda: The United States Government, Nazi Scientists, and Project Paperclip 1945–1990*. New York: St. Martin's Press, 1991.

Ignatieff, Alexander. *A Canadian Research Heritage*. Ottawa: Canadian Government Publication Centre, Supply and Service, 1981.

International Conference on Bituminous Coal, Proceedings, 1926, 1928, 1931.

Ipatiev, Vladimir. *Catalytic Reactions at High Pressures and Temperatures*. New York: Macmillan Publishers Ltd., 1936.

Ipatiev, Vladimir. *The Life of a Chemist*. Stanford: Stanford Univ. Press, 1946.

Ishida, Ryoichi. *Story of Coal Liquefaction*. Translated by Takeshi Kobayishi. Tokyo: Chûô Shuppan Insatsu, 1990.

Johnson, Arthur M. *The Challenge of Change: The Sun Oil Company, 1945–1977*. Columbus, Ohio: The Ohio State University Press, 1983.

Johnson, Benjamin. *Making Ammonia: Fritz Haber, Walther Nernst, and the Nature of Scientific Discovery*. Berlin/Heidelberg: Springer Verlag, 2002.

Johnson, Jeffery A. *The Kaiser's Chemists*. Chapel Hill: University of North Carolina Press, 1990.

Judt, Matthias and Giesla Burghard, eds. *Technology Transfer out of Germany*. Amsterdam: Harwood Academic Publishers, 1996.

Katzenellenbogen, Simon. "The Politics of Uneconomic Technology, South African Coal-to-Oil Production, 1932–87." University of Liverpool: Manchester-Liverpool Economic Historians Meeting, 22 May 1992.

Klein, Burton. *Germany's Economic Preparations for War*. Cambridge, MA: Harvard University Press, 1959.

Knott, David. "Synthetic Fuels Aiding South Africa to Advance its World Energy Role." *Oil and Gas Journal*, (17 March 1997): 23–28.

Krammer, Arnold. "Fueling the Third Reich." *Technology and Culture*, 19 (1978): 394–422.

Krönig, Walther. *Die katalysche Druckhydrierung von Kohlen, Teeren und Mineralölen*. Berlin: Springer-Verlag, 1950.

Kuczynski, Jurgen. *Germany: Economic and Labor Conditions under Fascism*. New York: Greenwood Press, 1968.

Landsberg, Hans H. and Sam H. Schurr. *Energy in the United States*. New York: Random House, 1968.

Langer, William C. *An Encyclopedia of World History*, 5th ed. Boston: Houghton Mifflin Company, 1972.

Larson, Henrietta M.; Evelyn H. Knowlton; and Charles Popple. *History of Standard Oil Company (New Jersey): New Horizons 1927–1950*. New York: Harper & Row Publishers, 1971.

Lasby, Clarence G. *Project Paperclip*. New York: Atheneum, 1971.

Lewes, V. B. *The Carbonisation of Coal*. London: Allan & Co., 1912.

Linde, Carl von. *Lowest Temperature in Industry*. Munich: Gësellschaft für Lindes Eismaschinen AG, 1912.

Marchese, Joseph. "Government and Energy: The Demonstration Program for Synthetic Liquid Fuels." PhD. diss., Princeton University, 1983.

Mason, T. W. "Labour in the Third Reich." *Past and Present*, 33 (1966): 112–41.

Mernitz, Kenneth. "Progress at a Price." PhD. diss., University of Missouri, 1983.

Meyer, R. F., C. T. Steele, and J. C. Olson, eds. *The Future of Heavy Crude and Oil Sands*. New York: McGraw-Hill, 1981.

Miller, Albert E. "The Story of the Technical Oil Mission, Preprint to the Twenty-fifth Annual Meeting of the American Petroleum Institute." Chicago, Illinois, 14 November 1945. Miller's paper has a list of the 26 TOM members who went to Germany and their affiliations.

Milward, Alan, S. *The Germany Economy at War*. London: University of London, Athlone Press, 1965.

Mines Branch Report. Ottawa, Canada: Government Printing Bureau, 1909–60s. Particularly those by T. E. Warren, K. W. Bowles, and R. E. Gilmore.

Mittasch, Alwin and W. Frankenburg. "Early Studies of Multicomponent Catalysis." *Advances in Catalysis*, Volume 2. New York: Academic Press Inc., 1950, pp. 81–104.

Mittasch, Alwin. *Geschichte der Ammoniaksynthese*. Weinheim: Verlag Chemie, 1951.

Molony, Barbara. *Technology and Investment: The Prewar Japanese Chemical Industry*. Council on East Asian Studies, Harvard University, 1990.

Morris, Peter. "The Development of Acetylene Chemistry and Synthetic Rubber by IG Farben Industries Aktiengesellschaft: 1926–1945." PhD. diss., University of Oxford, 1982.

"Multilateral: German Industries." *United States Treaties and other International Agreements*, Part I, 1951, 35 Vols. Washington, DC: Government Printing Office, 1950–84, Vol. 2 (1952), pp. 962–72, on p. 963.

"Multilateral: Incorporation of Germany into European Community of Nations." *United States Treaties and other International Agreements*, Part II, 1952, 35 Vols. Washington, DC: Government Printing Office, 1958–84, Vol. 3 (1954), pp. 2714–2722, on p. 2716.

Nagel, Alfred von. *Methanol Treibstoffe*. Ludwigshafen: Schriftenreihe des Firmenarchivs der Badischen Anilin- und Soda-Fabik AG, 5, 1970.

Nagel, Alfred von. *Sticktsoff.* Ludwigshafen: Schriftenreihe des Firmenarchivs der Badischen Anilin- und Soda-Fabrik AG, 3, 1970.

Nolte, Ernst. "Big Business and German Politics: A Comment." *American Historical Review*, 75 (1969): 71–78.

O'Connor, Richard. *The Oil Barons*. Boston: Little, Brown and Company, 1971.

Operational Archives Branch, US Naval History Division. Washington, DC, published a complete microfilm edition of all the reports in 1974. Information on Japan's synthetic fuel industry also is available in *Materials about Five Year Plan*, Part 2, Volume 3 (Liquid Fuel), edited by the Commerce and Industry Section of SMR, microfilm YD/324, section MOJ 275, National Diet Library, Tokyo; and in GHQ/SCAP, Record Group 331 (Records of General Headquarters/Supreme Commander for the Allied Forces), National Archives and Records Administration, Washington, DC.

Partington, J. R. and L. H. Parker. *The Nitrogen Industry*. London: Constable and Company Limited, 1922.

Payton-Smith, D. J. *Oil*. London: Her Majesty's Stationary Office, 1971.

Petroleum Facilities of Germany. Prepared by the Enemy Oil Committee for the Fuels and Lubricants Division, Office of the Quartermaster General, March, 1945.

Petroleum Press Service, 1947–55.

Petzholdt, George Paul Alexander. *Beitrag zur Kenntnis der Steinkohlenbildung nebst Kritik des Werkes von P. F. Reinsch*. Leipzig: Weigel, 1882.

Popple, C. S. *Standard Oil Company (New Jersey) in World War II*. New York: Standard Oil Company (New Jersey), 1952.

Potonié, Henry. *Die Enstehung der Steinkohle*. Berlin: Gebrüder Borntraeger, 1910.

Potter, E. B., ed. *Sea Power: A Naval History*. Englewood Cliffs, New Jersey: Prentice Hall, 1960.

Powell, Fred Wilbur. *The Bureau of Mines*. New York: D. Appleton and Company, 1922.

"Principles to Govern the Treatment of Germany in the Initial Control Period," No. 848, The Conference of Berlin (The Potsdam Conference), 1945. *Foreign Relations of the United States, Diplomatic Papers*, 17 Vols. Washington, DC: Government Printing Office, 1949–68, Vol. 2 (1960), pp. 750–53, on p. 752.

Proceedings of the Royal Society, Series A, 96. London: The Royal Society, 1919.

Rasch, Manfred. *Geschichte des Kaiser-Wilhelm-Instituts für Kohlenforschung 1913–943*. Weinheim: VCH, 1989.

Rasch, Manfred. *Kohlechemie im Revier. Zur Geschichte der Ruhrchemie AG 1927–1966*. Münster, 2018.

Reader, William J. *Imperial Chemical Industries: A History*, Volume II. London: Oxford University Press, 1975.

Reid, William T. "Low Temperature Carbonization of Coal in Japan." *US Bureau of Mines Information Circular 7430* (February 1948).

Report of the Fuel Research Board. London, 1930s–40s.

Report on the Petroleum and Synthetic Oil Industry of Germany. The Ministry of Fuel and Power. London: His Majesty's Stationery Office, 1947.

Review of Petroleum Technology, 1937–55.

Rich, Norman. *Hitler's War Aims*. New York: W. W. Norton & Company, Inc., 1973.

Richardson, Francis W. *Oil from Coal*. Park Ridge, New Jersey: Noyes Data Corporation, 1975.

Ritter, Gerhard. *The Schlieffen Plan: Critique of a Myth*. New York: Praeger Press, 1958.

Roberts, Glyn. *The Most Powerful Man in the World: The Life of Sir Henri Deterding*. Westport, CN: Hyperion Press Inc., 1976.

Roggersdorf, Wilhelm. *In the Realm of Chemistry*. Düsseldorf: Econ-Verlag, 1965.

Russell, Robert P. and Robert J. Haslam. "Hydrogenation of Petroleum." *Industrial and Engineering Chemistry*, 22 (1930): 1030–37.

Sabatier, Paul. *La catalyse en chimie organique*. Paris: Librairie Polytechnique, 1913. Translated by E. Emmett Reid. *Catalysis in Organic Chemistry*. New York: D. Van Nostrand, 1922.

Sasol 50 Years of Innovation, Sasol.com.

Sasuly, Richard. *I.G. Farben*. New York: Boni & Baer, 1947.

Schmidt-Pauli, Edgar von. *Friedrich Bergius*. Berlin: E. S. Mittler & Sohn, 1943.

Schuster, Curt. *Wissenschaft und Technik*. Ludwigshafen: Schriftenreihe des Firmenarchivs der Badischen Anilin- und Soda-Fabrik AG, 14, 1976.

Schweitzer, Arthur. *Big Business in the Third Reich*. Bloomington, Indiana: Indiana University Press, 1964.

Sheppard, Deri. *Robert Le Rossignol: Engineer of the Haber Process*. Cham, Switzerland: Springer Nature, 2000.

Shirer, William L. *Berlin Diary*. New York: Alfred A. Knopf, 1934.

Simon, Leslie. *German Research in World War II*. New York: John Wiley & Sons, Inc., 1947.

Soldberg, Carl. *Oil Power*. New York: New American Library, 1976.

Speer, Albert. *Inside the Third Reich*. New York: Macmillan Publishing Co., Inc., 1970.

Spitz, Peter. *Petrochemicals: The Rise of an Industry*. New York: John Wiley, 1988.

Statistical Abstract of the United States 1939. Washington, DC: Government Printing Office, 1939, p. 208.

Stokes, Raymond. *Divide and Prosper: The Heirs of IG Farben Under Allied Authority*. Berkeley: University of California Press, 1988.

Stokes, Raymond. *Opting for Oil*. New York: Cambridge University Press, 1994.

Storch, H. H. "Hydrogenation of Coal and Tar," in *Chemistry of Coal Utilization*. Edited by H. H. Lowry, 3 Vols. New York: John Wiley & Sons, 1945 and 1963, Vol. 2, pp. 1750–96.

Storch, Henry H., Norma Golumbic, and Robert Anderson. *The Fischer-Tropsch and Related Syntheses*. New York: John Wiley and Sons, 1951.

Stranges, Anthony N. "Canada's Mines Branch and its Synthetic Fuel Program for Energy Independence." *Technology and Culture*, 32 (1991): 521–54.

Stranges, Anthony N. "Friedrich Bergius and the Rise of the German Synthetic Fuel Industry." *Isis*, 75 (1984): 643–67.

Stranges, Anthony N. "From Birmingham to Billingham: High-Pressure Coal Hydrogenation in Great Britain." *Technology and Culture*, 20 (1985): 726–57.

Stranges, Anthony N. "Germany's Synthetic Fuel Industry 1930–45." Working paper 5.64. Center for German and European Studies, University of California, Berkeley (March 1997): 1–94.

Stranges, Anthony N. "Synthetic Fuel Production in Japan: A Case Study in Technological Failure." *Annals of Science*, 50 (1993): 229–65.

Stranges, Anthony N. "Synthetic Petroleum from Coal Hydrogenation: Its History and Present State of Development in the United States." *Journal of Chemical Education*, 60, 8 (April 1983): 617–25.

Stranges, Anthony N. "The US Bureau of Mines' Synthetic Fuel Programme." *Annals of Science*, 54 (1997): 29–68.

Tammen, Helmuth. *Die I.G. Farbenindustrie Aktiengesellschaft (1925–1933)*. Berlin: by the author, 1978.

Taylor, F. Sherwood. *A History of Industrial Chemistry*. New York: Arno Press, 1972.

Taylor, R. "High-Pressure Technique." *Chemistry and Industry*, 56 (1937): 41–50T.

Technical Oil Mission. Interviews of its members and of German scientists. Synthetic Fuels Collection, Texas A&M University, Archives.

Technical Oil Mission Reports. Microfilm and documents. Synthetic Fuels Collection, Texas A&M University Archives.

Ter Meer, Fritz. *Die IG Farben*. Dusseldorf: Econ-Verlag, 1953.

Thumen, Karl-Heindrich von. *Jahrbuch der deutschen Mineralolwirtschaft*. Frankfurt am Main: Fritz Knapp, 1939.

Transactions of the Fuel Conference: World Power Conference, 2 Vols. London: Percy Lund, Humphries & Co., 1928.

Travis, Anthony S. *Farbenindustrie*. Alperton, Wembley, Middlesex: Brent Schools & Industry Project, 1984.

Travis, Anthony S. *Nitrogen Capture: The Growth of an International Industry (1900–1940)*. Berlin/Heidelberg: Springer Verlag, 2018.

Trials of War Criminals Before the Nuremberg Military Tribunals Under Control Council Law No. 10. 15 Vols. Washington, DC, 1949–53.

Turner, Henry Ashby Jr., *German Big Business and the Rise of Hitler*. New York: Oxford University Press, 1985.

Turner, Henry Ashby Jr. "Big Business and the Rise of Hitler." *American Historical Review*, 75 (1969): 56–70.

United States Bureau of Mines, *Bulletins*. Washington: United States Government Printing Office, 1968, 1975.

United States Bureau of Mines, *Information Circulars 7375* (August 1946); *7446* (January 1948). Washington: United States Government Printing Office.

United States Bureau of Mines, *Report of Investigations*. Washington: United States Government Printing Office, 1950, 1954, 1959.

United States Bureau of Mines, *Statement*. Washington: United States Government Printing Office, 1943.

United States Bureau of Mines, *Technical Papers*. Washington: United States Government Printing Office, 1926, 1941, 1942, 1943, 1944.

United States Naval Technical Mission to Japan. Microfilm, Synthetic Fuels Collection, Texas A&M University Archives.

United States Strategic Bombing Survey, *The German Oil Industry Ministerial Report Team 78*, Oil Division, 1st ed. 5 September 1945, 2nd ed. January 1947, pp. 19–20. *The United States Strategic Bombing Survey*, Summary Report (European War), 30 September 1945, 18 pages.

Urteil im I.G. Farben-Prozess, Das. Offenbach A. M.: Bollwerk-Verlag Karl Drott, 1948.

Wagenfuhr, Rolf. *Die deutsche Industrie in Kriege 1939–1945*. Berlin: Duncker & Humblot, 1954.

Webster, Charles and Frankland, Noble. *The Strategic Air Offensive Against Germany*. London: H. M. Stationery Office, 1961.

Weil, B. H. and J. C. Lain. *Synthetic Petroleum from the Synthine Process*. Remsen Press Division, Chemical Pub. Co., 1948.

Williamson, Harold F. *The American Petroleum Industry*, 2 Vols. Evanston, IL: Northwestern University Press, 1959, Vol. 2.

Williamson, Harold F., Ralph L. Andreano, Arnold R. Daum, and Gilbert C. Klose. *The American Petroleum Industry: The Age of Energy 1899–1959*. Evanston, IL: Northwestern University Press, 1963.

Witschakowski, Walter. *Hochdrucktechnik*. Ludwigshafen: Schriftenrelhe des Firmenarchivs der Badischen Anilin- & Soda-Fabrik AG, 12, 1974.

Wolf, Gerhart. *BASF: Development of a World-wide Enterprise*. Ludwigshafen: Schriftenreihe des Firmenarchivs der Badischen Anilin- & Soda-Fabrik AG, 6, 1970.

World Petroleum Congress, *Proceedings*, 1933, 1937, 1951, 1955.

Wu, W. R. K. and Storch, H. H. "Hydrogenation of Coal and Tar." *United States Bureau of Mines Bulletin 633*. Washington: US Government Printing Office, 1968.

Index

www.ingramcontent.com/pod-product-compliance
Lightning Source LLC
LaVergne TN
LVHW010541300926
849162LV00002B/200
9789004544116